Concrete Mixture Proportioning

Modern Concrete Technology Series
Series Editors

Arnon Bentur
National Building Research Institute
Technion–Israel Institute of Technology
Technion City
Haifa 32 000
Israel

Sydney Mindess
Department of Civil Engineering
University of British Columbia
2324 Main Mall
Vancouver
British Columbia
Canada V6T 1W5

1. **Fibre Reinforced Cementitious Composites**
   A. Bentur and S. Mindess

2. **Concrete in the Marine Environment**
   P.K. Mehta

3. **Concrete in Hot Environments**
   I. Soroka

4. **Durability of Concrete in Cold Environments**
   M. Pigeon and R. Pleau

5. **High-Performance Concrete**
   P.-C. Aïtcin

6. **Steel Corrosion in Concrete**
   Fundamentals and civil engineering practice
   A. Bentur, S. Diamond and N. Berke

7. **Optimization Methods for Material Design of Cement-based Composites**
   Edited by A. Brandt

8. **Special Inorganic Cements**
   I. Odler (forthcoming)

9. **Concrete Mixture Proportioning**
   F. de Larrard

10. **Permeability and Poresize in Cement-based Materials**
    K. Aligizaki and P.D. Cady (forthcoming)

11. **Sulfate Attack on Concrete**
    J.P. Skalny, P.W. Brown, V. Johansen and J. Marchand (forthcoming)

12. **Fundamentals of Durable Reinforced Concrete**
    M.G. Richardson (forthcoming)

Concrete Mixture Proportioning
A scientific approach

François de Larrard

Routledge
Taylor & Francis Group

LONDON AND NEW YORK

First published 1999 by E & FN Spon

2 Park Square, Milton Park, Abingdon, Oxfordshire OX14 4RN
52 Vanderbilt Avenue, New York, NY 10017

Routledge is an imprint of the Taylor & Francis Group, an informa business

First issued in paperback 2019

Copyright © 1999 François de Larrard

British Library Cataloguing in Publication Data
A catalogue record for this book is available
from the British library

Library of Congress Cataloging in Publication Data
Larrard, François de
    Concrete mixture proportioning : a scientific approach / François
de Larrard.
        p.   cm. -- (Modern concrete technology)
    Includes bibliographical references and index.
    ISBN 0-419-23500-0
    1. Concrete--Mixing.   I. Title.   II. Series: Modern concrete
technology series (E. & F.N. Spon)
    TA439.L32   1999
    666'.893--dc21   98-39234
                                                                   CIP

ISBN 978-0-419-23500-2 (hbk)
ISBN 978-0-367-86356-2 (pbk)

Contents

Foreword

To write a foreword to this book is a pleasure but the task also presents some difficulties. The pleasure arises from the great importance of the work, but I feel I should first dispose of my 'difficulties'.

As François de Larrard acknowledges in his letter inviting me to write this foreword, my approach to concrete is pragmatic, coupled with doubts about ready-to-use mathematical models for mix design. This is true; in my book *Properties of Concrete* (4th edition), published in 1995, I say: 'an exact determination of mix proportions by means of tables or computer data is generally not possible: the materials used are essentially variable and many of their properties cannot be assessed truly quantitatively. For example, aggregate grading, shape and texture cannot be defined in a fully satisfactory manner.'

My views arose from consideration of commercial models, developed in Australia, the United Kingdom and the United States, each based on a limited range of experimental data, and yet each claiming universal validity. The critical feature of those models is that they are more-or-less statistical fits to experimental data, without the necessary explanation in terms of actual physical phenomena. Correlation of variables should not necessarily be taken to imply causation. For example, over a period of ten years, in a given state, the consumption of alcohol increased and, at the same time, the salaries of teachers increased too. It was concluded that the more the teachers are paid the more alcohol is drunk. This may be an apocryphal story, but *si non è vero è ben trovato*.

The fact is that some engineers fall into that kind of trap. Not so, de Larrard. His models are physical models of which the parameters, in so far as possible, correspond to measured quantities. Herein lies one considerable virtue of his work. There is no denying that, if observed phenomena cannot be expressed in mathematical terms, generalized use of the observed relations is not possible. At the same time, if the mathematical terms are not related to physical phenomena in a manner that stands up to a logical interpretation, there is a great risk of using a model whose limits of validity are not known.

Having read de Larrard's book, I fully accept that he is on the right path. This path must be followed in order to bring the use of concrete and, above all, the mix design, into line with the use of other materials, such as polymers and metals. In other words, de Larrard's approach will make concrete a truly purpose-designed manmade material, instead of just a reconstituted rock.

My use of the future tense might give the impression that the problem has not been fully solved. This is no criticism of de Larrard's work: nobody could achieve more today, if only because we are still unable to quantify all the properties of aggregate, as I mentioned earlier. And this is a formidable task. The physical means to achieve it are available without further basic research; the difficulty lies in the fact that no commercial part of the 'concrete system' has an interest in financing the necessary development.

As for de Larrard's models, I am impressed by the way in which he has validated them by independent assemblies of coherent data, without ever contradicting practical experience. His approach is the best way to progress from today's most common building material, that is, bad concrete, to concrete which, in every case, will be purpose-designed and tailored to the expected use.

There are a few, more general, comments which I would like to make. François de Larrard, a Frenchman in the great tradition of École Polytechnique, has chosen to write this book, not in what the French love to call *la langue de Molière*, which is also *la langue de Descartes*, quoted by de Larrard, but in what has indubitably become the worldwide language of technology, science, and much more. Yet he has retained the Cartesian approach and, for bringing this rigour to concrete, we should be very grateful to him.

Indeed, up to now, there has been very little interaction between the French, very mathematical, approach to concrete on the one hand and, on the other, the British and American pragmatic approach. French books on concrete are rarely translated into English; likewise, it is only in 1998 that, for example, my book *Properties of Concrete* appeared in a French translation (having, over the years, been translated into twelve other languages). An yet, it is the same well-designed concrete mix that is needed on both sides of the Channel and on both sides of the Atlantic, and the marrying of the French and Anglo-American approaches cannot but be highly beneficial for all concrete users. So, this 'French' book in English is very welcome. Its title includes the words 'a scientific approach'. It is a scientific approach by a civil engineer, and this is the way forward.

Adam Neville CBE, FEng
London
15 April 1998

Preface

... diviser chacune des difficultés que j'examinerais en autant de parcelles qu'il se pourrait, et qu'il serait requis pour les mieux résoudre...
... conduire par ordre mes pensées, en commençant par les objets les plus simples et les plus aisés à connaître, pour remonter peu à peu, comme par degrés, jusques à la connaissance des plus composés....
... faire partour des dénombrements si entiers et des revues si générales, que je fusse assuré de ne rien omettre.

... to divide every difficulty that I should examine into as many parts as possible and as necessary to better solve it...
... to drive my thoughts in order, starting by the most simple objects, the easiest to know, so as to rise gradually, step by step, up to the knowledge of the most compound ones...
... to make such complete accounts and so general reviews, that I might be assured of missing no thing.

(René Descartes, *Discours de la méthode*, 1637)

The aim of this book is to build a consistent, rational and scientifically based approach to designing concrete mixtures for most civil engineering applications. An attempt has been made to consider the following facts, which change the nature of the problem, as compared with the situation faced by our predecessors (from René Féret and Duff Abrams to more recent contributors):

- Concretes are no longer made of aggregate, Portland cement and water only. Often, if not always, they incorporate at least one of the following products: organic admixtures, supplementary cementitious materials, fibres.
- Nowadays concretes must meet a comprehensive list of requirements, which are not limited to the final compressive strength, but include rheological properties, early-age characteristics, deformability properties and durability aspects.

- As for the desired properties, the range of attainable values has displayed a dramatic increase in the past few years. For instance, no-slump to self-compacting concrete can be used. The compressive strength at 28 days may be as low as 10 MPa for some dam mass concretes, or as high as 200 MPa or more for some special precast products.
- A purely experimental and empirical optimization is less and less likely to succeed, because of the high number of parameters involved (both input and output), the high labour expenses of such studies, and the economic and time constraints that are characteristics of the industrial world.

But besides these negative aspects (with regard to the ease of optimizing a concrete), there are, fortunately, positive ones:

- Concrete technology is no longer a young technology; since the beginning of this century a huge amount of experimental data has been published, which can be exploited and synthesized. The latest edition of Adam Neville's book provides a unique survey of this literature (Neville, 1995).
- The development of computers helps the researcher to manage large amounts of experimental data. From these data, he may discover the physical laws underlying the behaviour of the concrete system, and translate them into semi-empirical mathematical models, linked with databases (Kaëtzel and Clifton, 1995).
- Once those models are incorporated in user-friendly software, practitioners may readily use them, even with limited knowledge of the vast complexity of the system. Thus optimal responses to industrial problems may be given quickly. At the present state of development, experiments are still necessary, but the software allows the formulator to cut the number of tests drastically, to focus the experimental programmes better and take maximum advantage of the new data generated.

On the basis of these statements, the objectives of this book are twofold: (i) to develop simplified models linking concrete composition and properties, so that an analytical solution of optimization problems becomes possible, allowing one to demonstrate the main rules and trends observed, and therefore understand the concrete system better; (ii) to build more comprehensive and complex models, which may be easily implemented in software and be used in practice for designing concretes with real materials for real applications.

Designing a concrete is first of all a packing problem. All existing methods implicitly recognize this statement, either by suggesting measurement of the packing parameters of some components (ACI

211), or by approximating an 'ideal' grading curve that is assumed to lead to maximum compactness with real materials. It is noteworthy that all authors propose different curves (or families of curves), which raises doubts as the soundness of this concept. Chapter 1 of this book presents in detail a theory developed by the author and his colleagues after a 12-year research effort. It is the first one, to the author's knowledge, to solve the question of the packing density of dry mixtures in all its general extent, with sufficient accuracy for practical application. The theory agrees with most classical results, and shows that the ideal proportions of a given set of aggregate fractions depend not only on the grading curves, but also on the packing abilities of each grain fraction, including the fine ones.

Chapter 2 is devoted to an analysis of the relationships between the composition of concrete and its properties. The models developed often refer to packing concepts, considering either the whole range of solid materials in fresh concrete or the aggregate skeleton in hardened concrete. Most engineering properties are dealt with, but with less emphasis on durability. This is not because the subject is not important, but simply because at present few data are available that both cover a wide range of concretes and relate to true durability-related material properties.

Chapter 3 lists the constituent properties that control concrete properties in the previously developed models. It is noteworthy that this list appears limited compared with the huge collections of parameters that encumber some concrete studies.

Chapter 4 shows how to use the models presented in Chapter 2. It is the core of the book. With the help of simplified models, optimization problems are first solved analytically. This approach qualitatively highlights the most important features of the concrete system. Then a more refined solution is worked out by using the complete models together with a spreadsheet package containing an optimization module (*solver*). General questions dealing with mixture proportioning are discussed in the light of numerical simulations, and some existing empirical methods are critically reviewed.

Chapter 5 presents a number of applications of the proposed approach to a series of industrial problems. It is shown here that the book provides a conceptual framework that applies to any cementitious granular material, including mortars, roller-compacted concrete and sprayed concrete. A special emphasis is put on high-performance concrete, but it is demonstrated that the approach applies just as well to very common 'commodity' concrete. Real concrete formulae, with measured properties, are presented, and are compared with the model predictions obtained for the same specifications.

In the conclusion, research needs for improving the models are stressed. Clearly, the area of durability is one that deserves the highest effort. Engineers must eventually be able to design a concrete for a given structure life span, in a given environment. Today we are lagging far behind this objective. It is also recognized that some models are too empirical, because the scientific basis for some problems is still lacking, making extrapolations sometimes hazardous. However, the author's defence is that an engineer must solve a problem with the most appropriate means that are available to him. Nevertheless, future research should aim to decrease as much as possible the part that empiricism plays.

In regard to the practical use of this book, a flowchart is provided in Appendix 1, which makes it possible to implement the various models in software. Also, it is briefly shown that the concepts presented may be used not only for designing new mixtures but also for quality control purposes. In an industrial production process, as raw materials are continuously changing, the concrete system should remain alive, adapting itself to give an output that is as constant as possible (Day, 1995). This is probably one of the greatest challenges of today's concrete industry, equal to the development of new and 'exotic' formulae.

This book is not a state of the art: that is, a compilation of existing and published knowledge about mixture proportioning. It contains essentially original findings, acquired by the author and his colleagues at the Laboratoire Central des Ponts et Chaussées (Central Laboratory for Roads and Bridges), Paris. Here the author had the chance to spend 12 years in an exceptional scientific environment, while being in contact with the 'real concrete world' through the network of the regional 'Ponts et Chausseés' laboratories. The approach tries to be ahead of the current technology. However, the views presented herein are often personal, and it is the author's hope that this book will help in sharing them with the international scientific and technical community.

A last remark addressed to civil engineers: this book contains very little (if any) chemistry.[1] This is partly because the author is not a chemist, but also because the level at which mix design of concrete is envisaged is that of assembling the components. Facts related to the phases are taken as hypotheses; only their consequences in concrete are studied in this book. However, the author is conscious that more chemistry would be required, if a comparable approach were to be extended to all durability aspects.

F. de Larrard
Bouguenais, January 1998

---

[1] This remark may be considered either as negative or positive, depending on the reader's background.

Acknowledgements

My deepest gratitude goes to various people without whom this work would not have been possible:

- to the two institutions in which I had the fortune to serve as a researcher (Laboratoire Central des Ponts et Chaussées, Paris/ Nantes, and National Institute of Standards and Technology, Gaithersburg, Maryland, who kindly welcomed me as a guest researcher from September 1996 to June 1997);
- to all the people who supported me in the project of writing such a book by their interest regarding our research: our industrial partners, our students and the practitioners who attended the numerous concrete training sessions to which I have contributed for about 10 years;
- to all the friends who participated as co-researchers in the scientific adventure related in this book: Paul Acker, Daniel Angot, Chiara F. Ferraris, Chong Hu, Ali Kheirbeck, Pierre Laplante, Robert Le Roy, Jacques Marchand, Thierry Sedran, Tannie Stovall, Patrick Tondat, Vincent Waller;
- to all my colleagues who were kind enough to communicate precious unpublished data for model calibration/validation: François Cussigh, Pierre Laplante, Gilbert Peiffer, Paul Poitevin, Thierry Sedran, Erik Sellevold;
- to all my colleagues who spent time correcting some parts of the manuscript: Mony Ben-Bassat, Dale Bentz, Nicholas Carino, Gilles Chanvillard, Jim Clifton, Clarissa Ferraris, Edward Garboczi, Geneviève Girouy, Robert Le Roy, Thierry Sedran, Vincent Waller;
- and last but not least to my wife and my children, who suffered to see their husband/father glued to his computer for months instead of doing everything they could normally expect from him.

Finally, this book is dedicated to Albert Belloc, Senior Technician, who spent 39 years at LCPC managing the concrete production team. By his courage and rigour he was a unique example for generations of concrete technologists, and played a major role in many original experiments related in this book. I am also extending this dedication to all experimentalists in LCPC and elsewhere, without whom no models or theories can be conceived.

Note to readers: when using the proportioning method readers are advised to check material properties on trial batches.

1 Packing density and homogeneity of granular mixes

The packing density of a granular mix is defined as the solid volume Φ in a unit total volume. Alternatively, the compaction may be described by the porosity:

$$\pi = 1 - \Phi \tag{1.1}$$

or by the voids index:

$$e = \frac{\pi}{\Phi} = \frac{1}{\Phi} - 1 \tag{1.2}$$

The prediction of the packing density of a granular mixture is a problem of extreme relevance for concrete (Johansen and Andersen, 1991), but it is also relevant in many other industrial sectors (Guyon and Troadec, 1994). Many composite materials, like concrete, are made up of granular inclusions embedded in a binding matrix. The aim is often to combine grains in order to minimize the porosity, which allows the use of the least possible amount of binder.

The packing density of a polydisperse grain mixture depends on three main parameters:

- the size of the grains considered (described by the grading curves);
- the shape of the grains;
- the method of processing the packing.

In the past, the design strategy has generally been to proportion the different grains to obtain a grading curve close to an 'ideal' grading curve, which is supposed to produce the maximum packing density. The final optimization is achieved by trial and error. Empirical models exist to describe the variation of the packing density of a given mixture with some parameters describing the compaction process. But few models were available (to the author's knowledge) that provided sufficient accuracy to design granular mixtures (accounting for the three parameters listed above). In particular, most available models (Johansen and Andersen, 1991; Dewar, 1993) either deal with a limited number of aggregate fractions, or assume a simplified grading distribution of each individual granular class. By contrast, the *compressible packing model* (CPM), the theoretical content of which is described in this chapter, covers combinations of any number of individual aggregate fractions, having any type of size distribution.

The CPM is a refined version of a previous model, the linear packing density model for grain mixtures (Stovall *et al.*, 1986; de Larrard, 1988), which was later transformed into the solid suspension model (de Larrard *et al.*, 1994a,b). This work has been carried out independently of another model, similar to the linear model, which has been proposed by Lee (1970). In section 1.1, the fundamentals of the linear model are presented. Here the virtual packing density is dealt with, defined as the maximum packing density attainable with a given grain mixture (with maximum compaction). From this virtual packing density the actual packing density is deduced, by reference to a value of a compaction index. The calculation of the actual packing density, together with a series of experimental validations, is given in section 1.2. Then the question of the influence of the boundary conditions is examined and modelled in section 1.3 (namely the wall effect exerted by the container, and the perturbation due to inclusion of fibres). In section 1.4 the model developed is used for the research of optimal grain size distributions. Finally, the limited knowledge of segregation is briefly reviewed in section 1.5, and the definition of two new concepts, namely the filling diagram and segregation potential, is proposed, with a view to characterizing mixtures that are likely to segregate.

1.1 VIRTUAL PACKING DENSITY OF A GRANULAR MIX

The *virtual packing density* is defined as the maximum packing density achievable with a given mixture, each particle keeping its original shape and being placed one by one. For instance, the virtual packing density of a mix of monosize spheres is equal to 0.74 (or $\pi/3\sqrt{2}$), the packing density of a face-centred cubic lattice of touching spheres, while the physical packing density that can be measured in a *random* mix is close to 0.60/0.64 (see e.g. Cumberland and Crawford, 1987), depending on the compaction. Throughout this section, the packing density dealt with is the virtual one.

1.1.1 Binary mix without interaction

Let us consider a mixture of grains 1 and 2, the diameters of which are d_1

and d_2. The two grain classes are said to be *without interaction* if

$$d_1 \gg d_2 \tag{1.3}$$

'Without interaction' means that the local arrangement of an assembly of grains of one size is not perturbed by the vicinity of a grain of the second size.

Let us calculate the (virtual) packing density of a mixture of these grains. First, we must know the result of the packing for each class on its own. The *residual* packing density of each class is denoted by β_1 and β_2 for grains 1 and 2 respectively. Second, the mutual volume fractions are y_1 and y_2, with

$$y_1 + y_2 = 1 \tag{1.4}$$

The *partial volumes* Φ_1 and Φ_2 are the volumes occupied by each class in a unit bulk volume of the granular mix. We have

$$y_i = \frac{\Phi_i}{\Phi_1 + \Phi_2} \tag{1.5}$$

and the packing density is

$$\gamma = \Phi_1 + \Phi_2 \tag{1.6}$$

We now make a distinction between two situations: *dominant coarse grains* and *dominant fine grains*.

When the coarse grains are dominant, they fill the available volume as if no fine grains were present (Fig. 1.1) so that

$$\Phi_1 = \beta_1 \tag{1.7}$$

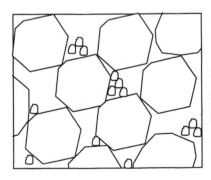

Figure 1.1 Binary mixture without interaction. Coarse grains dominant.

Therefore the packing density may be calculated as follows:

$$\gamma = \Phi_1 + \Phi_2 = \beta_1 + \Phi_2$$
$$= \beta_1 + (\Phi_1 + \Phi_2) y_2$$
$$= \beta_1 + \gamma y_2$$

then

$$\gamma = \gamma_1 = \frac{\beta_1}{1 - y_2} \tag{1.8}$$

When the small grains are dominant, they are fully packed in the porosity of the coarse grains (Fig. 1.2), so that

$$\Phi_2 = \beta_2(1 - \Phi_1) \tag{1.9}$$

With a similar approach, it is deduced that

$$\gamma = \gamma_2 = \frac{\beta_2}{1 - (1 - \beta_2)y_1} \tag{1.10}$$

Note that γ_1 and γ_2 may be calculated, whatever the dominant class. Hence in any case we can state

$$\gamma \leqslant \gamma_1$$

$$\Leftrightarrow \Phi_1 + \Phi_2 \leqslant \frac{\beta_1}{1 - \dfrac{\Phi_2}{\Phi_1 + \Phi_2}}$$

$$\Leftrightarrow \Phi_1 \leqslant \beta_1 \tag{1.11}$$

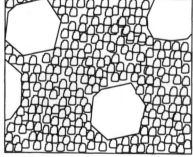

Figure 1.2 Binary mixture without interaction. Fine grains dominant.

in which we recognize equation (1.7). This last inequality is called the *impenetrability constraint* relative to class 1. Similarly, we always have

$$\gamma \leqslant \gamma_2$$

$$\Leftrightarrow \Phi_1 + \Phi_2 \leqslant \frac{\beta_2}{1 - (1 - \beta_2)\Phi_1/(\Phi_1 + \Phi_2)}$$

$$\Leftrightarrow \Phi_2 \leqslant \beta_2(1 - \Phi_1) \tag{1.12}$$

because of the validity of the impenetrability constraint relative to class 2. As either class 1 or 2 is dominant,[1] we can write, with no more concern about which is the dominant class:

$$\gamma = \text{Min}(\gamma_1, \gamma_2) \tag{1.13}$$

In Fig. 1.3 we can see the evolution of the packing density, from pure coarse-grain packing on the left-hand side to pure fine-grain packing on the right. Going from $y_2 = 0$ towards the peak, the packing density increases since a part of the coarse grain interstices is filled by fine grains. At the peak, the fine grains just fill all the space that is made available by the coarse grains. For larger values of y_2, coarse grains are replaced by an equivalent bulk volume of fine grains, decreasing the overall solid volume.

1.1.2   Binary mix with total interaction

Two grain populations are said to have total interaction when

$$d_1 = d_2 \tag{1.14}$$

while the β_i are generally different. For calculating the packing density of such a mix we state that total segregation does not change the mean

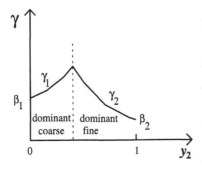

Figure 1.3 Evolution of the packing density vs. fine grain proportion, for a binary mix without interaction.

compactness, so that it is possible to consider that part of the container is filled with only class 1 grains, while the rest is filled with class 2 grains (Fig. 1.4). Therefore

$$\frac{\Phi_1}{\beta_1} + \frac{\Phi_2}{\beta_2} = 1 \tag{1.15}$$

The packing density is calculated by replacing in equation (1.6) one of the partial volumes by its expression as a function of the other one, taken from equation (1.15). In order to keep the same formalism as in section 1.1.1 we may write

$$\gamma_1 = \frac{\beta_1}{1 - (1 - \beta_1/\beta_2)y_2}$$
$$\gamma_2 = \frac{\beta_2}{1 - (1 - \beta_2/\beta_1)y_1} \tag{1.16}$$

so that equation (1.13) still applies. By using the relation $y_1 + y_2 = 1$, it is easy to see that in this particular case $\gamma = \gamma_1 = \gamma_2$ (Fig. 1.5).

1.1.3   Binary mix with partial interaction

We now consider a partial interaction between the classes, defined by the following inequality:

$$d_1 \geqslant d_2 \tag{1.17}$$

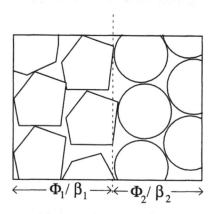

$$\longleftarrow \Phi_1/\beta_1 \longrightarrow\!\!\!\ast\!\!\!\longleftarrow \Phi_2/\beta_2 \longrightarrow$$

Figure 1.4 Calculation of the packing density in the case of total interaction.

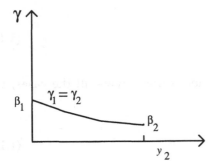

Figure 1.5 Evolution of the packing density vs. fine grain proportion, for a binary mix with total interaction.

We shall begin by describing two physical effects that can be found in binary mixes. Then we shall build general equations, incorporating these effects consistently with the two previous and ideal cases.

If a class 2 grain is inserted in the porosity of a coarse-grain packing (coarse grains dominant), and if it is no longer able to fit in a void, there is locally a decrease of volume of class 1 grains (loosening effect, Fig. 1.6). If each fine grain is sufficiently far from the next, this effect can be considered as a linear function of the volume of class 2 grains, so that

$$\gamma = \Phi_1 + \Phi_2 = \beta_1(1 - \lambda_{2 \to 1}\Phi_2) + \Phi_2$$
$$= \beta_1 + (\Phi_1 + \Phi_2)(1 - \beta_1\lambda_{2 \to 1})y_2$$
$$= \beta_1 + \gamma(1 - \beta_1\lambda_{2 \to 1})y_2$$

where $\lambda_{2 \to 1}$ is a constant that depends on the characteristics of both

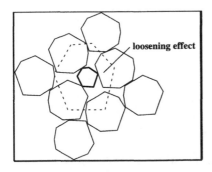

Figure 1.6 Loosening effect exerted by a fine grain in a coarse grain packing.

grain populations. The packing density is then

$$\gamma = \gamma_1 = \frac{\beta_1}{1 - (1 - \beta_1 \lambda_{2 \to 1}) y_2} \tag{1.18}$$

However, to adopt a continuous model that covers all the cases, we prefer the following form:

$$\gamma_1 = \frac{\beta_1}{1 - (1 - a_{12} \beta_1 / \beta_2) y_2} \tag{1.19}$$

where a_{12} is the *loosening effect coefficient*. When $d_1 \geqslant d_2$ (no interaction), $a_{12} = 0$, while when $d_1 = d_2$ (total interaction), $a_{12} = 1$.

When some isolated coarse grains are immersed in a sea of fine grains (which are dominant), there is a further amount of voids in the packing of class 2 grains located in the interface vicinity (wall effect, Fig. 1.7). If the coarse grains are sufficiently far from each other this loss of solid volume can be considered as proportional to $\Phi_1/(1 - \Phi_1)$, so that we can write:

$$\gamma = \Phi_1 + \Phi_2$$

$$= \Phi_1 + \beta_2 \left(1 - \frac{\Phi_1}{1 - \Phi_1} \lambda_{1 \to 2} \right) (1 - \Phi_1)$$

$$= \beta_2 + \gamma [1 - \beta_2 (1 + \lambda_{1 \to 2})] y_1$$

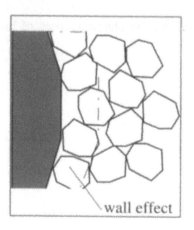

wall effect

Figure 1.7 Wall effect exerted by a coarse grain on a fine grain packing.

where $\lambda_{1 \to 2}$ is another constant, depending on the characteristics of both grain populations. The packing density is then

$$\gamma = \gamma_2 = \frac{\beta_2}{1 - [1 - \beta_2(1 + \lambda_{1 \to 2})]y_1} \tag{1.20}$$

which we prefer to write as follows:

$$\gamma_2 = \frac{\beta_2}{1 - [1 - \beta_2 + b_{21}\beta_2(1 - 1/\beta_1)]y_1} \tag{1.21}$$

b_{21} is the *wall effect coefficient*. When $d_1 \gg d_2$ (no interaction), $b_{21} = 0$, while when $d_1 = d_2$ (total interaction), $b_{21} = 1$.

As for the binary mix without interaction, it is easy to demonstrate that for any set of proportions y_i we have

$$\gamma = \text{Min}(\gamma_1, \gamma_2)$$

1.1.4 Polydisperse mix without interaction

Let us now consider a mix with n classes of grains ($n > 2$), with

$$d_1 \gg d_2 \ldots \gg d_n \tag{1.22}$$

Class i grains are dominant if

$$\Phi_i = \beta_i(1 - \Phi_1 - \Phi_2 - \ldots - \Phi_{i-1}) \tag{1.23}$$

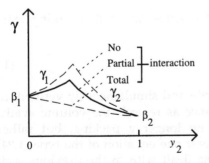

Figure 1.8 Evolution of the packing density of a binary mix vs. fine grain proportion. General case.

In that case, the packing density is calculated as follows:

$$\gamma = \sum_{j=1}^{n} \Phi_j$$

$$= \sum_{\substack{j=1 \\ j \neq i}}^{n} \Phi_j + \beta_i \left(1 - \sum_{j=1}^{i-1} \Phi_j \right)$$

$$= \beta_i + (1 - \beta_i) \sum_{j=1}^{i-1} \Phi_j + \sum_{j=i+1}^{n} \Phi_j$$

$$= \beta_i + \gamma \left[(1 - \beta_i) \sum_{j=1}^{i-1} y_j + \sum_{j=i+1}^{n} y_j \right]$$

Then

$$\gamma = \gamma_i = \frac{\beta_i}{1 - (1 - \beta_i) \sum_{j=1}^{i-1} y_j - \sum_{j=i+1}^{n} y_j} \tag{1.24}$$

Now, let us demonstrate that there is always at least one dominant class. If class 1 is not dominant, then we have

$$\Phi_1 < \beta_1 \tag{1.25}$$

Focusing on the interstitial medium of class 1 grains (that is, the packing of finer particles and the voids volume), if class 2 is not dominant, we can state

$$\Phi_2 < \beta_2(1 - \Phi_1) \tag{1.26}$$

By considering smaller and smaller scales, and still assuming that no class is dominant, we have finally:

$$\Phi_n < \beta_n(1 - \Phi_1 - \dots \Phi_{n-1}) \tag{1.27}$$

When these n inequalities are strictly and simultaneously verified, each class of grain has a certain clearance as regards the volume available (Fig. 1.9). Therefore the mix is no longer a packing, but rather a suspension. We conclude that at least one equation of the type (1.24) is verified. Otherwise, as for the case dealt with in the previous section dealing with binary mixes, the impenetrability constraint relative to the

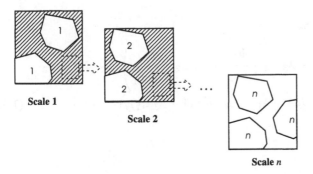

Figure 1.9 A polydisperse mix examined at various scales, where no class is dominant.

ith class is equivalent to

$$\gamma \leqslant \gamma_i \tag{1.28}$$

so that

$$\gamma = \underset{1 \leqslant i \leqslant n}{\text{Min}} \ \gamma_i \tag{1.29}$$

1.1.5   Polydisperse mix: general case

Let us first consider the case of a ternary mix, in which

$$d_1 \geqslant d_2 \geqslant d_3 \tag{1.30}$$

Let us assume that class 2 is dominant. Here, class 2 grains are submitted to a loosening effect exerted by class 3 grains, plus a wall effect exerted by class 1 grains (Fig. 1.10). Therefore the packing density

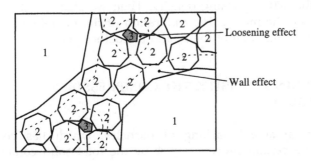

Figure 1.10 Perturbations exerted on the intermediate class by coarse and fine grains in a ternary mix.

of the mix is

$$\gamma = \Phi_1 + \Phi_2 + \Phi_3$$

$$= \Phi_1 + \beta_2 \left(1 - \lambda_{3\to2} \frac{\Phi_3}{1 - \Phi_1} - \lambda_{1\to2} \frac{\Phi_1}{1 - \Phi_1} \right)(1 - \Phi_1) + \Phi_3$$

$$= \beta_2 + \gamma[1 - \beta_2(1 + \lambda_{1\to2})]y_1 + \gamma(1 - \lambda_{3\to2})y_3 \qquad (1.31)$$

and

$$\gamma = \gamma_2$$

$$= \frac{\beta_2}{1 - [1 - \beta_2(1 + \lambda_{1\to2})]y_1 - (1 - \lambda_{3\to2})y_3}$$

$$= \frac{\beta_2}{1 - [1 - \beta_2 + b_{21}\beta_2(1 - 1/\beta_1)]y_1 - (1 - a_{23}\beta_2/\beta_3)y_3} \qquad (1.32)$$

The linear formulation ensures the additivity of all interactions suffered by one class. This derivation can be easily generalized for n classes of grains with interactions. Also, equation (1.29) still applies, and the equation defining the packing density when class i is dominant is

$$\gamma_i = \frac{\beta_i}{1 - \sum_{j=1}^{i-1} [1 - \beta_i + b_{ij}\beta_i(1 - 1/\beta_j)]y_j - \sum_{j=i+1}^{n} [1 - a_{ij}\beta_i/\beta_j]y_j} \qquad (1.33)$$

giving the most general formulation for the virtual packing density of a granular mix.

As the virtual packing density is, by definition, non-accessible by experiments, we are going first to continue with the theory to show how we can calculate the actual packing density. Then we shall be able to calibrate and validate the model. The calibration of the model consists essentially in the determination of the a_{ij} and b_{ij} coefficients.

1.2   ACTUAL PACKING DENSITY: THE COMPRESSIBLE PACKING MODEL

We now consider an actual packing of particles, which has been physically built by a certain process. Let Φ be the total solid content, with $\Phi < \gamma$.

1.2.1 Compaction index and actual packing density

We are looking for a scalar index K, which would take a value that depends only on the process of building the packing. By analogy with some viscosity models (Mooney, 1951), we assume that this index is of the following form:

$$K = \sum_{i=1}^{n} K_i \tag{1.34}$$

with $K_i = H\left(\dfrac{\Phi_i}{\Phi_i^*}\right)$

where Φ_i is the actual solid volume of class i, while Φ_i^* is the maximum volume that particles i may occupy, given the presence of the other particles. In other words, the n classes of grain with partial volumes equal to $\Phi_0, \Phi_1, ..., \Phi_{i-1}, \Phi_i^*, \Phi_{i+1}, ..., \Phi_n$ would form a virtual packing.

Function H can be calculated by considering only the self-consistency of the model. Let us deal with a binary mix, the two classes of which are identical (that is, $d_1 = d_2$; $\beta_1 = \beta_2 = \beta$). The only impenetrability constraint is

$$\Phi_1 + \Phi_2 \leqslant \beta \tag{1.35}$$

For calculating the compaction index of the mix, we can write

$$K = H\left(\frac{\Phi_1}{\beta - \Phi_2}\right) + H\left(\frac{\Phi_2}{\beta - \Phi_1}\right) = H\left(\frac{\Phi_1 + \Phi_2}{\beta}\right) \tag{1.36}$$

which corresponds to the following functional equation:

$$H\left(\frac{x}{1-y}\right) + H\left(\frac{y}{1-x}\right) = H(x+y) \tag{1.37}$$

with $x = \dfrac{\Phi_1}{\beta}$; $y = \dfrac{\Phi_2}{\beta}$

Let us show that the only functions for which equation (1.37) is verified are of the form

$$H(u) = k\,\frac{u}{1-u} \tag{1.38}$$

Define the function $k(u)$ such that

$$H(u) = k(u) \left(\frac{u}{1-u} \right) \tag{1.39}$$

Replacing H in the functional equation (1.37), it becomes

$$(x+y) \cdot k(x+y) = x \cdot k \left(\frac{x}{1-y} \right) + y \cdot k \left(\frac{y}{1-x} \right) \tag{1.40}$$

If $x = y = u/2$ then

$$k(u) = k \left(\frac{u/2}{1-u/2} \right) = k \left[\frac{u/4}{1 - \left(\frac{1}{2} + \frac{1}{4} \right) u} \right]$$

$$= \dots k \left[\frac{u/2^n}{1 - \left(1 - \frac{1}{2^n} \right) u} \right] = \dots k(0) \tag{1.41}$$

Therefore $k(u)$ is a constant for $u \in [0, 1[$, which is the domain of variation of Φ_i/Φ_i^*. For the sake of simplicity, we take

$$k = 1 \tag{1.42}$$

Thus the compaction index becomes

$$K = \sum_{i=1}^{n} \left(\frac{\Phi_i/\Phi_i^*}{1 - \Phi_i/\Phi_i^*} \right) \tag{1.43}$$

Φ_i^* is equal to Φ_i when class i is dominant, so that we can use the same approach as in equation (1.31), replacing Φ_i^* by the expression

(generalized to n classes, i is dominant)

$$\Phi_i^* = \beta_i \left[1 - \sum_{j=1}^{i-1} \left(1 - b_{ij} \left[1 - \frac{1}{\beta_j} \right] \right) \Phi_j - \sum_{j=i+1}^{n} \frac{a_{ij}}{\beta_j} \Phi_j \right] \qquad (1.44)$$

which gives for K:

$$K = \sum_{i=1}^{n} \frac{y_i/\beta_i}{\frac{1}{\Phi} - \left\{ \sum_{j=1}^{i-1} \left[1 - b_{ij} \left(1 - \frac{1}{\beta_j} \right) \right] y_j - \sum_{j=i+1}^{n} \frac{a_{ij}}{\beta_j} y_j + \frac{y_i}{\beta_i} \right\}} \qquad (1.45)$$

However, replacing 1 by $\sum_{i=1}^{n} y_i$ in equation (1.33), we recognize that

$$\sum_{j=1}^{i-1} \left[1 - b_{ij} \left(1 - \frac{1}{\beta_j} \right) \right] y_j - \sum_{j=i+1}^{n} \frac{a_{ij}}{\beta_j} y_j + \frac{y_i}{\beta_i} = \frac{1}{\gamma_i} \qquad (1.46)$$

so we reach the final expression for the compaction index

$$K = \sum_{i=1}^{n} K_i = \sum_{i=1}^{n} \frac{y_i/\beta_i}{1/\Phi - 1/\gamma_i} \qquad (1.47)$$

As K is a characteristic of the packing process, the packing density is then the value of Φ defined implicitly by equation (1.47). Actually, K is a strictly increasing function of Φ, as the sum of such functions, so that there is a unique value of Φ satisfying this equation for any positive K value (Fig. 1.11).

The y_i are the control parameters of the experiment, β_i are characteristics of the grain classes, the γ_i are given by equation (1.33), and the value of K depends on the process of making the mixture. For a

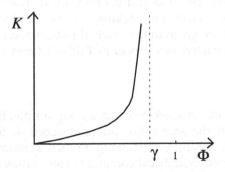

Figure 1.11 Variation of K vs. Φ.

monodisperse packing, it follows that

$$K = \frac{1}{\beta/\Phi - 1}$$

(1.48)

1.2.2 Calibration of the model with binary data

Tests related in this section have been carried out at the Laboratoire Régional des Ponts et Chaussées de Blois (de Larrard *et al.*, 1994b). The aim was first to prepare a series of binary mixes, for calibration of the interaction coefficients. The individual fractions were selected in order to be as monodisperse as possible.

Materials

Two families of aggregate were selected and sieved:

- rounded aggregate from the Loire (Decize quarry), with nearly spherical shapes;
- crushed angular aggregate from the Pont de Colonne quarry at Arnay le Duc.

These materials were expected to cover the range of civil engineering materials, from smooth quasi-spherical grains (such as those of fly ash) to angular, flat and elongated ones (such as certain crushed aggregate).

For each family, five monosize classes were prepared, limited by two adjacent sieves in the Renard series. This French standard series has diameters in geometrical progression, with a ratio of $\sqrt[10]{10}$ ($\approx$1.26). The mean sizes of the classes were chosen for obtaining size ratios of 1/2, 1/4, 1/8 and 1/16. Extreme diameters were limited on the high side, for container dimension purposes, and on the low side, to avoid materials that were too humidity sensitive. For polydisperse mixtures (see section 1.2.3) a sixth class was added, ranging between 0.08 and 0.5 mm, for increasing both the grading span and the maximum packing values obtained. Photographs of the particles used are given in the original publication (de Larrard *et al.*, 1994b), and their characteristics appear in Tables 1.1 and 1.2.

Packing process used

After weighing the granular classes, selected to obtain a 7 kg sample, the mixes were homogenized. When the size ratio did not exceed 4, the grains were poured in a Deval machine (following French standard P 18 577), equipped with a modified cylindrical container. This container was then rotated around an oblique axis (with regard to the symmetry

Table 1.1 Characteristics of rounded aggregates used in packing experiments. As the same grains were used for a series of experiments, a certain attrition took place, increasing the β_i values. Corrected packing density values have been obtained by linear regressions in the binary mixes.

Names	d_{min} (mm)	d_{max} (mm)	Packing density	Corrected packing density
R < 05	0.08	0.5	0.593	–
R05	0.5	0.63	0.592	0.594
R1	1	1.25	0.609	0.613
R2	2	2.5	0.616	0.620
R4	4	5	0.6195	0.629
R8	8	10	0.628	0.632

Table 1.2 Characteristics of crushed aggregate used in the packing experiments.

Names	d_{min} (mm)	d_{max} (mm)	Packing density	Corrected packing density
C < 05	0.08	0.5	0.630	–
C05	0.5	0.63	0.516	0.523
C1	1	1.25	0.507	0.528
C2	2	2.5	0.529	0.525
C4	4	5	0.537	0.557
C8	8	10	0.572	0.585

axis), for 2 min or 66 revolutions. The cylinder has a diameter of 160 mm and a height of 320 mm, and supported another 160 × 160 mm cylinder, used for pouring the uncompacted mixture. After removal, the container served for measuring the packing density.

For the other binary mixes that were prone to segregation, a manual homogenization was carried out: aggregates were poured after mixing in horizontal layers, and then removed vertically with a shovel and cast by successive layers in the cylindrical mould. Both techniques of preparation were used for polydisperse mixtures.

Then the cylinder containing aggregates was closed with a 20 kg steel piston, applying a mean compression of 10 kPa on the top of the sample. The whole set was put on a vibrating table and submitted to the following vibration sequence: 2 min at a 0.4 mm amplitude, 40 s at 0.2 mm and 1 min at 0.08 mm. The height of the sample was recorded continuously by an ultrasound telemeter having a precision of 0.001 mm. The vibration process was fixed for having a comparable response with all granular classes, while keeping the total process within a reasonable time. Hence no definite stabilization of the height appears in such a process, which confirms that actual packing density is not a material property, but rather depends on the mixture *and* the process. The packing density of the mixture was calculated by dividing the mass of the sample by the mean density of the aggregate and by the total volume of the specimen. Each experimental value obtained for the model was the mean of two successive measurements carried out on the same aggregate sample. Between the two measurements, the cylinder was emptied and reconstructed using the entire process.

Packing density of monosize classes

Actual packing densities are given in Tables 1.1 and 1.2. For the two families, they increase with the diameter of the particles (Fig. 1.12). This could be due to differences in the shape of the grains, but is mostly due to the fact that in spite of our efforts, vibration is more efficient in compacting coarse grains than fine ones, because fewer contacts are present per unit volume of the mix. At equal size, rounded aggregates are more compact than crushed ones. Incidentally, no general law seem to govern the relationship between size and packing density of aggregates.

Calibration of the model

The interaction coefficients, a_{ij} and b_{ij}, were expected to depend mainly on the ratio between the particle diameters d_i and d_j. However, to assess the soundness of this hypothesis, an attempt has been made to duplicate, when possible, the binary series of a given size ratio, for both coarse grain and fine grain. For the smallest ratio (1/16), only a single series could be produced. For each class combination, the variation of packing density was expected to be steeper on the dominant coarse grain side than on the dominant fine grain side (Fig. 1.8). This is why the fine grain proportion was incremented by 5% steps between 0 and 30%, and by 10% steps afterwards. Obtained packing density values are summarized in Tables 1.3 and 1.4, and Figs 1.17 and 1.18. Here we note that, while the size ratio

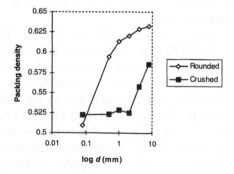

Figure 1.12 Experimental packing density of monodisperse classes vs. size of particles.

Table 1.3 Packing densities of binary mixtures: rounded grains.

% fine	R8R05 exp	R8R05 theo	R8R1 exp	R8R1 theo	R4R05 exp	R4R05 theo	R8R2 exp	R8R2 theo	R2R05 exp	R2R05 theo	R8R4 exp	R8R4 theo	R1R05 exp	R1R05 theo
0	0.628	0.632	0.628	0.632	0.6195	0.629	0.628	0.632	0.616	0.62	0.628	0.632	0.609	0.613
5	0.657	0.6557	0.6545	0.6526	0.645	0.6492	0.653	0.6481	0.635	0.6354	0.6375	0.6416	0.624	0.6217
10	0.6865	0.6808	0.6795	0.6743	0.6715	0.6703	0.682	0.6646	0.663	0.6512	0.643	0.6511	0.633	0.6304
15	0.71	0.7069	0.707	0.6966	0.689	0.692	0.697	0.6813	0.678	0.6671	0.654	0.6603	0.64	0.6386
20	0.729	0.733	0.724	0.7189	0.706	0.7136	0.714	0.6976	0.692	0.6826	0.66	0.6689	0.656	0.6462
25	0.754	0.7558	0.742	0.7391	0.7265	0.7326	0.7235	0.7122	0.708	0.6964	0.663	0.6762	0.666	0.6526
30	0.758	0.7677	0.748	0.7528	0.7485	0.7446	0.728	0.7229	0.708	0.7063	0.6595	0.6817	0.6705	0.6572
40	0.753	0.7544	0.7285	0.7496	0.736	0.7379	0.723	0.7251	0.693	0.7067	0.6565	0.6853	0.6635	0.659
50	0.7385	0.7256	0.7095	0.7277	0.725	0.7138	0.705	0.7111	0.67	0.6909	0.6535	0.6806	0.6545	0.6525
60	0.7165	0.6959	0.6965	0.7029	0.7	0.6875	0.689	0.6927	0.656	0.6709	0.649	0.6719	0.644	0.6421
70	0.68	0.6677	0.677	0.6786	0.6745	0.662	0.671	0.6737	0.633	0.6506	0.6445	0.6616	0.636	0.6303
80	0.652	0.6414	0.6585	0.6554	0.648	0.6379	0.646	0.6551	0.613	0.6308	0.638	0.6508	0.6215	0.6181
90	0.6195	0.6168	0.635	0.6336	0.614	0.6152	0.632	0.6371		0.6119	0.629	0.6398	0.61	0.6059
100	0.592	0.594	0.609	0.613	0.592	0.594	0.616	0.62	0.592	0.594	0.6195	0.629	0.592	0.594

Table 1.4 Packing densities of binary mixtures: crushed grains.

% fine	C8C05 exp	C8C05 theo	C8C1 exp	C8C1 theo	C4C05 exp	C4C05 theo	C8C2 exp	C8C2 theo	C2C05 exp	C2C05 theo	C8C4 exp	C8C4 theo	C1C05 exp	C1C05 theo
0	0.572	0.585	0.572	0.585	0.537	0.557	0.572	0.585	0.529	0.525	0.572	0.585	0.507	0.528
5	0.62	0.6066	0.613	0.6034	0.591	0.575	0.597	0.5986	0.54	0.5388	0.5825	0.5931	0.527	0.5362
10	0.642	0.6295	0.646	0.6226	0.6185	0.594	0.611	0.6125	0.552	0.5531	0.5875	0.6011	0.532	0.5444
15	0.676	0.6535	0.6755	0.6425	0.638	0.6137	0.625	0.6264	0.5515	0.5679	0.588	0.6087	0.545	0.5524
20	0.705	0.6779	0.699	0.6624	0.669	0.634	0.634	0.6398	0.566	0.583	0.592	0.6158	0.552	0.5602
25	0.731	0.7001	0.7215	0.6806	0.693	0.6536	0.643	0.6516	0.573	0.5979	0.5955	0.6217	0.5485	0.5673
30	0.7365	0.7137	0.7245	0.693	0.711	0.67	0.651	0.6594	0.594	0.6115	0.594	0.6259	0.555	0.5733
40	0.723	0.6998	0.7025	0.6861	0.691	0.6741	0.643	0.6554	0.588	0.626	0.5875	0.6271	0.556	0.5792
50	0.6941	0.6666	0.6705	0.6586	0.667	0.6502	0.6335	0.6349	0.582	0.617	0.587	0.6198	0.549	0.576
60	0.6585	0.6331	0.638	0.629	0.64	0.6219	0.6245	0.6111	0.579	0.5987	0.587	0.6084	0.546	0.5674
70	0.616	0.6019	0.611	0.6008	0.603	0.5945	0.5975	0.5877	0.568	0.5788	0.572	0.5956	0.5425	0.5567
80	0.583	0.5732	0.5965	0.5746	0.571	0.5688	0.5695	0.5654	0.5555	0.5593	0.564	0.5825	0.537	0.5455
90	0.5655	0.547	0.5435	0.5504	0.545	0.545	0.5435	0.5445	0.534	0.5406	0.553	0.5696	0.53	0.5341
100	0.516	0.523	0.507	0.528	0.516	0.523	0.529	0.525	0.516	0.523	0.537	0.557	0.516	0.523

appears to be the main parameter controlling the behaviour of the binary mix, significant differences appear for pairs of equal size ratio.

Each binary mixture series displays one experimental point for the two coefficients a and b. The relationship between the voids index and the fine grain proportions appears to be two straight lines linked by a curved part. The slope of these lines directly expresses the granular interaction between classes (Powers, 1968) (Fig. 1.13).

In the case of an infinite value for the compaction index (that is, when dealing with virtual packing density), one can easily show using equation (1.33) that

$$
a_{12} = \beta_2 \left(\left| \frac{\partial e}{\partial y_2} \right|_{y_2 = 0} + \frac{1}{\beta_1} \right)
$$

$$
b_{21} = \frac{1/\beta_2 - 1 - \left| \dfrac{\partial e}{\partial y_2} \right|_{y_2 = 1}}{1/\beta_1 - 1}
$$

(1.49)

Actually, for sufficiently high values of K, the real packing density curves have approximately the same tangent as the virtual one (Fig. 1.14). One may therefore apply the previous equation to the present data, replacing the β_i by the α_i (actual residual packing densities of monodisperse classes). Obtained values for a and b are given in Table 1.5.

In Fig. 1.16, the experimental points for a and b have been plotted against the size ratios. These coefficients increase with the ratio d_2/d_1, which matches the theory. At first sight, other parameters should also play a role. However, no systematic trend appears between rounded and crushed aggregate, or between coarse and fine pairs (as shown by the

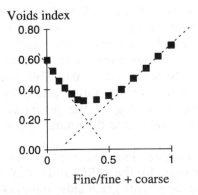

Figure 1.13 Voids index vs. fine grain proportion, for the R8/R01 mixture.

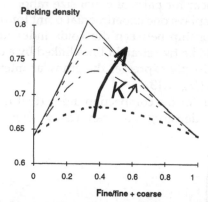

Figure 1.14 Packing density of binary mix of grains with a size ratio of 1/8 after the compressible packing model. Actual residual packing densities of the two classes are assumed to be equal to 0.64, and the different curves stand for low to high **K** values.

Table 1.5 Experimental values for the interaction functions deducted from binary mix experiments.

	d_1/d_2						
	16	8	8	4	4	2	2
Rounded	R8R05	R8R1	R4R05	R8R2	R2R05	R8R4	R1R05
a	0.26	0.31	0.50	0.63	0.46	0.71	0.66
b	−0.05	0.30	0.08	0.33	0.33	0.65	0.56
Crushed	C8C05	C8C1	C4C05	C8C2	C2C05	C8C4	C1C05
a	0.21	0.14	0.31	0.50	0.67	0.77	0.72
b	−0.03	0.28	0.05	0.11	0.47	0.68	0.70

comparison of the R8R4 and R1R05, or C8C4 and C1C05 mix series. However, the sieves used in civil engineering often suffer some flaws (either initial ones, or those that result from the wear exerted by aggregates). Thus the actual ratio of a fraction couple with a theoretical ratio of 1/2 may probably range between 0.4 and 0.6. This would correspond to a horizontal shifting of the corresponding points in the figures. One may conclude that the scattering of the points is not necessarily the sign of another influence, related to the grains' shape, for example.

This hypothesis of inaccurate sieve size does not explain the two *negative* values of the b coefficient, obtained for the R8R05 and C8C05 mix series. For those couples, on the fine dominant grain side, the packing density evolves faster than in the case without interaction. In other terms, the fine grains suffered an *anti-wall effect*. This observation is related to the effect of vibration on compaction. Coarse grains probably act as internal vibrators in the mixture of fine grains (Aïtcin and Albinger, 1989). In smoothing the results, this phenomenon has been neglected, as the model is expected to apply in general, including non-vibrated mixtures (see section 1.2.3).

The regression function used to fit the experimental values has to satisfy the following conditions:

- Continuity with the case of binary mixture without interaction $(d_2/d_1 = 0)$: $a = b = 0$.
- Continuity with the case of binary mixture with total interaction $(d_2/d_1 = 1)$: $a = b = 1$.
- Moreover, if one considers the case of a binary mixture in which y_1 is small, d_2 is fixed, and d_1 varies around d_2 (Fig. 1.15), one should have

when $d_1 \geqslant d_2$:

$$\gamma = \frac{\beta_2}{1 - (1 - \beta_2 + b_{21}\beta_2(1 - 1/\beta_1)y_1)} \tag{1.50}$$

when $d_1 \leqslant d_2$:

$$\gamma = \frac{\beta_2}{1 - (1 - a_{21}\beta_2/\beta_1)y_1} \tag{1.51}$$

and the derived function should be also continuous in $d_2 = d_1 = d$,

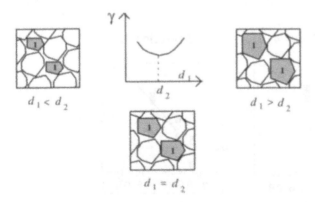

Figure 1.15 Continuity between dominant coarse grain and dominant fine grain cases.

which gives

$$-\left(1 - \frac{1}{\beta_1}\right)\left[\frac{\partial b}{\partial x}\right]_{x=1} + \frac{1}{\beta_1}\left[\frac{\partial a}{\partial x}\right]_{x=1} = 0 \tag{1.52}$$

where x is the ratio of fine to coarse grain diameters.

As the two factors in this equation are positive, this implies for the derived values of a and b:

$$\left[\frac{\partial a}{\partial x}\right]_{x=1} = \left[\frac{\partial b}{\partial x}\right]_{x=1} = 0 \tag{1.53}$$

The following functions verify the three listed conditions, while giving a reasonable approximation of the experimental points:

$$\begin{aligned} a_{ij} &= \sqrt{1 - (1 - d_j/d_i)^{1.02}} \\ b_{ij} &= 1 - (1 - d_i/d_j)^{1.50} \end{aligned} \tag{1.54}$$

For the calibration of the model, the compaction index remains to be fixed. As shown in Fig. 1.14, the higher the value of K, the sharper the binary curves. With the above interaction functions and a K value of 9, the experimental values of packing density obtained are best smoothed with a mean error in absolute value equal to 0.77% for the rounded grains and 1.71% for the crushed grains (Figs 1.17, 1.18). By comparison, the means of within-test standard deviations, describing the repeatability of the measurements, are 0.0026 and 0.0078 for rounded and crushed aggregates respectively. Therefore one may estimate that the better predictions

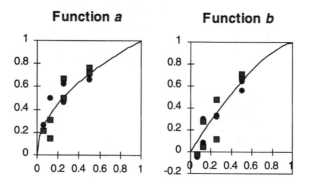

Figure 1.16 Fitting of interaction functions a or b vs. size ratio. Squares and circles stand for crushed and rounded aggregates respectively.

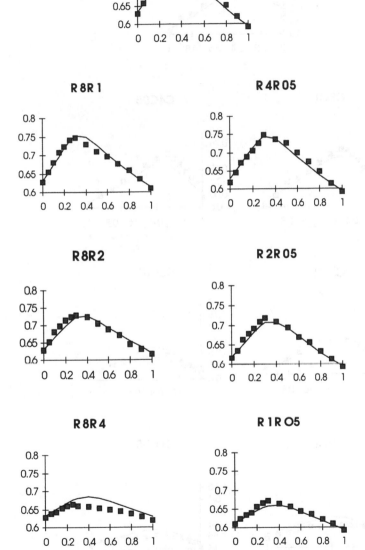

Figure 1.17 Binary mixes of rounded grains. Packing density vs. fine grain proportion. The dots stand for experimental points, while the curves are the model smoothing.

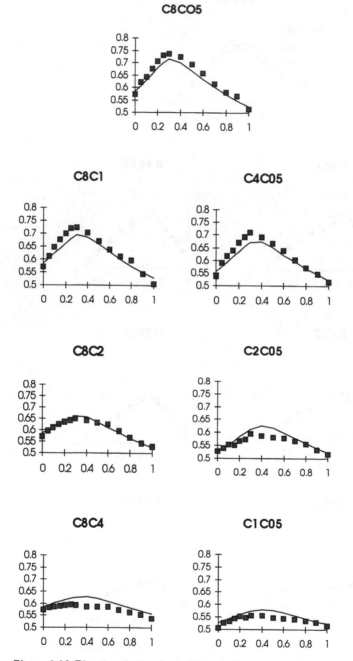

Figure 1.18 Binary mixes of crushed grains. Packing density vs. fine grain proportion.

obtained for the rounded aggregates come at least partially from the better soundness of the aggregates. In fact the crushed aggregates exhibited significant wear and dust production during the tests.

Lecomte and Zennir (1997) carried out a similar experimental programme with different fractions of limestone. They obtained similar values for the interaction coefficients. From their experimental results they reached interaction functions slightly different from ours. However, we did not consider their data because the soft nature of their aggregate may have impaired the precision of their tests.

1.2.3 Validation with data of various origins

In this section the compressible packing model is validated by comparison between experimental results and predictions of the theory. Both original data and literature data are used for this purpose. For the latter case, since the process of packing is different, the compaction index K is adjusted to obtain the best fit. Finally, the values of K found for each process are summarized.

Preliminary binary mixtures (from Laboratoire Régional de Blois)

Previously to the binary mixtures already dealt with in section 1.2.2, mixes were prepared to optimize the process of packing density measurement (Cintré, 1988). For these data (Table 1.6 and Fig. 1.19) the process of mixing and pouring was the one adopted for the calibration measurements. For each family (namely rounded and crushed aggregates) four binary mixture series were prepared with three aggregate fractions. For each tested mix, a first measurement was taken prior to vibration. Then a vibration process was applied, slightly different than the one for our calibration measurements (one minute more for the first sequence). After vibration, a second measurement was performed, characterizing the compacted sample. The latter measurements were simulated with the model, with the same value of the K index (that is, 9), although the vibration time was slightly longer than in the calibration measurements. Except for some erratic points for the R8R05 mix, which could be attributed to segregation, the agreement between model and experiments is quite satisfactory, including the crushed aggregates (mean error in absolute value equal to 1.03%).

For the non-compacted mixes, the same β_i values were kept, and a lower K value was adopted (equal to 4.1). The accuracy of the prediction remains reasonable (mean error in absolute value equal to 1.65%), which shows that the model may apply to loosely compacted mixtures. However, the precision is lower because of the higher scatter of the measurements.

Table 1.6 Binary mixture measurements performed by Cintré (1988). Experimental data and predictions of the model.

	% fine	Vibrated		Non vibrated	
		C_{theo}	C_{exp}	C_{theo}	C_{exp}
C8C2	0	0.593	0.593	–	–
	10	0.6225	0.633	0.5541	0.535
	20	0.6524	0.66	0.577	0.542
	25	0.6658	0.692	–	–
	30	0.6758	0.665	0.5935	0.546
	40	0.6769	0.66	0.5975	0.544
	50	0.66	0.65	0.5887	0.5415
	60	0.6386	0.638	0.5728	0.542
	70	0.6171	0.62	0.5542	0.536
	80	0.5963	0.601	0.535	0.522
	90	0.5766	0.588	0.5163	0.5085
	100	0.558	0.558	–	–
C8C05	0	0.593	0.593	–	–
	10	0.6392	0.651	0.5679	0.557
	20	0.6899	0.701	0.6066	0.612
	30	0.7312	0.717	0.6361	0.649
	35	0.7339	0.739	0.6419	0.638
	40	0.7248	0.721	0.6405	0.66
	50	0.6955	0.71	0.6231	0.631
	60	0.6645	0.682	0.5979	0.604
	70	0.635	0.648	0.5713	0.571
	80	0.6077	0.615	0.5457	0.535
	90	0.5824	0.577	0.5217	0.508
	100	0.559	0.559	–	–
R8R2	0	0.637	0.637	0.569	0.584
	10	0.6689	0.676	0.5952	0.58
	20	0.6991	0.705	0.6194	0.598
	30	0.7233	0.715	0.6362	0.607
	40	0.722	0.714	0.64	0.607
	50	0.7058	0.697	0.6316	0.61
	60	0.6861	0.676	0.6165	0.599
	70	0.666	0.661	0.5988	0.59
	80	0.6465	0.641	0.5805	0.572
	90	0.6278	0.623	0.5624	0.5525
	100	0.61	0.61	0.5449	0.544
R8R05	0	0.637	0.637	0.569	0.583
	10	0.6861	0.676	0.6106	0.611
	20	0.7425	0.707	0.6528	0.66
	30	0.7849	0.719	0.6854	0.709
	40	0.7739	0.721	0.6933	0.728
	50	0.7538	0.742	0.6803	0.699
	60	0.7294	0.714	0.6591	0.672
	70	0.7047	0.7	0.6357	0.645
	80	0.6809	0.674	0.6125	0.619
	90	0.6583	0.656	0.5902	0.593
	100	0.637	0.637	0.569	0.567

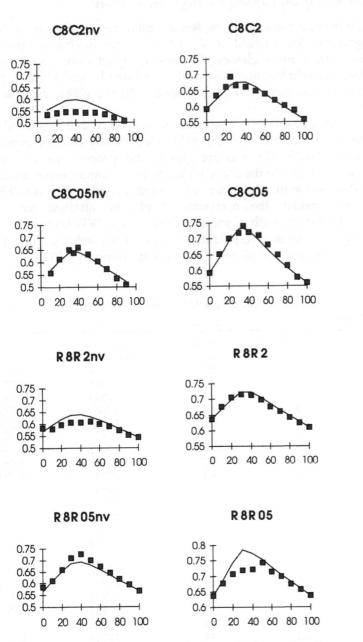

Figure 1.19 Cintré's binary mixtures. Packing density vs. fine grain proportion. nv = means non-vibrated.

Ternary mixtures (from Laboratoire Régional de Blois)

After the calibration measurements, ternary mixtures were tested in the same programme (de Larrard *et al.*, 1994b). After having arbitrarily selected the <05, 1 and 8 classes, the most compact combination was experimentally sought by successive trials (see Table 1.7 and 1.8, and Fig. 1.20). These measurements were simulated with the CPM, using a K value of 9.

For the rounded grain series, the mean error is equal to 0.76%, a value comparable to that obtained in the calibration binary mixture experiments. The 0.22/0.25/0.53 mixture (respective proportions of each fraction, from the finest to the coarsest) led to the optimum experimental value of 0.7945, while the theoretical value of this combination is 0.7897. Following the model, the optimum should be attained for the 0.20/0.19/0.61 mixture, with a packing density of 0.7959 (experimental value obtained by linear interpolation: 0.7764). Then, relying upon the model would have led one to 'miss' the real optimum by 0.0179.

Table 1.7 Packing density measurements for rounded grains, and predictions of the model.

% R < 05	% R1	% R8	C_{theo}	C_{exp}
25	0	75	0.7624	0.752
30	0	70	0.7746	0.753
22	10	68	0.7865	0.7715
27	10	63	0.7883	0.777
32	10	58	0.7785	0.764
37	10	53	0.764	0.756
42	10	48	0.7481	0.738
12	20	68	0.7814	0.767
17	20	63	0.7947	0.7745
22	20	58	0.7946	0.779
27	20	53	0.7849	0.7815
32	20	48	0.7708	0.7735
37	20	43	0.755	0.76
42	20	38	0.7388	0.7455
17	25	58	0.7936	0.791
22	25	53	0.7897	0.7945
27	25	48	0.779	0.7845
12	30	58	0.7842	0.781
17	30	53	0.7862	0.7825
22	30	48	0.7816	0.7835
27	30	43	0.7714	0.776
32	30	38	0.758	0.761
37	30	33	0.7431	0.7465
39	31.5	29.5	0.7351	0.7355
17	35	48	0.7755	0.7715
22	35	43	0.7714	0.769

Table 1.8 Packing density measurements for crushed grains, and predictions of the model.

% C < 05	% C1	% C8	C_{theo}	C_{exp}
22	10	68	0.7553	0.7748
27	10	63	0.7784	0.7755
32	10	58	0.7862	0.7973
37	10	53	0.7811	0.7897
42	10	48	0.7702	0.7871
12	20	68	0.7341	0.7335
17	20	63	0.7607	0.744
22	20	58	0.7789	0.7522
27	20	53	0.7854	0.7765
32	20	48	0.7817	0.7876
37	20	43	0.7723	0.7862
42	20	38	0.7603	0.7812
17	30	53	0.7532	0.7698
22	30	48	0.7615	0.769
27	30	43	0.7646	0.7745
32	30	38	0.7623	0.7784
37	30	33	0.7553	0.7691
42	30	28	0.7453	0.7654
32	15	53	0.7864	0.7993

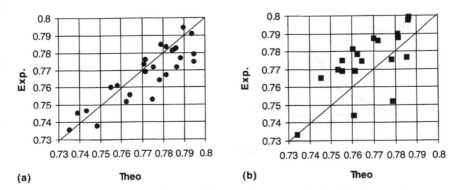

Figure 1.20 Comparison between predictions and experiments for the ternary mixtures: (a) rounded grains; (b) crushed grains.

As for crushed aggregates, the mean error is equal to 1.31%, still giving the same trends as in the calibration experiments. Here the experimental optimum is the 0.32/0.15/0.53 mixture, with an experimental packing density of 0.7993 (theoretical value 0.7854). The theoretical optimum is obtained with the proportions of 0.29/0.15/0.56, which give a theoretical packing density of 0.7879, and an experimental one of 0.7826 (calculated by linear interpolation). Here the error is equal to 0.0167. Note that these calculations are pessimistic: because of the concave shape of the packing density surface, the linear interpolations are less than reality. Moreover, these series of measurements, as the following, suffer from the fact that the real gradings of the finest aggregate fractions (R<05 and C<05), which were wider than the coarser ones, have unfortunately not been measured.

Dreux-type gradings (from Laboratoire Régional de Blois)

One of the most popular French concrete mix design methods (Dreux, 1970) is based upon ideal grading curves of the type given in Fig. 1.21. The A coefficient (proportion of grains finer than half the maximum size) is tabulated for different types of aggregate and compaction energies. In other words, Dreux's main statement is to consider that an ideal aggregate would be a binary mixture of two basic fractions (coarse and fine), each one being evenly distributed in a logarithmic diagram. No theory was referred to for this statement. Dreux only collected real grading curves of 'fair' concretes that had been used on site, and smoothed those curves using this model.

Such grading curves were reconstituted in the experimental programme, and the optimal value of A was sought experimentally for each aggregate family (see Table 1.9 and Fig. 1.22). The agreement between

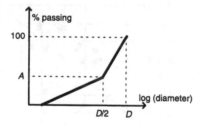

Figure 1.21 Ideal grading curve (after Dreux).

Table 1.9 Dreux-type continuous gradings. Theoretical and experimental packing densities.

A	$X < 05$ (%)	$X05$ (%)	$X1$ (%)	$X2$ (%)	$X4$ (%)	$X8$ (%)	Rounded		Crushed	
							C_{theo}	C_{exp}	C_{theo}	C_{exp}
0.3	13.4	1.9	4.7	5	5	70.0	0.7641	0.766	0.7142	0.7312
0.35	15.5	2.2	5.5	5.9	5.9	65.0	0.7839	0.7865	0.7379	0.7543
0.4	17.8	2.5	6.3	6.7	6.7	60.0	0.7969	0.792	0.7593	0.7643
0.45	20.0	2.9	7.1	7.5	7.5	55.0	0.8016	0.8015	0.7765	0.7713
0.5	22.1	3.17	7.9	8.4	8.4	50.0	0.7996	0.792	0.7866	0.7673

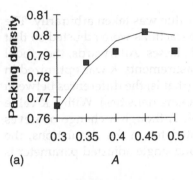 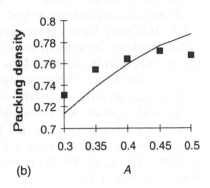

Figure 1.22 Effect of A (proportion of grains finer than half the maximum size) on the packing density, in Dreux-type graded aggregates: (a) rounded; (b) crushed.

model and experience is excellent for the rounded aggregates (mean error of 0.34%, same optimum), and fair for the crushed ones (mean error of 1.26%).

Joisel's data

With the view of building a concrete mix-design method, Joisel (1952) carried out packing density measurements on dry mixtures. Thus he provided an outstanding set of numerous and consistent experimental data. From a unique rounded aggregate source, six fractions were sieved, with limit sizes forming a geometrical progression with a ratio of 2. The extreme sizes were 0.2 and 12 mm, and the fractions were called a–f, from the finest to the coarsest. The samples were manually mixed in a plate, then poured in a $14 \times 14 \times 14$ cm cube, submitted to vibration (without pressure), trowelled and weighed.

Looking for an ideal distribution, Joisel assumed that a succession of optimum mixtures, mixed together, should eventually produce this ideal distribution. Joisel first investigated elementary binary mixtures x–y, from which he found the optimum, called xy. A second generation of binary mixtures xy–zt (which were in fact quaternary mixtures) led to a local optimum called xy,zt. At a third step, new mixtures of the type $x_1 y_1, z_1 t_1 - x_2 y_2, z_2 t_2$ were produced, leading to a grading curve assumed to be close to the optimal one. The whole strategy was followed from different combinations of basic fractions. Finally, a mean of all third-generation gradings was supposed to give *the* optimal grading. Strangely enough, the packing density of this final grading is not provided in the original publication. Nevertheless, this process has the merit of having produced 28 binary curves, corresponding to 165 experimental values (each one measured three times).

For simulating these experiments, a K value was taken arbitrarily. The virtual packing densities of the aggregate fractions were adjusted so that the model could be calibrated on the a–f classes. Afterwards, the model could be run on the whole package of measurements. K was optimized in order to give a zero mean of the residues (that is, the differences between model and experimental values of the packing densities). With a K value of 4.75 the mean error obtained is 0.70%. Although a change of sieves during the experiments was not accounted for in the simulations, the quality of fitting for the 165 points with one single adjusted parameter is noticeable (Fig. 1.23).

Kantha Rao and Krishnamoothy's data

In a more recent paper, Kantha Rao and Krishnamoothy (1993) investigated the problem of optimal grading for polymer concretes. They performed a large number of experiments, dealing with binary mixtures of sand and coarse aggregate. Packing density measurements were performed following an Indian standard, which suggests compacting the mixtures by dry rodding.

The material grading curves are given in the paper, together with optimal binary mixtures (proportions and porosity values). However, the porosities of elementary materials were not given. Some have been obtained directly from the authors (de Larrard *et al.*, 1994b), which allowed us to simulate 20 binary mixture series using 13 different materials. The K value used in the simulation is equal to 4.5, leading to a mean error in predicting the optima porosities of 0.79% (Fig. 1.24). Moreover, the fit between actual and theoretical optima is particularly good (Fig. 1.25). This is encouraging for later applications of the model to concrete grading optimization.

Summary

The model has been validated on binary, ternary, gap-graded six-fraction, continuous six-fraction and continuous, wide-range binary mixtures. The precision obtained is at least as good with literature data as with our original data. Table 1.10 compares the K values obtained in the different optimization processes.

Note that within the various types of dry packing process the ranking is logical: that is, the more energy is given to the mixture, the higher the compaction index. We shall see later that the model applies as well for mixtures of wet fine materials, having the minimal water dosage to behave as a thick smooth paste (see section 2.1.2 and Fig. 2.10). In the

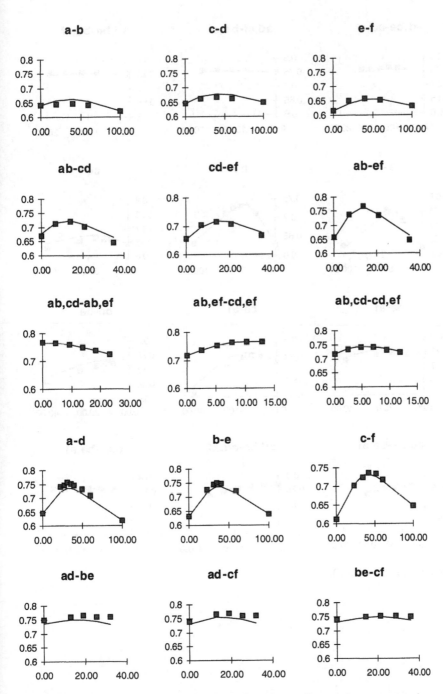

Figure 1.23 Joisel's experiments (packing density vs. fine grain percentage).

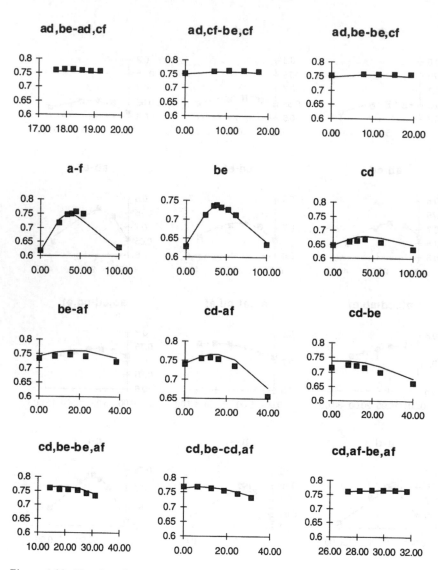

Figure 1.23 (Continued).

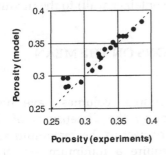

Figure 1.24 Comparison between experimental and theoretical porosities, for Kantha Rao and Krishnamoothy's optimal mixtures.

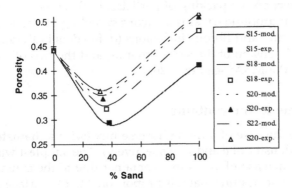

Figure 1.25 Binary mixtures of different sands with the same coarse aggregate (12 mm dia. glass marbles). Comparison between theoretical predictions and experimental data from Kantha Rao and Krishnamoothy.

Table 1.10 Summary of K values (compaction index) for different packing processes.

Dry packing					Wet packing
Pouring (Cintré, 1988)	Sticking with a rod (Kantha Rao and Krishnamoothy, 1993)	Vibration (Joisel, 1952)	Vibration + compression 10 kPa (de Larrard et al., 1994b)		Smooth, thick paste (see section 2.1.2)
4.1	4.5	4.75	9		6.7

following section, an attempt is made to account for the presence of other bodies, either internal (fibres) or external (container wall) to the mixture.

1.3   EFFECT OF BOUNDARY CONDITIONS ON THE MEAN PACKING DENSITY

Up to now, only mixtures placed in infinite volumes have been considered. However, in concrete technology the dimensions of the structures are not necessarily large with respect to the maximum size of aggregate (MSA). Some standards require a minimum structure dimension/MSA ratio of 5, which is not always observed (for example in some small precast pieces). Moreover, in a reinforced concrete piece the container to be considered is the inner volume of the form minus the reinforcement. Also, the distance between two bars may be as low as 1.5 times the MSA. Finally, fibres can be included in the mixture, which have a strong disruptive effect on the packing of particles. In this section, two basic cases are studied: a boundary surface (corresponding to the wall effect exerted by the form), and fibrous inclusions (defined as individual volumes having one dimension of the same order as, and the two others small compared with, the MSA; see Fig. 1.26).

1.3.1   Wall effect due to the container

Let us first investigate the case of a monodisperse mix (with a diameter and a virtual packing density of d and β, respectively). The simplest way of calculating Δv, the increase of voids in a container due to the contact with a plane surface S, is to assume that it is proportional to the surface of the plane and to the size of particles (Caquot, 1937):

$$\Delta v \propto Sd \tag{1.55}$$

However, this formula does not account for the effect of the surface curvature.

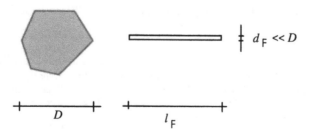

Figure 1.26 Definition of a fibrous inclusion.

Ben-Aïm (1970) improved this approach. He postulated the existence of a perturbed volume V_p (in a unit total volume of mixture), in which the distance between the wall and each point is less than $d/2$ (Fig. 1.27). In this volume the mean packing density is $k_w\alpha$, with $k_w < 1$, while it remains α in the rest of the container. The mean value for the whole container is then

$$\bar{\alpha} = (1 - V_p)\alpha + V_p k_w \alpha$$
$$= [1 - (1 - k_w)V_p]\alpha \qquad (1.56)$$

Unlike Caquot's model, this model accounts for the surface curvature. With the hypothesis that the grains are spherical, Ben-Aïm calculated a value of $k_w = 11/16$, which matched his experimental data well.

In order to check whether the model would apply to aggregates, the Laboratoire Régional de Blois carried out packing measurements on monodisperse grain fractions, placed in cylinders of constant height but variable diameter (from 20 to 80 mm; see Tables 1.11 and 1.12; de Larrard *et al.*, 1994b). The process of pouring, compacting and vibrating the samples was the same as that adopted for the calibration measurements (see section 1.2.2). As expected, the packing density increased with the diameter of the container, mainly for the coarsest fractions. A calibration of Ben-Aïm's model on the data obtained provided two different values of k_w:

- for rounded grains, $k_w = 0.88$ (mean error 0.6%);
- for crushed grains, $k_w = 0.73$ (mean error 0.6%).

Figure 1.28 shows how the model agrees with the experimental data.

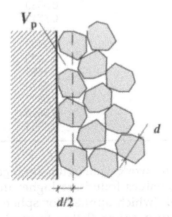

Figure 1.27 Ben-Aïm's model for calculating the mean packing density in a volume submitted to a wall effect.

Table 1.11 Packing measurements to investigate the wall effect due to the container (rounded grains).

d (mm)	Ø (mm)	Experimental packing density	Model packing density
8	80	0.6263	0.6248
8	40	0.6119	0.6119
8	20	0.523	0.5907
4	80	0.6189	0.6196
4	40	0.6126	0.6126
4	20	0.606	0.5997
2	80	0.6155	0.6214
2	40	0.6194	0.6177
2	20	0.616	0.6106
1	80	0.6114	0.6124
1	40	0.6089	0.6105
1	20	0.613	0.6069
0.5	80	0.5854	0.5929
0.5	40	0.592	0.5920
0.5	20	0.604	0.5902

d, grain size; $\varnothing$, cylinder diameter.

Table 1.12 Packing measurements to investigate the wall effect due to the container (crushed grains).

d (mm)	Ø (mm)	Experimental packing density	Model packing density
8	80	0.5945	0.5945
8	40	0.577	0.5671
8	20	0.5	0.5220
4	80	0.5755	0.5768
4	40	0.5765	0.5624
4	20	0.536	0.5360
2	80	0.563	0.5719
2	40	0.5645	0.5645
2	20	0.551	0.5503
1	80	0.56	0.5630
1	40	0.56	0.5593
1	20	0.552	0.5520
0.5	80	0.555	0.5639
0.5	40	0.562	0.5620
0.5	20	0.579	0.5583

Therefore Ben-Aïm's model appears to work well with aggregate having random shapes. However, the k_w values found are higher than Ben-Aïm's original value of $11/16 \approx 0.687$ (which applied for spheres). This means that, statistically speaking, most grains flatten themselves against the wall, so that the increase of porosity is lessened.

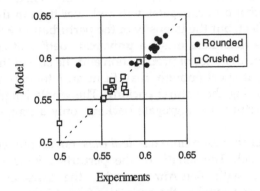

Figure 1.28 Calibration of Ben-Aïm's model for aggregate. Theoretical vs. experimental packing densities.

The introduction of the container wall effect in the compressible packing model is natural. First, equation (1.56) may be applied to the *virtual* packing densities of monodisperse fractions, as it has been demonstrated that, for such a mixture, virtual and actual packing values are proportional to each other (see equation (1.48)). Then it can be noted that in a polydisperse mix the probability of a particle's being in the perturbed volume (in Ben-Aïm's sense) is still v_p (Fig. 1.29). Therefore, the basic equations of the model remain the same (that is, equations (1.33) and (1.47)), replacing the β_i values by $\bar{\beta}_i$ as given by equation (1.56).

1.3.2 Effect of fibrous inclusions

The case of fibres is treated similarly. The main difference from the previous case lies in the curvature of the interface (the surface of the

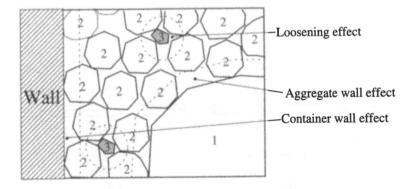

Figure 1.29 Container wall effect in a polydisperse mix. Example of a ternary mix where the intermediate class is dominant.

fibre), the radius of which is small with respect to the MSA. Then the distance at which the perturbation propagates will be lower than in the case of the container wall effect, but the intensity of the perturbation will be higher. We shall assume that there exists a universal coefficient k_F, which is the ratio between the distance of propagation and the size of the particles. Moreover, a fibre, if short enough, may fit in an interstice of coarse grains without disturbing the natural packing. The whole length of the fibre is not able to perturb the aggregate packing, only a limited one (Fig. 1.30).

We shall therefore consider that each fibre has *dead ends*, not accounted for in the perturbation model. The length of the perturbed zone v_p is $l_F - d/2 \geqslant 0$. By analogy with Ben-Aïm's model, the transverse dimensions of v_p will be proportional to the grain size (Fig. 1.31).

The mean packing density for a monodisperse packing mixed with a volume of fibres Φ_F is:

$$\bar{\alpha} = (1 - \Phi_F - N_F v_p)\alpha \qquad (1.57)$$

Figure 1.30 A fibre that is short enough can fit in the packing of particles without creating any significant perturbation.

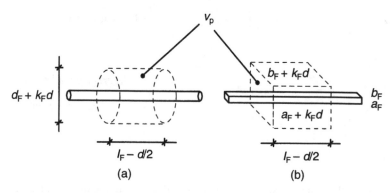

Figure 1.31 Perturbed volumes for (a) cylindrical and (b) prismatic fibres.

where N_F is the number of fibres per unit volume. Here we have assumed that there is no aggregate in the perturbed zone.

Only one parameter must be adjusted (k_F) to calibrate the model. In the data used hereafter (Bartos and Hoy, 1996), dry mixtures of fibres with aggregate were carried out. The perturbating effect of two sorts of fibres – a cylindrical one (Dramix) and a prismatic one (Harex) – on the packing of two aggregate fractions (a coarse gravel and a sand) was investigated. The mixtures of fibres and particles were dry rodded, in accordance with BS 812 Part 2. The evolution of packing density with the proportion of fibres was systematically measured for the two kinds of fibre. In addition, the effect of the dimensions of the fibres at constant fibre volume was investigated. The experimental results are plotted in Figs 1.32 and 1.33.

All results have been simulated with the model. The aggregate fractions used were not monodisperse. An equivalent size has been used in the calculation for each aggregate fraction (0.5 and 16 mm for the sand and the gravel respectively), corresponding to the sieve size at which half of the mass of particles passed. By optimization, a value of 0.065 was found for k_F. The mean error given by the model was then equal to 1.1%. The only significant disagreement between theory and experiment is found with the right-hand point in Fig. 1.33. Here the model over-estimates the packing density by 5.6%. The reason probably lies in the length of the fibres (60 mm), in which an additional container wall effect originates. However, both experiments and model reflect the fact that at constant volume and transverse dimensions the perturbation exerted by fibres increases with their length (Rossi, 1998).

Finally it should be mentioned that this model does not work if *flexible* fibres are dealt with (such as Fibraflex fibres: Bartos and Hoy, 1996;

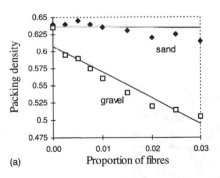
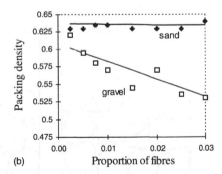

(a) (b)

Figure 1.32 Packing density of dry aggregate fractions against the percentage of fibres. (a) 'Dramix' fibres are cylindrical, with a diameter of 0.5 mm and a length of 30 mm. (b) 'Harex' fibres are roughly prismatic (length 32 mm, width 3.8 mm, depth 0.25 mm). Dots are experimental points, and the solid lines stand for the model predictions.

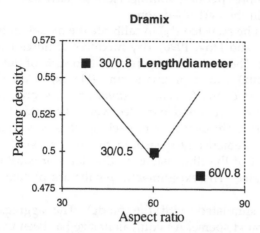

Figure 1.33 Effect of the fibre aspect ratio on the packing density of gravel
(fibre dosage 3%).

Chaudorge, 1990). These fibres have the ability to deform when
submitted to the pressure exerted by coarse aggregates, which decreases
their perturbation effect (Fig. 1.34). Therefore the model is not expected to
give good predictions with other types of fibre, such as glass or
polypropylene. Fortunately, most fibres used in structural concretes are
generally stiff steel.

As for the container wall effect, corrected values of β_i have to be
introduced in the CPM equations to predict the packing density of
polydisperse granular mixes containing fibres.

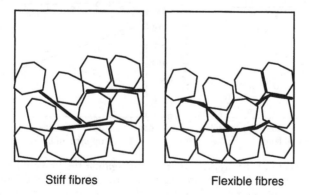

Figure 1.34 Difference in perturbation effects exerted by stiff and flexible
fibres.

1.4 GRANULAR MIXES OF MAXIMUM PACKING DENSITY

In this section the compressible packing model is used to determine the mixture that leads to the maximum packing density, in a number of cases (from the simplest to the most general). Numerical optimization is performed with spreadsheet software. In order to reach a deeper understanding of the problem, a simplified analytical model is also used, which highlights some of the most important trends. These include the effects of grading span, residual packing densities and compaction index on optimal grading curves and packing density value.

1.4.1 An appolonian model for a simplified approach

A simplified model is proposed hereafter, with the view of covering mixtures having a wide grading span and a high compaction index. It refers to an intuitive way of building an optimal mix, first proposed by the ancient Greek Appolonios de Perga, cited in Guyon and Troadec (1994) (Fig. 1.35).

Let us consider a sequence of diameters in geometrical progression, so that

$$d_1 = D$$
$$d_2 = \lambda D$$
$$d_3 = \lambda^2 D \qquad\qquad (1.58)$$
$$\cdots$$
$$d_n = \lambda^{n-1} D$$

with $\lambda \ll 1$, so that the granular interactions can be neglected.

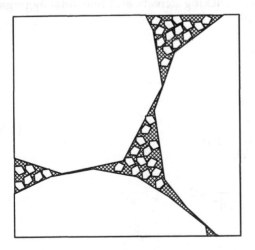

Figure 1.35 The appolonian model for building an optimal mix.

The combination leading to the highest *virtual* packing density is easy to calculate (see section 1.1.4):

$$\Phi_1 = \beta_1$$
$$\Phi_2 = \beta_2 (1 - \beta_1)$$
$$\dots \tag{1.59}$$
$$\Phi_n = \beta_n (1 - \Phi_1 - \Phi_2 \cdots - \Phi_{n-1})$$

and the optimal value of virtual packing density is then

$$\gamma = 1 - \prod_{i=1}^{n} (1 - \beta_i) \tag{1.60}$$

We shall then call an *appolonian mixture* a polydisperse mixture of grain fraction, the size of which corresponds to such a geometrical sequence with low λ value. An *optimized* appolonian mixture will have the partial volumes defined by equation (1.59). Let us now review the case of real mixtures with an increasing number of aggregate fractions. We shall look first at the predictions of the appolonian model, and then we shall compare these predictions with some simulation results.

1.4.2   Binary mixtures

Optimum calculation by the appolonian model

A number of real cases have been studied in sections 1.2.2 and 1.2.3. Let us calculate the optimal virtual packing density and fine/total aggregate ratio in the appolonian model:

$$\gamma_{max} = 1 - (1 - \beta_1)(1 - \beta_2)$$

$$y_2 = \frac{\beta_2(1 - \beta_1)}{1 - (1 - \beta_1)(1 - \beta_2)}$$

$$\frac{\partial y_2}{\partial \beta_2} = \frac{\beta_1 (1 - \beta_1)}{[1 - (1 - \beta_1)(1 - \beta_2)]^2} > 0 \tag{1.61}$$

$$\frac{\partial y_2}{\partial \beta_1} = - \frac{\beta_2}{[1 - (1 - \beta_1)(1 - \beta_2)]^2} < 0$$

From these formulae, the following trends can be then predicted:

- The packing density of the optimal binary mix is an increasing function of each fraction of the residual packing density.
- The proportion of fine aggregate at the optimum increases with the packing density of the fine aggregate, and decreases with that of the coarse aggregate.

Simulations (actual packing density predicted by the CPM)

Let us deal first with the effect of fine/coarse size ratio on the actual packing density. In Fig. 1.36 we see that the optimal packing density increases when the ratio decreases (as expected, because of the progressive disappearance of granular interactions). Also, y_2 at the optimum tends to decrease, which is logical: the less the sizes are comparable, the less the fractions are interchangeable.

As for the effect of each fraction of packing density, it follows that predicted by the appolonian model (Fig. 1.37): the optimum packing density increases when β_1 or β_2 increases. The shift of the peak abscissa is noticeable.

Finally, the effect of the compaction index can be seen in Fig. 1.38. The higher it is, the higher the packing density values and the sharper the curve. There is also a significant effect on y_2 at the optimum. The more compacted the mix, the less room there is in the coarse grain voids to place the fine grains, so that y_2 at the optimum decreases.

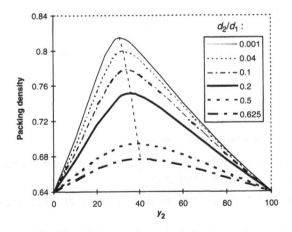

Figure 1.36 Effect of fine/coarse size ratio on the binary mix behaviour. $K = 9$, $\beta_1 = \beta_2 = 0.71$.

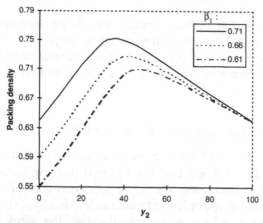

Figure 1.37 Effect of changes of the coarse aggregate packing density on the binary mix behaviour. $K = 9$, $\beta_2 = 0.71$, $d_2/d_1 = 0.2$.

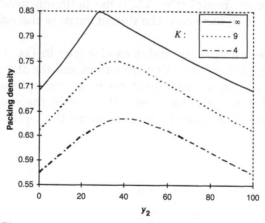

Figure 1.38 Effect of changes of the K value on the binary mix behaviour. $\beta_1 = \beta_2 = 0.71$, $d_2/d_1 = 0.2$.

1.4.3   Ternary mixtures

Optimum calculation by the appolonian model

The calculations are very similar to the ones performed in the binary case. It can accordingly be demonstrated that:

- the packing density of the optimal ternary mix is an increasing function of the β_i;
- for any aggregate fraction, its proportion at the optimum increases with its packing density, but decreases when any other residual packing density increases.

Simulations (actual packing density predicted by the CPM)

Let us now look at the predictions of the compressible packing model. With a given grading span (that is, d_1 and d_3 fixed, and d_2 variable), it can be demonstrated (de Larrard, 1988) that the intermediate diameter that leads to the highest virtual packing density is the geometrical mean of the extreme diameters: that is, $\sqrt{d_1 d_3}$.

When the grain classes have the same residual packing density, and when the ratio between two successive diameters is not too close to 1, the ternary optimum requires the use of the three fractions. The wider the grading span (or the lower the ratio d_3/d_1), the higher the optimal packing density, and the higher the coarse material proportion at the optimum (Fig. 1.39). It is also verified in the simulation that the optimal intermediate diameter for the actual packing density is, as theoretically

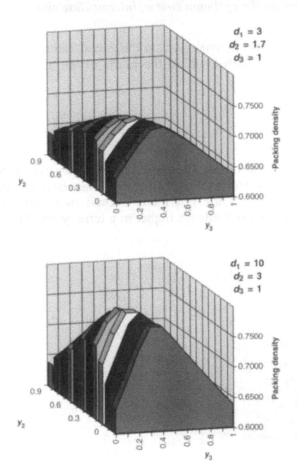

Figure 1.39 Ternary diagram. Effect of the grading span ($d_3/d_1 = 1/3$ and 1/10).

predicted for the virtual one, the geometrical mean of the extreme diameters.

Each aggregate proportion at the optimum is influenced by the residual packing density of the fraction concerned (β_i). When the width of the grading span is low, and β_2 is significantly lower than β_1 and β_3, it may happen that the optimum is a binary mixture of the extreme sizes (Fig. 1.40).

Finally, when investigating the effect of the compaction index on the ternary diagrams, the same trend as that dealing with binary mixes is found: when K increases, the maximum packing density and the proportion of coarse grains at the optimum also increase, and the 'mountain' becomes sharper (Fig. 1.41).

Effect of the finer class on the optimum coarse/intermediate size proportion ratio

The Φ_1/Φ_2 ratio at the optimum, analogous to the gravel/sand ratio in a concrete, can be easily calculated within the frame of the appolonian model:

$$\frac{\Phi_1}{\Phi_2} = \frac{\alpha_1}{\alpha_2(1-\alpha_1)} \tag{1.62}$$

According to this simplified model, this ratio does not depend on the presence of the finest class. However, in reality the finest class exerts a loosening effect, which affects the intermediate class more than the coarse one. Therefore the ratio is always higher in a ternary mix than in

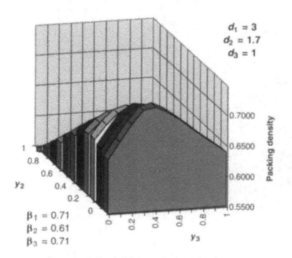

Figure 1.40 Ternary diagram in which the optimum is a binary mix.

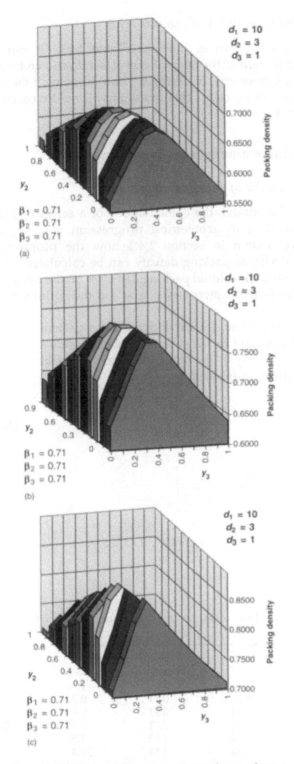

Figure 1.41 Effect of the compaction index on the ternary diagrams: (a) $K = 4$; (b) $K = 9$; (c) $K = \infty$.

the binary mix of the two coarse fractions (Table 1.13). It can be concluded that determination of the optimal gravel/sand ratio on binary dry mixtures (Toralles-Carbonari *et al.*, 1996) can be misleading if the aim is to minimize the *concrete* porosity. This question will be discussed in more detail in section 4.2.

1.4.4 Optimal mixtures among a given grading span

Optimum calculation by the appolonian model

Within the frame of this model, the combinations of a series of grain fractions having diameters in geometrical progression $\{d_i\}_{1 < i < n}$ are considered. We have shown in section 1.4.1 how the proportions leading to the optimal virtual packing density can be calculated. Here again the effect of individual residual packing densities on the optimum may be easily inferred. When β_i increases, ϕ_i and γ at the optimum also increase.

Let us solve the same problem dealing with the *actual* packing density. We shall restrict our attention to the case of uniform residual packing densities (so that $\beta_i = \beta$ for any i). The optimal mix should be the solution of the following optimization problem:

$$\text{Max}\left(\sum_{i=1}^{n} \Phi_i \right) \tag{1.63}$$

$$\text{with } K = \sum_{i=1}^{n} H\left[\frac{\Phi_i}{\beta(1 - \Phi_1 - \cdots - \Phi_{i-1})} \right]$$

Table 1.13 Influence of the fine class on the optimum Φ_1/Φ_2 ratio (CPM simulations).

d_1	10	10	10	3
d_2	3	1.4	3	1.4
d_3	1	1	1	1
β_1	0.71	0.71	0.71	0.71
β_2	0.71	0.71	0.61	0.61
β_3	0.71	0.71	0.71	0.71
$[\Phi_1/\Phi_2]_{\text{opt.}}$ (binary)	1.92	1.92	1.92	1.68
$[\Phi_1/\Phi_2]_{\text{opt.}}$ (ternary)	3.14	4.34	21.4	∞

Making a change of variables, we define the Ψ_i as follows:

$$1 - \Psi_1 = \Phi_1$$

$$1 - \Psi_2 = \frac{\Phi_2}{1 - \Phi_1}$$

(1.64)

$$\cdots$$

$$1 - \Psi_n = \frac{\Phi_n}{1 - \Phi_1 - \cdots - \Phi_{n-1}}$$

Then the optimization problem becomes

$$\text{Min}\left(\prod_{i=1}^{n} \Psi_i\right)$$

(1.65)

$$\text{with } K = \sum_{i=1}^{n} H\left(\frac{1 - \Psi_i}{\beta}\right)$$

Using the Lagrange method it can be derived that all Ψ_i will have the same value at the optimum. Also, all the partial compaction indices K_i are equal. If α is the common value of the $1 - \Psi_i$, we therefore have

$$\Phi_1 = \alpha$$

$$\Phi_2 = \alpha(1 - \alpha)$$

(1.66)

$$\cdots$$

$$\Phi_n = \alpha(1 - \alpha)^{n-1}$$

with

$$\alpha = \frac{\beta}{1 + n/K}$$

(1.67)

The mixture leading to the optimum actual packing density is the same as that giving the optimum virtual packing density, provided that β is replaced by α. The minimal virtual and actual porosities are respectively

$$\pi_{min} = (1 - \beta)^n$$

$$p_{min} = \left(1 - \frac{\beta}{1 + n/K}\right)^n \tag{1.68}$$

For calculating the theoretical grading distribution, the cumulated y_i value must be plotted against the logarithm of the diameter. An example of such a distribution is given in Fig. 1.42.

Simulations: convergence towards the optimal distribution

When performing numerical optimizations with the CPM, with a fixed grading span and uniform residual packing densities, it is numerically found that the series of diameters still form a geometrical progression. Then the question arises whether there is a finite number of classes in the optimal mix, or whether the aggregate size distribution tends to be continuous. As far as the virtual packing density is concerned it has already been found that the optimal mixture has a discrete support: that is, it is gap-graded (de Larrard, 1988). However, with the actual packing density, and keeping a realistic value of K, a limit of this type has not been found. Such an ideal discrete distribution could exist, but would require more powerful calculation tools for it to be found. Meanwhile, it must be kept in mind that in concrete technology only a finite number of aggregate fractions can be handled, so that the search for optimal gradations with very close successive diameters has no practical relevance.

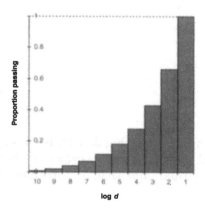

Figure 1.42 Theoretical grading curve of an optimal appolonian mixture. $n = 10$, $\beta = 0.71$, $K = 9$.

Here the optimal mixture having an increasing number of classes has been determined. An example of the evolution of the grading curve is shown in Fig. 1.43. Apparently, the curve tends to be continuous, except at both extremities of the grading span, where a gap tends to appear. In other words the partial volumes of extreme classes tend towards finite values. As analysed by Caquot (1937), these extreme classes have to supplement the lack of coarser or finer particles: the coarsest fraction is the only one that is not submitted to a wall effect,[2] while the finest one is the only one not perturbed by the loosening effect. Thus one can deduce that in an optimized real mixture the coarser and the finer classes will be more represented than their closest neighbours.

If another type of plotting is chosen, the curve may be linearized. Caquot proposed using the fifth root of the diameter as the abscissa parameter, which seems in the present case to give a fairly straight broken line, matching the optimal curves suggested by Faury (1944) (Fig. 1.44).

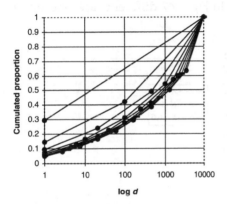

Figure 1.43 Evolution of the optimal grading curve, when more aggregate fractions are added. $d/D = 10^{-4}$.

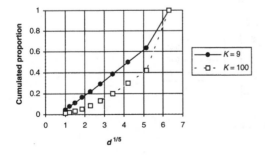

Figure 1.44 Optimal grading curve with 10 classes of grain, in Caquot's coordinates. $\beta = 0.71$.

In fact, within the frame of the appolonian model, it can be demonstrated that the curve given in Fig. 1.42 may be linearized with d^q as the abscissa parameter. However, the value of q depends on β and K. This is why the curve is no longer made of two straight lines when K is increased (Fig. 1.44).

Effect of the input parameters on the optimal grading curve

In Fig. 1.45 the effect of the K index on the optimal 10-class mix is highlighted. This effect appears to be significant, which is a first argument against the concept of a 'universal' ideal curve[3]. Similarly to the binary and ternary cases, an increase in K tends to promote a higher proportion of coarse grains.

The effect of change in the residual virtual packing density has been investigated. As long as this value is uniform within the grading span, the changes in the optimal grading curve are minor (Fig. 1.46).

However, when different β_i values are taken within the grading span, the curve changes significantly. In Fig. 1.47 different cases are simulated.

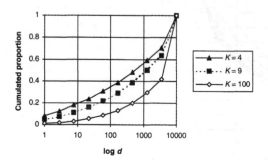

Figure 1.45 Effect of the compaction index on the optimal grading curve. $\beta = 0.71$.

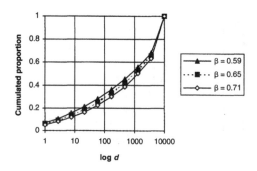

Figure 1.46 Effect of change in the value of β on the optimal grading curve. $K = 9$.

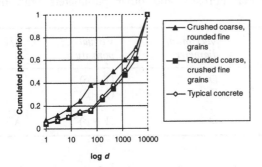

Figure 1.47 Optimal grading curves in various cases. Non-uniform β; $K = 9$.

Having crushed grains in the superior part of the grading span and rounded grains in the inferior one tends to increase the proportion of fine elements. The reverse is found in the opposite situation. Finally, the third curve deals with the optimal curve when 'concrete-like' materials are used (that is, crushed rough grains at the cement level, round grains at the sand level, and crushed grains again at the coarse aggregate level).

Minimum porosity vs. grading span

Caquot (1937) proposed an empirical 'law' to predict the minimum porosity of a granular mix, the grading span of which is contained within the (d, D) limits:

$$p_{\min} \propto \sqrt[5]{\frac{d}{D}} \tag{1.69}$$

According to this formula, the minimum porosity of an optimized packing of particles would tend towards zero for an infinite grading span.

First, let us show that this formula may be demonstrated within the appolonian model, *but only for virtual porosity.* In Table 1.14, the evolution of the different quantities is calculated when further classes are added in the appolonian model.

We then have:

$$\pi_{\min} = (1 - \beta)^n$$
$$= (1 - \beta)^{1 + [\log(d/D)/\log(\lambda)]}$$
$$= (1 - \beta)\left(\frac{d}{D}\right)^{[\log(1 - \beta)/\log(\lambda)]} \tag{1.70}$$

Table 1.14 Calculation of minimal porosity and other parameters with the appolonian model.

Number of classes	Minimum diameter	Maximum diameter	Virtual porosity	d/D
1	D	D	$(1 - \beta)$	1
2	λD	D	$(1 - \beta)^2$	λ
...	...	...	...	...
n	$\lambda^{n-1}D$	D	$(1 - \beta)^n$	λ^{n-1}

The last formula has the same shape as Caquot's law. However, with the same simplified model, one may also calculate the minimal *actual* porosity. In equation (1.68) n appears twice, so that p_{min} cannot be reduced to a simple power law of the d/D ratio. Moreover, if we look at the limit of p_{min} when d/D tends towards 0, we have

$$p_{min} = \exp\left[n \log \left(1 - \frac{\beta}{1 + \dfrac{n}{K}} \right) \right]$$

$$\approx \exp\left(-\frac{n\beta}{1 + \dfrac{n}{K}} \right) \tag{1.71}$$

then

$$p_{min} \longrightarrow \exp(-\beta K) \tag{1.72}$$

when $n \rightarrow +\infty$

so that there appears to be a finite inferior limit for p_{min}, which can be 0 only for infinite compaction (corresponding to virtual packing).

This result is fully confirmed by CPM simulations. There is a difficulty here in that, for a given grading span, the computer does not generally attain the optimal actual porosity with a finite number of classes. Therefore the following method was used. For each grading span (ranging from 10^{-1} to 10^{-4}) the sequence of the minimal actual porosities was plotted against the number of aggregate fractions. For n between 5 and 10 the values fitted very well with the following empirical model:

$$p_{\min, n} = u + \frac{v}{n^w} \tag{1.73}$$

where u, v and w were adjustable parameters ($w > 0$). Extrapolating this equation, the minimal actual porosity was taken as u. The overall results are given in Fig. 1.48. It can be seen that the virtual porosity found by the CPM exactly matches a Caquot-type power law, while the actual porosity tends towards a non-zero limit.

These simulations are verified by experience, as can be seen in Fig. 1.49. Here, optimal mixtures (single size, binary and six-class) have been collected from Joisel's data and Blois' data (see section 1.2.3), and the porosity plotted against the d/D ratio.

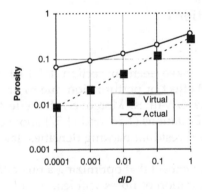

Figure 1.48 Minimal virtual and actual porosity for a polydisperse mix, after CPM simulations. $\beta = 0.71$, $K = 9$.

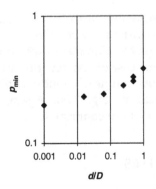

Figure 1.49 Experimental minimal porosity of rounded aggregate mixtures, against the extreme size ratio.

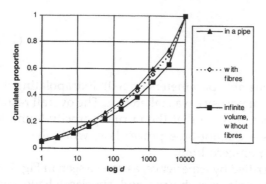

Figure 1.50 Effect of boundary conditions on the optimal grading curve. $d/D = 10^{-4}$, $n = 10$, $\beta = 0.71$.

1.4.5 Effect of boundary conditions

Analysis by the appolonian model

Within the frame of the CPM, we have seen in section 1.3 that the boundary conditions (container wall effect or perturbation due to fibres) are accounted for by decreasing the β_i values. Moreover, this effect is significant only for the coarse sizes. Equations (1.56) and (1.57) allow one to calculate the values of the corrected residual packing densities, taking account of the boundary conditions.

From the appolonian model we can expect that optimizing a mix either in a 'small' container or with incorporation of fibres will lead to a lower packing density, with more fine elements at the optimum.

Simulations

To illustrate this statement, the optimization already performed ($d/D = 10^{-4}$, $n = 10$, $\beta = 0.71$) has been repeated twice. First it was assumed that the mixture was placed in a pipe the diameter of which was twice the MSA (this corresponds to an extreme case of concrete pumping). Another calculation was then done with 2% by volume of cylindrical fibres ($l_F/\text{MSA} = 3$, $l_F/\varnothing = 60$). The results appear in Fig. 1.50. Again, the qualitative predictions of the appolonian model are confirmed.

1.5 SEGREGATION OF GRANULAR MIXES

An optimized size distribution must not only lead to a high packing density. Between mixing and placement, granular mixes are submitted to accelerations (due to gravity, external shocks and/or vibration). This may result in a separation of the mixture, which is generally detrimental to the function of the final product. In concrete technology a significant segregation creates spatial variations in final properties, such as visual appearance, strength, elastic modulus, and shrinkage. This last feature may in turn provoke cracking patterns that impair the durability of the structure. The aim is therefore to design granular mixes that are not prone to segregation.

Few studies of segregation, either experimental or theoretical, are available in the literature, especially if wide grading range mixtures are dealt with. Here we shall review some experimental facts that are well known by the engineering community. Then we shall try to build some simple conceptual tools that will at least have the merit of summarizing these experimental facts, and which could eventually provide a way to design stable mixtures.

1.5.1 Some experimental facts

The following facts are based on Powers (1968), Dreux and Gorisse (1970), Popovics (1982), and Neville (1995).

Fact 1: For any granular mix, compaction decreases the likelihood of later segregation

When a granular mix is compacted (by compression and/or vibration), the action exerted on the mixture may create a certain segregation. For instance, in a binary mixture with a fine/coarse aggregate size ratio lower than 1/4, and when the coarse fraction is dominant, the fine grains settle in the bottom of the container (Cintré, 1988). However, if the compaction is effective, each particle tends to find an equilibrium position, and no further segregation may appear afterwards.

Fact 2: For a given set of aggregate fractions and a given compaction index, the higher the whole mixture packing density, the lower the proneness to segregate

In other words, when the proportions of a mixture are close to the optimum, the grain fractions are blocking each other. This can be understood in the case of the binary mixture without interaction (Fig. 1.51).

Fact 3: Within optimized mixes having the same compaction index and the same grading span, continuously graded mixtures will exhibit less segregation than gap-graded ones

This is a classical argument in favour of continuously graded concretes. In a binary packing, there is a critical particle diameter for which the fine

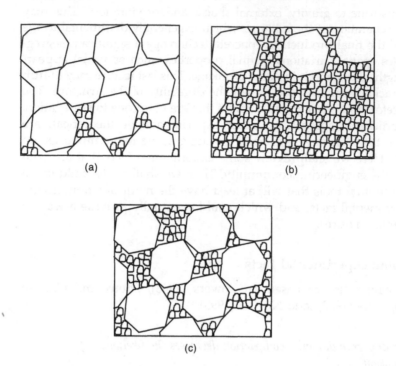

Figure 1.51 Binary mixtures without interactions, subject to gravity. When coarse
grains are dominant (a), the fine grains settle in the coarse grain
porosity. When fine grains are dominant (b), the coarse grains tend to
accumulate in a region of the container. In wet mixes, experience
shows that the coarse grains sink in the bottom of the container.
However, in vibrated dry mixtures, the coarse fraction rises up
(Rosato *et al.*, 1987). Finally, at the optimum (c), no more segregation
is possible.

grains may circulate through the coarse grain porosity, which is of the
order of $D/6$.[4] Then, if all the fine materials are finer than this critical size,
the separation of the mix is easier.

*Fact 4: Within optimized mixes having the same compaction index and
the same type of granular distribution, a wider grading span is a factor
for increasing the segregation*

It is generally recognized that the higher the maximum size of aggregate
(MSA), the higher the tendency to segregate. This constitutes one of the
positive arguments for the use of sand concrete (see section 5.5.3), which
is basically concrete with no aggregate coarser than 5 mm. Unlike usual
mortars, a part of the fine materials consists of supplementary
cementitious materials (such as fly ash or filler). A physical justification

of the MSA effect could be related to Stokes' law describing the fall of a spherical particle in a viscous liquid. Following this law, the speed of fall is proportional to the second power of the particle diameter.

It is also believed that adding silica fume to a concrete tends to reduce the likelihood of segregation. However, silica fume particles are colloidal (that is, their size is less than 1 μm), and so they may form a gel, which prevents the circulation of water in fresh concrete. Therefore the mechanism of segregation reduction is in this case outside the scope of dry packing theories.

Fact 5: For minimizing the risk of segregation, the amount of fine elements required is higher in a loose packing than in a compacted one

First let us refer to an analogy between the compaction index in a dry mix and the workability of fresh concrete. A high K index would correspond to a stiff concrete mixture, while a low K would correspond to a wet mix.

In his original concrete mix-design method, Day (1995) calculates his mix suitability factor (MSF) by referring to the specific surface of particles, with various corrections. This empirical parameter is tabulated for various concrete consistencies: the more fluid the mix, the higher the MSF. But the higher the MSF, the more fine particles have to be used. Then, following Day, a loose packing needs a greater amount of fine particles to be stable. Other methods, such as Dreux' (Dreux, 1970), also recognize the need for more fine elements in softer fresh mixtures to prevent segregation.

Fact 6: Differences in absolute density among the aggregate fractions increase the rate of segregation

Quite obviously, a mix in which coarse aggregates have an absolute density either higher or lower than the rest of the aggregate fractions has an increased tendency to segregate. This is the case for lightweight aggregates (see section 5.5.1), which tend to rise in a lightweight concrete during vibration. Conversely, in heavy concretes, such as the ones used in nuclear engineering, coarse aggregates tend to sink.

1.5.2   Quantitative indicators: filling diagram and segregation potential

Definition of filling diagram and segregation potential within the frame of the appolonian model

The appolonian model, presented in section 1.4.1, provides a simplified view of a granular mix, in which the concept of segregation ability comes naturally. In an appolonian mix, let the *i*th class be a non-dominant class (see section 1.1.4). The i grains may pack in a volume that is the room left available by the other fractions. Let us assume that the particles tend to fall in the bottom of this volume. Then the whole mixture will have a maximum content of i particles in the bottom part of the container, the maximum height of which is Φ_i/Φ_i^* (see Fig. 1.52), where Φ_i^* was defined in section 1.2.1. In the upper part of the container, no i grains will be present. The proportion of this segregated volume would then be

$$S_i = 1 - \frac{\Phi_i}{\Phi_i^*} \tag{1.74}$$

Let us define the filling diagram as giving the Φ_i/Φ_i^* (called the *filling proportion* or filling ratio of the i class) against the size of the grain fractions (see Fig. 1.53). The segregation potential will correspond to the

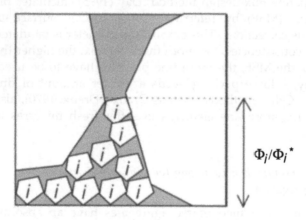

Φ_i/Φ_i^*

Figure 1.52 Segregation of class i in a mixture.

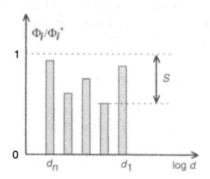

Figure 1.53 Filling diagram and segregation potential.

biggest heterogeneous volume in the container, which is

$$S = \text{Max } S_i \qquad (1.75)$$

for $1 \leqslant i \leqslant n$

Each value of Φ_i/Φ_i^* indicates the extent to which the ith class fills the voids of the coarser ones. The higher the S value, the lower the proportion of segregated volume if maximum segregation occurs. Hence a high S value does not necessarily indicate that much segregation will occur. For instance, coming back to the simple binary case (Fig. 1.51), we can guess that the rate of segregation in a real case will be much higher when coarse aggregates are dominant (Fig. 1.51a) than when fine aggregates are dominant (Fig. 1.52b). If $S = 0$ it means that each fraction is fully packed: that is, we have an optimal virtual packing. Here no segregation is possible. As, in reality, the virtual packing density is never achieved, S will always have a non-zero value.

Significance of the segregation potential with regard to the main experimental facts

Let us now use the appolonian model to demonstrate that most of the experimental facts listed in section 1.5.1 are summarized by the segregation potential concept. We can first find a link between the compaction index and the segregation potential. From equation (1.34), we have

$$K_i = \frac{\Phi_i/\Phi_i^*}{1 - \Phi_i/\Phi_i^*}$$

so that

$$S = 1 - \underset{1 \leqslant i \leqslant n}{\text{Min}} \left(\frac{K_i}{1 + K_i} \right) \qquad (1.76)$$

When the compaction index increases (fact 1), each Φ_i tends to approach Φ_i^* and then all S_i decrease. So does S, and the segregation hazard is lower.

When a mixture is optimized at a given K value (fact 2), we have at the same time (see section 1.4.4):

$$K = \sum_{i=1}^{n} K_i \qquad (1.77)$$

and

$$K_i = \frac{K}{n}$$

Then the lowest K_i takes its maximum possible value (see Fig. 1.54), which maximises the quantity $\text{Min}_{1 < n < i}[(K_i/(1 + K_i)]$, and then minimizes S.

The question of gap-graded mixes (fact 3) may also be covered. Although all appolonian mixtures are virtually gap-graded, we shall consider that discontinuous distributions in reality correspond to appolonian mixtures in which at least one intermediate fraction is missing. Then there exists an i for which

$$\Phi_i = 0$$
$$\Rightarrow S_i = S = 1 \tag{1.78}$$

The segregation potential is maximum.

For demonstrating fact 4 within the framework of this theory, let us assume that the mixture is optimized for a given K value. We then have

$$K_i = \frac{K}{n}$$

$$S = 1 - \frac{K}{n + K} \tag{1.79}$$

Thus S is an increasing function of n, and the likelihood of segregation increases with the grading span. Hence, as we assumed that the

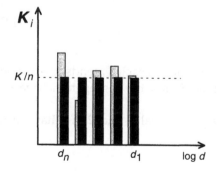

Figure 1.54 Effect of granular optimization on the K_i distribution. Grey bars: before optimization. Black bars: after optimization.

compaction index is a cumulative parameter, the more grain fractions for a given K value, the less each fraction is packed inside the mix.

Finally, fact 5 is simply a consequence of the effect of K on optimal grading (see section 1.4.4). When K decreases, the proportion of fine elements at the optimum increases.

Fact 6 (effect of differences in absolute density) is the only one that is not displayed by the concept of segregation potential. It deals with the *rate* of segregation more than with its maximum amount. An optimum virtual packing will not segregate, whatever the densities of each grain fraction.

Extension of the concept for real granular mixtures: critical percolation ratio

Let us consider now a real granular mixture, for which two successive diameters d_i and d_{i+1} may be close to each other. Let us calculate the maximal ratio $\lambda^* = d_{i+1}/d_i$ for which the $i+1$ grains may percolate through the i ones. If the i grains are fully packed spheres, we have, following Fig. 1.55:

$$\lambda^* = \frac{2}{\sqrt{3}} - 1 \approx \frac{1}{6.46} \tag{1.80}$$

However, we have seen that for finite K values no grain fraction is fully packed (as in a virtual packing). If the packing is optimized, we will admit that we have

$$K_i = \frac{K}{n} \tag{1.81}$$

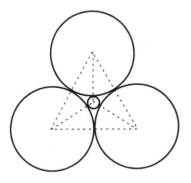

Figure 1.55 Binary virtual packing pattern, in which the fine grain may percolate through the coarse grain packing.

The pattern of Fig. 1.55 may be kept, reducing the diameter of the coarse particles by a factor equal to the cube root of the volume reduction coefficient. From equation (1.67) we know that the latter coefficient is equal to $K_i/(K_i + 1) = K/(K + n)$. Thus the value of λ^* becomes (Fig. 1.56)

$$\lambda^* = \frac{2}{\sqrt{3}} - \sqrt[3]{\frac{K}{K + n}} \tag{1.82}$$

Now, we shall consider the real mixture as an appolonian one, except that the size ratio between two adjacent classes will be equal to λ^*. If the grading span is d/D, we shall have necessarily

$$\lambda^{*n-1} = \frac{d}{D} \tag{1.83}$$

Then λ^* is the implicit solution of the following equation:

$$\lambda^* = \frac{2}{\sqrt{3}} - \sqrt[3]{\frac{K}{K + 1 + \dfrac{\log(d/D)}{\log(\lambda^*)}}} \tag{1.84}$$

For $K = 9$ and $d/D = 10^{-4}$ we find numerically a value of $0.38 \approx 1/2.6$ for λ^*. In this case it implies that the filling diagram should be built for a series of 10 different fractions. The minimal segregation potential, deducted from equation (1.79), would then be equal to 0.53. The size distribution inside each $[d_{i+1}, d_i]$ interval will have no effect on the segregation ability of the mixture, while the mutual proportions of these clustered classes will be critical.

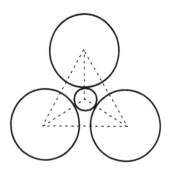

Figure 1.56 Increase of the critical percolation ratio λ^* due to the lower volume of coarse grains.

For calculating the S_i values of the clustered classes, we may use the fact that K is a cumulative parameter. We shall just sum the K_j within the $[d_{i+1}, d_i]$ interval, and then use equation (1.76) for calculating the corresponding S_i parameter.

1.5.3 Examples: some simulations with the CPM

In the following examples, several size distributions, all having the same grading span (10^{-4}) and the same K index (equal to 9) are compared. The grading curves are plotted in Fig. 1.57. The packing densities and segregation potentials are given in Table 1.15, and the filling diagrams are presented in Figs 1.58 to 1.64.

The reference distribution is the one leading to the maximum packing density, as already discussed in section 1.4.4. Its filling diagram is

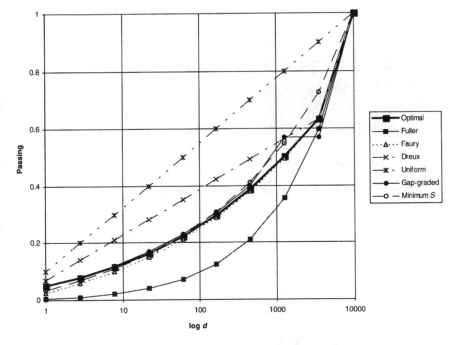

Figure 1.57 Simulated grading curves.

Table 1.15 Packing density and segregation potential of various distributions.

	Optimal	Fuller	Faury	Dreux	Uniform	Gap-graded	Minimum S
Φ_{opt}	**0.929**	0.869	0.927	0.914	0.891	0.928	0.926
S	0.59	0.96	0.59	0.80	0.85	1.00	**0.53**

uniform, except for the two extreme classes, which are somewhat over-represented as far as the filling proportion is concerned (Fig. 1.58).

Then the Fuller curve is plotted (Fig. 1.59), which corresponds to a uniform distribution in a diagram where the abscissa would be the square root of the particle diameter. The packing density is substantially lower than the optimum, and the filling diagram shows that a gradual lack of particles is found when we move towards the left-hand side of the grading span.

Faury's distribution – a broken line in a diagram having the fifth root of the diameter in abscissa – is, as already noted in section 1.4.4, close to the optimal one.[5] However, the filling diagram given in Fig. 1.60 shows a

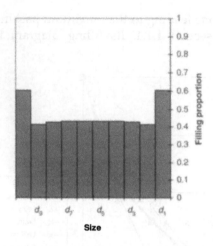

Figure 1.58 Filling diagram of the optimal distribution.

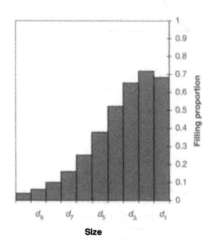

Figure 1.59 Filling diagram of Fuller's distribution.

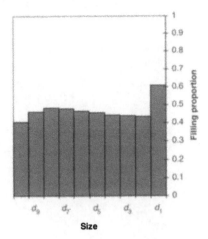

Figure 1.60 Filling diagram of Faury's distribution.

moderate risk of segregation caused by a lack of particles in the finest class.

As for Dreux' distribution (a broken line in a logarithmic *x*-scale diagram[6]), it achieves a reasonably satisfactory packing density, but an under-dosage of the intermediate classes on the coarse-grain side of the grading span (Fig. 1.61). Thus the segregation potential is high.

In contrast to the Fuller's case, a uniform distribution (in a semi-logarithmic diagram) produces an over-dosage of fine elements. The packing density is still low, and the segregation potential high (see Fig. 1.62).

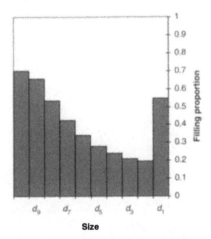

Figure 1.61 Filling diagram of Dreux' distribution.

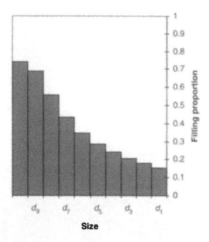

Figure 1.62 Filling diagram of the uniform distribution.

An interesting case is that of a gap-graded mix (Fig. 1.63), in which the second class (from the coarsest) has been removed ($\Phi_2 = 0$), and the rest of the mix has been optimized. The packing density is almost as high as the optimal, and the grading curve tends to approximate the reference continuous curve. As a fraction is totally lacking, the segregation potential is maximum.[7] It can be seen that, in the optimization process, the neighbour fractions tend to partially supplement the lack of the second class.

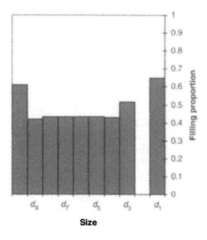

Figure 1.63 Filling diagram of the gap-graded distribution.

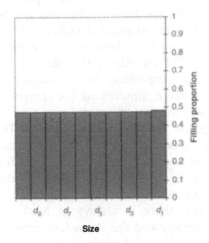

Figure 1.64 Filling diagram of the distribution minimizing the segregation potential.

Finally, the continuous mix has been re-optimized, with a different criterion. Instead of maximizing the packing density, the segregation potential has been minimized (still at constant K index).

The grading curve obtained is close to the one that maximizes the packing density. However, the overdosages of the extreme fractions have disappeared, and the filling diagram is uniform (Fig. 1.64). The segregation potential attains the value of 0.53, already calculated in the case of the appolonian model (see section 1.5.2).

Therefore it seems that the concepts developed are suitable for analysing grain mixtures. It is apparent that the grain size distribution is critical not only for obtaining a dense packing, but also to avoid the segregation of this packing. For high grading span, high packing density mixtures, the segregation potential is a more sensible tool to optimize a mix than the packing density. The filling diagram can be used to determine whether a certain size of particles is overdosed or underdosed in a mixture, as compared with the other ones. Of course, this diagram may be built taking the boundary conditions into account. A well-distributed granular mix, as considered in an infinite volume, may become overdosed in coarse particles when confined in small container.

1.6 SUMMARY

The questions of packing density and homogeneity of dry granular mixes have been investigated in this chapter.

A distinction was first made between *virtual* and *actual* packing density. The virtual packing density is the maximum value that can be achieved with a given mix, each particle being placed individually. It is a characteristic of the mix alone. The actual packing density is the result of a given placing process. It is then controlled not only by the mix, but also by the method of mixing, placement and compaction. At this stage, only mixtures placed in a container having large dimensions (as compared with the maximum size of the particles) were considered.

A model was built for calculating the virtual packing density. The parameters of the model are the grain size distribution and the virtual packing density of each monodisperse fraction (also called the *residual packing density*, the symbol for which is β_i). The main hypothesis of this model is that granular interactions in any mixture are essentially *binary* in nature. It was then demonstrated that there is always at least one *dominant* class, which ensures the continuity and the stress transmission in the dry packing. This class suffers two type of interactions from the other classes: a *wall effect* (from the coarser fractions) and a *loosening effect* (from the finer ones). These effects are thought to remain additive throughout the volume of the container.

From the virtual packing density calculation, a second model was proposed for calculating the actual packing density, called the *compressible packing model*. A relationship was developed between the virtual packing density and the compaction index K, a scalar parameter characterizing the process of building the packing. At this stage, the whole model needed two types of calibration, dealing with the interaction coefficients and the compaction index respectively. A series of binary mix experiments made it possible to derive equations for the interaction coefficients. As expected, these coefficients are mainly controlled by the size ratio of the two fractions considered. Then the model was extensively validated on several sets of independent experimental data. The compaction index values that are assigned to each type of packing experiment are logically ranked from low K values (for a loose packing) to high K values (for a dense compressed and vibrated packing). Knowing the compaction index corresponding to a given process, the model is then able to predict the packing density of any combination of monodisperse classes from the residual packing values. The precision achieved is generally better than 1% in absolute value.

A generalization of the model was then performed to include the effect of boundary conditions. The container wall creates a perturbed zone in the mix, in which the residual packing density of a monodisperse fraction is decreased by a certain factor. It is then easy to compute the mean value of the residual packing density within the volume of the container, for each grain fraction. All effects remain additive, and the formalism of the model is unchanged. Similarly, when rigid fibres are introduced in the mixture, each fibre is surrounded by a porous zone where the proportion of a given size is zero, thus lowering the mean value of the corresponding residual packing density. Finally, in a widely graded granular mixture, only the packing abilities of the coarsest fractions are altered by the boundary conditions.

The question of optimization of dry mixtures was then addressed. For a better understanding of the problem, a simplified model was proposed, called the *appolonian* model. It corresponds to the intuitive way of building a dense polydisperse packing: placing a maximum volume of coarse grains in a container, then adding a maximum volume of particles small enough to fit in the voids of the first fraction, and then repeating the operation until no more grain fraction is available. Both optimal virtual and actual packing density can be calculated within the frame of this model, which leads to particular size distributions. Yet it is realized that the proportions at the optimum depend strongly on the individual β_i values. Also, it is found that continuing the process until infinity does lead to a zero value for the virtual porosity, but not for the actual one.

These theoretical findings were checked with the CPM by numerical simulations. The evolution of virtual and actual porosity vs. grading span, already predicted by the appolonian model, was verified. Optimal distribution appears rather continuous, close to the classical Faury curves, but with an overdosage of extreme fractions. However, the effect of differences in the β_i values was confirmed. It is then clear that the concept of a *universal ideal grading curve* is valid only for particles having the same shape along the whole grading span, and for a given K index. This is of course a rather unusual case. Moreover, the ideal distribution for a given set of aggregate fractions, and for a given packing process, also depends on the boundary conditions. The model allows one to take the container wall and the presence of rigid fibres into account.

In the last section of the chapter an attempt was made to use the formalism of the CPM to assess the likelihood of segregation. Two simple tools were defined, which allow us to concentrate all the basic knowledge about segregation. The *filling diagram* is a diagram giving the relative height of a clustered fraction of grains (called the *filling ratio*) that would settle in the case of severe segregation. Clustered fractions include the elementary fractions that are too similar in size to percolate through each other. A well-distributed mixture would have even values in this diagram. If a size range is underdosed, as in gap-graded mixtures, the diagram would display a low filling ratio at this level. Finally, a *segregation potential* was defined from the lowest value within the grading span. It appears that this segregation potential could be a useful numerical index for optimizing mixtures for stability.

NOTES

1 If both inequalities (1.11) and (1.12) are not satisfied, the mix is no longer a packing, as the particles may be poured into a smaller volume.
2 Except if the mixture is confined in a container: see section 1.3.1.
3 However, good current methods (Faury, 1944; Dreux, 1970) acknowledge this fact, by changing the ordinate of the break point with the compaction energy.
4 For spheres, the ratio is equal to $2/\sqrt{3} - 1 \approx 1/6.46$.
5 Here, the proportion of the coarsest fraction has been taken equal to that of the reference optimal distribution.
6 Idem.
7 One could argue that this mixture is not likely to segregate, while having an S value of 1. This is actually a limit case of the theory. However, in reality, aggregate fractions contain some particles that are out of the nominal grading limits. In such a case, the second clustered class would be somewhat represented in the mixture.

2 Relationships between mix composition and properties of concrete

This chapter deals with models linking the mixture proportions and the most important concrete properties. While classical mix-design methods are generally limited to two concrete characteristics (slump, and the compressive strength at 28 days), the aim is to have a more global approach, considering the various properties that are of interest for engineers.

We shall logically start with fresh concrete properties, with a special emphasis on fresh concrete rheology. Then we shall deal with the heat of hydration of hardening concrete, an important factor for all massive structures. We shall continue with compressive strength, which remains the most critical concrete property. The tensile strength will then be investigated, as a critical parameter in some particular applications, such as unreinforced concrete. To complete the coverage of mechanical properties, models will be developed for concrete deformability, including elastic modulus, creep and shrinkage. Then, although no models are available for the durability-related properties, it will be shown how compressive strength and gas permeability can be correlated. A last section will summarize the packing concepts used throughout this chapter.

2.1 FRESH CONCRETE PROPERTIES

In this section, the main properties of fresh concrete are investigated. First, the laws governing the flow behaviour will be analysed from a mechanical viewpoint (which is the aim of rheology). Here, concrete is viewed as a *Herschel–Bulkley* material, characterized by three physical parameters. It will be shown that for practical purposes, however, the *Bingham* model is most often an acceptable approximation, reducing the number of parameters from three to two (the yield stress and the plastic viscosity). One of these two parameters (the yield stress) is well correlated with the Abrams cone slump. Models will be developed for predicting these parameters from the concrete mix design, in the light of the packing theory presented in Chapter 1.

An even more direct application of the packing concept is the definition of concrete *placeability*. After a transportation stage (during which concrete is sheared), the mixture has to be cast and consolidated in a given volume. For a given voids content, the granular mix is characterized by a compaction index. This index must be inferior to the one dealing with the type of placement.

The air content of fresh concrete after a standard consolidation will then be addressed. Finally, after consideration of segregation and bleeding, simplified models will be proposed for fresh concrete workability. Such models will allow an analytical treatment of the mixture-proportioning problem (in section 4.2).

2.1.1   The rheological behaviour of fresh concrete

Definition and measurement

Fresh concrete as a material is intermediate between a fluid and a humid packing of particles. Both types of material may flow and fill a volume of any shape, but the latter suffers limitations in terms of minimum dimensions (which must remain bigger than the maximum size of aggregate). Moreover, unlike fluids, granular mixes exhibit volume changes when they are sheared, and may loose their homogeneity (see section 1.5).

Fluids at low pressure, which can be considered homogeneous and incompressible, are described within the frame of rheology. The aim of this science is to establish relationships between stress and strain rate. If the sample volume remains approximately constant during the flow process, then the law of behaviour becomes a relationship between the shear stress and the shear strain rate (or strain gradient). Then such a law can be sought for fresh concrete, provided that the material meets two criteria:

- it does not segregate during flowing;
- its volume is almost constant during the shear process.

Concretes of dry/plastic consistency are bulky, and contain a significant proportion of entrapped air (5–30% or more) when they flow out of the mixer. They are therefore out of the scope of rheology, and can be characterized only with technological tests. Strictly speaking, only well-graded mixes that are sufficiently fluid may be assessed by the rheological tools. Experience shows that the condition that the *slump* be *higher than 100 mm* guarantees the applicability of rheology to concrete, provided that no excessive segregation takes place during the test (Hu *et al.*, 1995). As an illustration, the water content was changed from 183 to 211 l/m^3 in a mixture having a coarse/fine aggregate ratio of 1.56 and an aggregate/cement ratio of 4.36. The variations of several factors against the slump are given in Fig. 2.1. It can be seen that below a slump of 100 mm the mix exhibits a high dilatancy. Also, the apparent plastic viscosity – see definition below – suddenly drops, which means that the specimen is subject to a local failure, instead of being sheared throughout its whole volume.

A *rheometer* is an apparatus designed to measure the relationship between the shear stress and the strain gradient of a fluid. It can be of various types (Hu, 1995), the most popular being the coaxial, or 'Couette' viscometer. However, the constraint of having a gap between the cylinders that is at the same time small compared with the inner radius and large compared with the maximum size of aggregate leads to a gigantic device. Such a rheometer has been built by Coussot for coarse aggregate–mud suspensions (Coussot and Piau, 1995), and contains about 0.5 m³ of material. It was valuable for confirming the validity of measurements made by smaller devices (Hu *et al.*, 1996), but clearly has little practical relevance outside this research context.

Some researchers have developed smaller concrete coaxial viscometers in which the gap is no longer small with respect to the radius. As a result, the rotation at low speed creates a 'plug flow' phenomenon, which means that only a part of the concrete sample is sheared (Hu *et al.*, 1995). Here, it turns out that coarse aggregate particles tend to be expelled from the sheared zone towards the dead zone, changing the composition and the flowing properties of the sheared concrete, and leading to a systematic underestimation of concrete viscosity (Hu, 1995).

Parallel-plate geometry is more suitable for testing a coarse material such as concrete. Here, a torsional motion is imposed on the sample by means of two parallel planes, one of which is rotating around a perpendicular axis (Fig. 2.2). This principle has been adopted in the design of the BTRHEOM rheometer (de Larrard *et al.*, 1994e; Hu, 1995). While the strain field is not homogeneous as in a Couette viscometer, it is

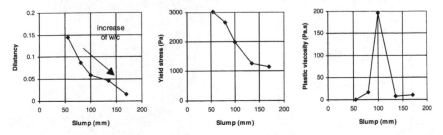

Figure 2.1 Effect of water dosage on the behaviour of a concrete mixture (unpublished test results). Here dilatancy is defined as the relative difference between the volume of fresh concrete sheared at low speed, and after consolidation (Hu, 1995). Unpublished measurements performed with the BTRHEOM rheometer.

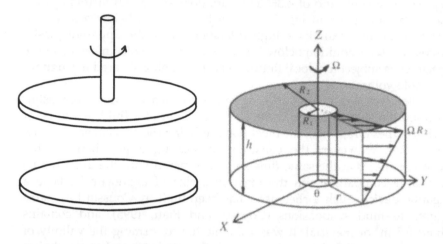

Figure 2.2 Concrete flow in a parallel-plate rheometer (de Larrard *et al.* 1994e, Hu 1995).

imposed by the geometry of the apparatus and governed by the speed of rotation of the top plate. By integration of this strain field one may calculate the law of behaviour of the tested material from the relationship between the torque and the rotation speed. The BTRHEOM rheometer is shown in Fig. 2.3.

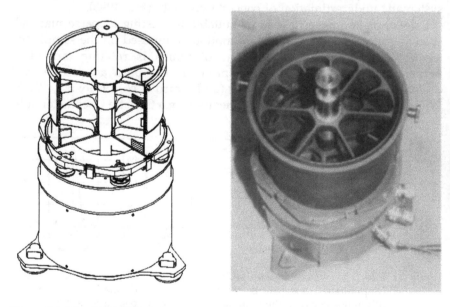

Figure 2.3 The BTRHEOM rheometer, developed at LCPC (Laboratoire Central des Ponts et Chaussées, Paris).

Once the container of the rheometer has been filled with fresh concrete, a light vibration is applied to consolidate the sample. Then the sample is sheared at the maximum rate (which generally corresponds to a rotation speed of 0.8 rev/s), and the speed decreases by successive steps down to 0.2 rev/s. At each step the torque is measured after a stabilization duration of about 20 s. The five couples (torque, rotation speed) allow the user to investigate the rheological behaviour of fresh concrete for a strain gradient ranging from 0.25 to 6 s^{-1}. It is believed that in most concreting operations the strain gradient of fresh concrete is contained within this range, with the exception of pumping.

The Herschel–Bulkley model

Since the work by Tatersall and his co-workers, fresh concrete is no longer considered as a material with a single parameter, called consistency or workability. Tatersall claimed for over 20 years that two parameters were needed to characterize the flow (Tatersall, 1991).

More recently, on the basis of a large experimental plan dealing with 78 different mixes made up with the same raw materials, it has been found that the behaviour of fresh concrete fits quite well with the Herschel–Bulkley model (Ferraris and de Larrard, 1998; de Larrard *et al.*, 1998), at least in the strain gradient range investigated. This model assumes a power-law relationship between shear stress and strain gradient increments:

$$\tau = \tau_0 + m\dot{\gamma}^n \qquad (2.1)$$

where τ is the shear stress applied to the specimen; $\dot{\gamma}$ is the strain gradient; and τ_0, m and n are three parameters characterizing the flow behaviour of the tested material.[1]

In a parallel-plate rheometer, an equation similar to equation (2.1) can be derived, in which the torque and rotation speed stand for the stress and strain rate respectively. An example of experimental results dealing with a mixture tested over an extended strain gradient range is shown in Fig. 2.4. From such a test, a power law can be calibrated by using the least-square method. Then the three material parameters are calculated from the calibration coefficients.

Simplified approach: the Bingham model

The Bingham model became popular in the concrete community following the works of Tatersall, summarized in Tatersall (1991). The Bingham model is a Herschel–Bulkley model with $n = 1$. The m parameter becomes μ, and is called the *plastic viscosity* (measured in Pa.s – pascal-seconds). In reality, the exponent in the Herschel–Bulkley

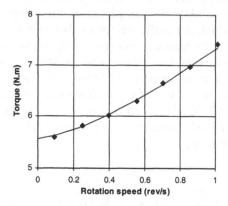

Figure 2.4 Relationship between torque and rotation speed in BTRHEOM rheometer, for a self-compacting concrete. Smoothing of experimental results with the Herschel–Bulkley model (Ferraris and de Larrard, 1998). In this test the strain gradient ranged from 0.13 to 7.6 s^{-1}.

model differs significantly from unity. For example, in the experimental plan carried out by Ferraris and de Larrard (1998), the mean value of n was 1.53 for non-superplasticized mixes and 1.36 for superplasticized mixes.

However, several difficulties arise from the use of a three-parameter model for fresh concrete:

- The uncertainty on the value of each parameter is higher than for a two-parameter model, as the number of experimental points is limited.
- The control of concrete flow properties by changing the mix-design ratios becomes very complicated.
- Practitioners may be reluctant to use rheological tools if the output is too complex, and not related to worker experience in the different construction stages.

This is why, in our opinion, the Bingham model can be kept as a basic approach for fresh concrete, provided that the rheometer tests are first exploited with the Herschel–Bulkley model, from which a linearization is carried out at the material level. This approach is summarized in Fig. 2.5. The Herschel–Bulkley rheogram (that is, the stress/strain gradient curve) is regressed in the $[0, \dot{\gamma}_{max}]$ interval by a straight line departing from $(0, \tau_0)$, using the least-square method. Here $\dot{\gamma}_{max}$ is the maximum strain gradient applied to the specimen during the rheological test (usually 6 s^{-1} in the rheometer test). The yield stress value is the

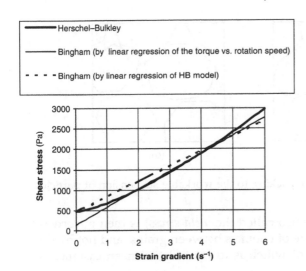

Figure 2.5 Linear approximation of the Herschel–Bulkley model by the Bingham model.

one obtained in the Herschel–Bulkley model, while the plastic viscosity is deduced from the other model parameters (Ferraris and de Larrard, 1998):

$$\mu = \frac{3m}{n+2}\, \dot{\gamma}_{max}^{n-1} \tag{2.2}$$

By contrast, if a linear regression is performed directly on the torque/rotation speed measurements (as proposed by Tatersall, 1991), negative values of yield stress can be encountered, especially with self-compacting concretes (Ferraris and de Larrard, 1998).

A physical explanation of the Bingham model

Let us consider fresh concrete as a granular mixture suspended in water. Here, the whole population of grains (including cement and ultrafine particles, if any) is included in the solid phase, and the air content is neglected. The minimum water volume corresponds to the porosity of the dry packing. By definition, the non-workable mixture is a dense packing, the porosity of which is just saturated with water (Fig. 2.6a).

An increase of the water dosage creates a certain degree of clearance; motions by sliding between particles become possible (Fig. 2.6b). If a shear stress is applied to the system a deformation may appear, provided that the stress is high enough to overcome the friction between particles.

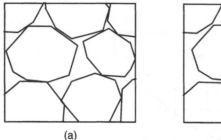
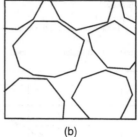

(a)   (b)

Figure 2.6 From dry packing to the workable suspension of particles.

The shear threshold (also called the *yield stress*) is then governed by the number and the nature of contacts between grains, and not by the liquid phase, the only role of which is to control the mean distance between grains.

   If a given strain gradient is now applied at a macroscopic scale, one may assume that the rate field will depend on one single parameter. In other words, if two experiments are performed on the same granular system with two different strain gradients, the particle motions will be statistically the same with the exception of the overall timescale. Then the fluid velocity field in the porosity of the granular mix will be complex, but still proportional to the strain gradient imposed on the system. With the hypothesis that the fluid is newtonian,[2] and remains in a laminar state, its contribution to the total stress will remain proportional to the mean strain gradient. Therefore, in the classical form of the Bingham model:

$$\tau = \tau_0 + \mu\dot{\gamma} \tag{2.3}$$

the τ_0 term appears as the solid phase contribution, and the $\mu\dot{\gamma}$ term as the liquid phase one (Fig. 2.7).

   Among a wide range of strain gradients, the hypothesis of a particle motion field depending on a unique parameter is only an approximation. The degree of deflocculation may vary with the macroscopic strain gradient, inducing changes in the circulation of water through the porous medium. Thus the plastic viscosity depends on the strain gradient, which produces a non-linear behaviour of the Herschel–Bulkley type.

Concrete rheology vs. time

Fresh concrete is by nature a system that is far from equilibrium (from a thermodynamic viewpoint). In any practical application (except in shotcrete technology) the evolution of behaviour over one or two hours

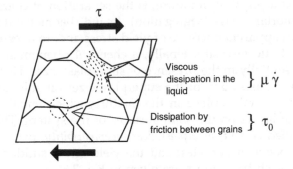

Figure 2.7 The solid and liquid contributions to the shear resistance of concrete.

is as important as the initial state. Rheological properties are first measured after the end of the mixing process, but it is the behaviour at the casting stage that really matters.

It is now possible to perform a monitoring of the fresh concrete behaviour with a single specimen. This procedure not only makes the test lighter, but also avoids a sampling error, which could overcome the significance of comparisons between tests results performed at different ages. Based on monitoring tests performed with the BTRHEOM rheometer we have proposed a classification of rheological behaviour, linked with the physico-chemistry of the cementitious system (de Larrard *et al.*, 1996c; Table 2.1). Delayed absorption of water by the aggregate skeleton is the first physical phenomenon that may take place in fresh concrete. On the basis of the physical models presented below, an

Table 2.1 Evolution of rheological parameters during the period of use of fresh concrete: interpretation and remedies.

Case	Yield stress	Plastic viscosity	Slump	Top surface of sample	Interpretation	Remedy
I	$\rightarrow$	$\rightarrow$	$\rightarrow$	–	Stable mix	–
II	$\uparrow$	$\rightarrow$	$\downarrow$	–	Chemical activity	Add a retarder or change of cement/SP system
III	$\uparrow$	$\uparrow$	$\downarrow$	–	Water absorption	Pre-saturate aggregates
IV	$\downarrow$	$\rightarrow/\downarrow$	$\rightarrow/\uparrow$	Coarse agg. rising	Segregation between mortar and coarse aggregate	Change the grading of aggregate or add a viscosity agent

Source: de Larrard *et al.* (1996c).

increase of both yield stress and plastic viscosity can be expected. A second phenomenon, still physical in nature, is the separation of coarse aggregates from the mortar phase (segregation). In the rheometer this phenomenon creates an apparent decrease of yield stress and, to a lesser extent, a decrease of plastic viscosity. Finally, a chemical phenomenon sometimes appears, essentially in the presence of superplasticizer. Here, because of a preferential adsorption of the superplasticizer on sulphate phases, an early hydration takes place in the mixture, which entails a precipitation of hydrate layers on the solid surfaces of the system. This phenomenon, often called *cement/superplasticizer incompatibility*, primarily affects the friction between particles, and the yield stress suddenly increases. Examples of such behaviours are given in Fig. 2.8.

Nowadays it is not possible to develop predictive models dealing with the evolution of rheological behaviour of fresh cementitious mixtures. The phenomena are too complex, and still poorly understood (especially the chemical ones). However, the initial behaviour can be linked with the mix composition, with the help of packing concepts. In the practical mix-design process, rheological tests have to be carried out during a period as long as the practical duration of use of the mix to be proportioned. Preliminary tests can be performed on the fine phase of fresh concrete (see section 3.4.2), and the suitability of the material combination tested on the whole mixture (see section 4.2.3).

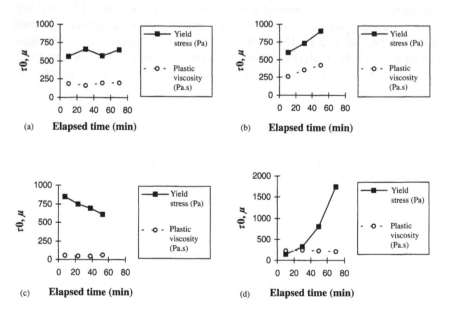

Figure 2.8 Typical behaviours of fresh concrete vs. time (de Larrard *et al.*, 1996c). (a) stable mixture; (b) water absorption by the aggregate; (c) segregation; (d) cement/SP incompatibility.

Keeping the physical analysis of fresh concrete in mind, and with the help of experimental data, we shall now develop specific models relating the two Bingham parameters (τ_0 and μ) to the concrete mix design just after mixing.

2.1.2 Plastic viscosity

Application of the compressive packing model to fresh concrete

As a preliminary step in the modelling of fresh concrete rheological properties it is necessary to predict the minimum water dosage that fills the voids. It is the overdose of water with respect to this minimum value that provides workability to the fresh material.

In Chapter 1 a comprehensive theory was presented that allows us to calculate the packing density of a *dry* granular mixture, from the knowledge of two types of material parameter:

- the grading curve of the mixture (describing the grain size distribution),
- the residual (virtual) packing density of each individual fraction present in the mix,

and from the compaction index (K), a characteristic of the packing process. The model has been validated with the help of several data sets, and its precision is generally better than 1% in absolute value.

To what extent can the model be used for granular mixtures with a very wide grading span, with a systematic agglomeration between fine particles? As shown in Fig. 2.9, the precision provided by the CPM for

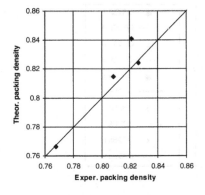

Figure 2.9 Validation of the CPM (compressible packing model) for dry concretes (Ferraris and de Larrard, 1998). All packing measurements were performed according to the procedure described in section 1.2.2. $K = 9$.

'dry[3] concrete' is still very satisfactory. Here, the packing density of components (coarse aggregate, fine aggregate, very fine sand, and cement) was measured, and the β_i coefficients were calculated with an assumption of uniformity within each component grading range.

A natural idea is to consider a 'humid concrete' having the minimum voids content as a packing of particles. However, does water play the same role in such a system as air (or voids) in a dry packing? Certainly, water changes the surface forces, which are important for fine particles. As an example, the experimental packing density of the cement in the experiments presented in Fig. 2.9 was 0.553. In the presence of water it was possible to produce a smooth (while thick) cement paste with a solid concentration of 0.565. Thus it appears that the mixing and shearing process used for preparing the cement paste, together with the capillary forces, is more efficient for compacting fine particles than the normal combination of vibration and pressure on the dry packing. Moreover, a thick cement paste still has the ability to be deformed by shear. Then, given the interactions that take place between particles, a more compact state is attainable, which means that there should be a finite compaction index corresponding to the state of thick paste. As we did for dry packing of particles, the relevant K coefficient has to be determined by simulation of binary mixtures: the higher K is, the sharper is the curve of packing density vs. proportion of fine grains (Fig. 1.14). From measurements dealing with the water demand of binary mixtures of powders (Fig. 2.10), a value of 6.7 has been found for the compaction index of a smooth thick cement paste.

Finally, the packing density of a concrete in the presence of water, which is the complement to 1 of the porosity, or minimum water dosage, can be calculated with the CPM. The β_i values for the aggregate are determined as

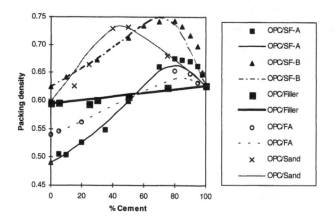

Figure 2.10 Solid concentration in water demand tests, for superplasticized binary mixtures of powders. Experimental data and predictions of the CPM (with $K = 6.7$). Data from Sedran (1999).

for a dry packing, and those for the binders come from the water demand measurements. If an organic admixture having plasticizing properties is to be used, this product must be added to the binder during the water-demand tests, at the same percentage (see sections 3.2.3 & 3.3.3).

In fact the introduction of a plasticizer/superplasticizer makes the rheology of cementitious materials far more complex than when no organic admixture is used. A first clarification consists in defining a *saturation dose* for a binder/admixture couple. This amount represents the amount above which no further plasticizing effect is experienced (see section 3.4.2). All practical percentages of organic admixture, P, will be found between 0 and the saturation dosage P^*. For predicting the packing density of concrete with any amount in this range, it is proposed to determine experimentally the β_i parameters of the binder, for $P = 0$ and $P = P^*$. Then it appears that the following empirical equation (parabolic model) leads to a fair prediction of β_i for intermediate amounts of plasticizer/superplasticizer:

$$\beta_i = \beta_i^{P=P^*} + \left(1 - \frac{P}{P^*}\right)^2 (\beta_i^{P=0} - \beta_i^{P=P^*}) \qquad (2.4)$$

A validation of this model, with three cement/superplasticizer systems, is displayed in Fig. 2.11.

As for the grading curve, it is clear that conventional measurements (such as LASER granulometry) give a picture that can be quite different from the real state of agglomeration of fine particles in the fresh concrete.

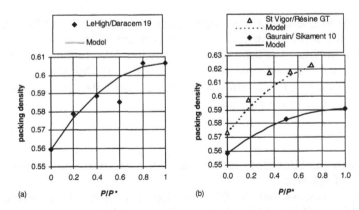

Figure 2.11 Prediction of packing density of binder at intermediate dosages of plasticizer/superplasticizer, with a parabolic model: (a) data taken from Ferraris and de Larrard (1998); (b) private communication from T. Sedran, LCPC (1997).

Fortunately this uncertainty has little effect on the proportioning calculations, for the following two reasons:

- The error is made twice (in the calculation of the β_i from the water demand test, and in the concrete calculations); to a certain extent the second error tends to overcome the effect of the first, if the state of agglomeration is comparable in the pure paste and in the concrete.
- Flocculation is of concern for the fine part of cement, which generally represents the finest part of the concrete-grading span. Therefore all corresponding interaction terms in the CPM equations deal with small d_i/d_j size ratios,[4] and have little weight in the final result (which is the packing density of the whole mixture).

Plastic viscosity and normalized solid concentration

Following the proposed physical analysis, the plastic viscosity of Bingham materials corresponds to the viscosity of Newtonian bodies. Various models can be found in the literature for the viscosity of suspensions. Most of them express the fact that viscosity increases with the solid concentration of the suspension, and tends towards infinity when the concentration is close to the packing value.

Such a model has been proposed by Krieger and Dougherty (1959), and applied to the apparent viscosity[5] of cement paste (Struble and Sun, 1995). The model is of the following form:

$$\frac{\eta}{\eta_c} = \left(1 - \frac{\Phi}{\Phi^*}\right)^{-[\eta]\Phi^*} \tag{2.5}$$

where η is the viscosity of the suspension, η_c is the viscosity of the fluid (water), Φ is the solid volume, Φ^* is the maximum solids concentration, and $[\eta]$ is the intrinsic viscosity of the suspension. We shall call the ratio Φ/Φ^* the *normalized solid concentration*. The Krieger–Dougherty model has the advantage of referring to packing concepts, and of being consistent with Einstein's equation:

$$\eta = 1 + 2.5\Phi \tag{2.6}$$

This equation applies for low solid concentration monodisperse suspensions of spheres.

However, it appears that the Φ^* term depends on the strain gradient (Struble and Sun, 1995), which is in contradiction with the fact that packing density is a *static* concept. Moreover, the predictions do not provide a good agreement with experimental values dealing with mortars and concrete (Ferraris and de Larrard, 1998).

Farris (1968) proposed a model applying for polydisperse suspensions, with the condition that all granular fractions are very different in size as compared with each other. This condition is unfortunately not verified for cementitious materials, the size distribution of which is rather continuous. There have been some attempts to apply Farris' model to concrete, by modelling the constituents as monodisperse fractions (Hu *et al.*, 1995; Hu, 1995). This oversimplification of the concrete size distribution can also be found in some packing models used for concrete mixture proportioning (Dewar, 1993). While these models offer good qualitative agreement with experience they cannot provide a quantitative prediction of concrete plastic viscosity, especially if the aim is to develop a model incorporating physical parameters that can be calibrated on each individual constituent. Moreover it is well known that the rheological behaviour of fresh concrete is sensitive to minor changes in the size distribution of the constituents. Clearly, such changes cannot be reflected by models that assume that the constituents are monodisperse.

For modelling the relationship between plastic viscosity and mix composition we shall refer to the fact that experimental values of plastic viscosity are ranked as the normalized solid concentration. This fact, found for binary mixtures of spheres (Chang and Powell, 1994), is quite well verified for mortar and concrete (Ferraris and de Larrard, 1998). In this large experimental programme, 78 mixtures produced with the same constituents were tested with the BTRHEOM rheometer (one test performed without vibration for each mixture). Thirty-three were mortars and concrete without admixture, while the other mixes contained a naphthalene-based superplasticizer at the saturation dosage (see section 3.4.2). The mixture proportions were chosen to cover virtually all feasible mixtures in terms of cement content, coarse/fine aggregate ratio, and consistency (Fig. 2.12). In addition to the two main series, some mixtures were designed with intermediate amounts of superplasticizer. Also, four concretes contained silica fume in the slurry form, with dosages ranging from 7.5 to 30% of the cement mass.

The relationship between experimental plastic viscosity (calculated from equation (2.2)) and the normalized solid concentration, Φ/Φ^*, appears in Fig. 2.13. Here the solid concentration is computed from the mix compositions, neglecting the air content (no entraining agent was used in this programme). As for the packing density, Φ^*, it was evaluated by using the CPM, with a compaction index of 9. This K value corresponds to the densest arrangement that can be obtained in a randomly produced packing: that is, not placed grain by grain (see section 1.2.3).

These data are smoothed by the following empirical model:

$$\mu = \exp\left[26.75\left(\frac{\Phi}{\Phi^*} - 0.7448\right)\right] \qquad \text{(mean error: 61 Pa.s)} \qquad (2.7)$$

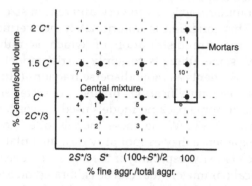

Figure 2.12 Experimental plan reported in Ferraris and de Larrard (1998). Each point represents a combination of dry materials, from which three mixtures were generated by changing the water/solid dosage.

The scatter around the model curve found in Fig. 2.13 can be explained as follows:

- Plastic viscosity is measured with an uncertainty margin of about ±10% in relative value (Hu, 1995), if the mixture does not segregate. However, in this programme the range of variations of mix proportions was so large that some mixtures exhibited some segregation (with an excess of coarse aggregate rising) and/or bleeding.
- The solid volume in the sheared specimen is less than the theoretical one calculated with the hypothesis of no air in the mix.
- Finally, the calculation of the packing density Φ^* is subject to an error of 0.005–0.01.

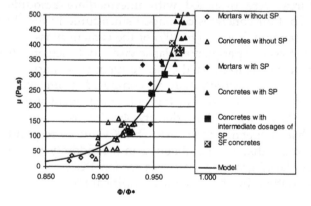

Figure 2.13 Relationship between plastic viscosity and normalized solid concentration, for a large series of mixtures produced with the same constituents (Ferraris and de Larrard, 1998).

This is why this curve supports the assumption that it is mainly the normalized concentration that controls the plastic viscosity.

Confirmation with other data

In another programme, 15 concretes were designed with various aggregate fractions (a crushed limestone from 'Le Boulonnais' mixed with a natural sand), a Portland cement, and, for some of them, fly ash, silica fume, air-entraining agent and superplasticizer (de Larrard *et al.*, 1996b). The target compressive strengths at 28 days ranged from 25 to 120 MPa. Two additional mixes were produced with a basalt aggregate (from 'Raon') and the same binders and admixtures. The mixture proportions and rheological properties appear in Table 2.2. All concretes were tested with the BTRHEOM rheometer, with the original procedure (Hu, 1995), slightly different from that adopted in the programme performed at NIST (Ferraris and de Larrard, 1998). In the present data the mixtures were tested first under vibration, and then without vibration. The relationship between the plastic viscosity and the normalized solid concentration appears in Fig. 2.14. Still, the normalized solid concentration provides a fair classification of the mixtures, in terms of plastic viscosity. The data are best fitted by the following model:

$$\mu = \exp\left[38.38 \left(\frac{\Phi}{\Phi^*} - 0.8385 \right) \right] \qquad \text{(mean error: 28 Pa.s)} \qquad (2.8)$$

The difference in the coefficient values, as compared with equation (2.7), is probably due to differences in the testing procedure. When a test under vibration is first carried out, some segregation of coarse aggregate[6] may take place, producing a lower apparent plastic viscosity. Otherwise, it can be stated from Fig. 2.14 that entrained air has little effect on plastic viscosity. The pattern is different for yield stress, as shown below.

Finally, other data have been taken from the LCPC rheology database. In the Grand Viaduc de Millau mix-design study, summarized in de Larrard *et al.* (1996a), four high-performance concretes were produced with different types of aggregate. In a previous work, another special HPC was designed for a nuclear power plant (de Larrard *et al.*, 1990). All these mixtures have been tested with the rheometer (following Hu's procedure), and their viscosity is plotted in Fig. 2.15 against their normalized solid concentration. The mean error given by equation (2.8) in predicting these results is equal to 53 Pa.s, which is of the same order of magnitude as the values found for the previous data. It can therefore be considered that this model is validated for plastic viscosities measured following the standard test (under vibration then without vibration) defined in Hu (1995).

Table 2.2 Mixture proportions and rheological properties of a series of concretes with target strength ranging from 25 to 120 MPa (de Larrard et al., 1996b).

Mix	Water (l/m³)	SP (kg/m³)	R (kg/m³)	AEA (kg/m³)	SF (kg/m³)	FA (kg/m³)	OPC (kg/m³)	River sand (kg/m³)	CL 0/5 (kg/m³)	CL 5/12.5 (kg/m³)	CL 12.5/20 (kg/m³)	Basalt 6/10 (kg/m³)	% air %	Slump (mm)	Plastic viscos. (Pa.s)
1	193	0.0	0.0	0	0	0	230	446	453	388	619	0	0.8	105	–
2	187	0.0	0.0	0	0	48	195	449	456	369	623	0	0.5	175	35
3	166	1.1	1.4	0	0	95	223	436	443	421	565	0	0.9	95	176
4	197	0.0	0.0	0	0	0	410	400	406	428	509	0	1.0	140	21
5	181	0.0	0.0	0	0	79	325	401	408	453	503	0	1.0	90	109
6	146	12.4	3.3	0	0	0	461	401	407	475	550	0	0.5	250	132
7	136	12.0	2.5	0	22	0	360	435	442	465	579	0	0.5	250	107
8	124	12.5	2.6	0	38	0	377	432	439	488	561	0	0.7	240	179
9	124	15.6	3.3	0	57	0	470	407	413	437	554	0	0.9	245	160
10	160	0.0	0.0	0.31	0	0	230	427	433	454	574	0	7.2	110	75
11	159	0.0	0.0	0.54	0	49	189	405	411	454	586	0	7.3	105	93
12	188	0.0	0.0	1.32	0	0	483	364	368	390	477	0	7.1	120	32
13	192	0.0	0.0	1.44	0	107	428	317	322	422	452	0	5.2	85	81
14	151	14.8	3.9	0.98	0	0	557	332	338	443	489	0	6.5	265	146
15	132	12.9	2.7	0.58	23	0	387	408	414	437	550	0	5.3	250	159
16	127	13.7	2.9	0	41	0	412	861	0	0	0	1037	0.9	245	308
17	122	15.5	3.2	0	55	0	461	837	0	0	0	1010	1.0	23	162

SP, superplasticizer; R, retarder; AEA, air-entraining agent; SF, silica fume; FA, fly ash; OPC, ordinary Portland cement; CL, crushed limestone.

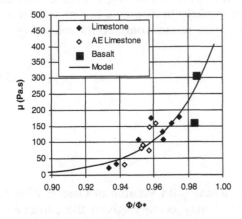

Figure 2.14 Relationship between plastic viscosity and normalized solid concentration (de Larrard *et al.* 1996b). AE: mixtures containing entrained air.

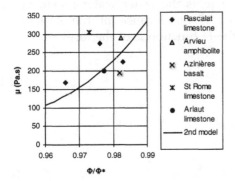

Figure 2.15 Relationship between plastic viscosity and normalized solid concentration. Experimental data and predictions from equation (2.8).

2.1.3 Yield stress

Yield stress of mixtures without organic admixture

As already stated in section 2.1.1, the yield stress is viewed as the macroscopic result of friction between the various granular fractions. Referring to the approach developed in section 1.2.1, it is tempting to look for a model similar to the compaction index equation, since the contributions of the grains are additive by nature. However, if we assume that the friction within a monodisperse packing is governed by the number of contacts between grains, we may expect the appearance of a *size effect*.

For two monodisperse assemblies of particles having the same geometry but different sizes, the finest one is likely to display a higher yield stress. Thus, a general shape of a yield stress model could be the following:

$$\tau_0 = f \left(\sum_{i=1}^{n} a_i K_i' \right) \tag{2.9}$$

$$\text{with } K_i' = H \left(\frac{\Phi_i}{\Phi_i^*} \right)$$

where f is a function; Φ_i is the actual solid volume of class i; Φ_i^* is the maximum volume that particles i may occupy, given the presence of other particles; a_i is a coefficient, which is supposed to increase when the diameter decreases; and K_i' is the contribution of the i fraction to the compaction index of the mixture.

Here the granular system is constituted by the de-aired concrete. In other words, the voids content equals the water/water + solid volume ratio (de Larrard and Ferraris, 1998). Because of the system self-consistency (see section 1.2.1), the H function is of the form

$$H(x) \propto \frac{x}{1-x} \tag{2.10}$$

Then we have for the yield stress:

$$\tau_0 = f \left(\sum_{i=1}^{n} a_i \frac{\Phi_i/\Phi_i^*}{1 - \Phi_i/\Phi_i^*} \right) = f \left(\sum_{i=1}^{n} a_i K_i' \right) \tag{2.11}$$

For calibration of the parameters a_i and function f, let us refer to the experimental plan already used for modelling the plastic viscosity (de Ferraris and Larrard, 1998; see Fig. 2.12). In this programme 23 mixtures without superplasticizer were characterized with the rheometer, leading to the following equation:

$$\tau_0 = \exp (2.537 + 0.540 K_g' + 0.854 K_s' + 1.134 K_c') \tag{2.12}$$

where K_g', K_s' and K_c' are the contributions of gravel, sand and cement to the compaction index respectively. The mean error provided by the model was 148 Pa (see Fig. 2.16).

To establish this equation it has been necessary to adopt mean values of a_i within each individual constituent. However, if the aim is to apply

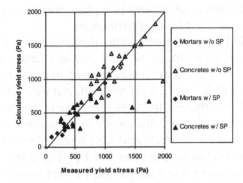

Figure 2.16 Comparison between experimental data and predictions of yield stress equations (2.12) and (2.15) (Ferraris and de Larrard, 1998).

the model to mixtures produced with other materials, it is necessary to derive for the a_i parameters an equation depending only on the size of the grain fractions. By plotting the three values against the mean size of the materials (that is, the size corresponding to a proportion of 50% passing), we obtain the diagram given in Fig. 2.17, which can be summarized by the following empirical equation:

$$a_i = 0.736 - 0.216 \log(d_i) \tag{2.13}$$

and

$$\tau_0 = \exp\left(2.537 + \sum_{i=1}^{n} a_i K_i'\right) \tag{2.14}$$

where d_i is the diameter of the grain fraction in mm. Then it is confirmed

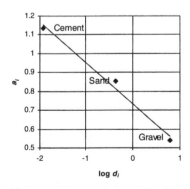

Figure 2.17 Smoothing of the a_i parameter as a function of the grain size.

that the contribution of a constituent to the yield stress increases when its size decreases. In particular, the effect of cement is higher than the sand contribution, which is superior to the gravel contribution.

Further validation is provided by the experiments already reported in Fig. 2.1. Here the mixtures were produced with a crushed limestone having a maximum size of 20 mm. The model has been used to predict the yield stress, by using equations (2.13) and (2.14) and without any fitting process. Although the model is applied for concretes made up with different materials, and in a range of lower consistency as compared with the original experimental programme (Ferraris and de Larrard, 1998), the agreement with experiments is still fair (mean error 161 Pa) (Fig. 2.18). As a result it is likely that yield stress measurements can be performed for slump lower than 100 mm, even if, in this range, the plastic viscosity obtained with the rheometer is no longer significant (Fig. 2.1).

Effect of superplasticizer (at the saturation point)

While plastic viscosity is moderately influenced by plasticizing admixtures (essentially through the virtual packing density of the fine particles), yield stress is fundamentally changed by plasticizers/ superplasticizers. Basically, the admixture exerts two microstructural effects on the cementitious system (Foissy, 1989):

- It creates a better deflocculation between fine particles, which increases the packing density of the system (geometrical effect).
- It lubricates the solid surfaces, decreasing the friction stresses between particles (mechanical effect).

A total of 25 mixtures containing a saturation dosage of superplasticizer were prepared and tested with the BTRHEOM rheometer (Ferraris and de Larrard, 1998). The variations of sand/total aggregate ratio and cement

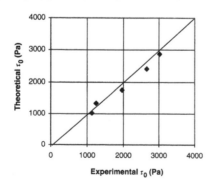

Figure 2.18 Comparison between predictions and measurements, for the yield stress of limestone concretes (unpublished results).

content adopted in this programme are illustrated in Fig. 2.12. The amount of superplasticizer was constant in terms of percentage of cement weight, and sufficed to make possible a maximum deflocculation of the cement in the fresh mixtures.

Here the model for yield stress was calibrated as follows:

$$\tau_0 = \exp (2.537 + 0.540K_g' + 0.854K_s' + 0.224K_c') \tag{2.15}$$

The mean error between predictions and experiments is equal to 177 Pa (Fig. 2.16). However, if the mixtures having at the same time a high coarse/fine aggregate ratio and a low cement content are removed from the data set[7] (number 4 in Fig. 2.12), the error drops to 109 Pa.

Comparing equations (2.12) and (2.15) it is clear that the contributions of aggregate fractions are the same, while that of cement is strongly reduced by the superplasticizer. As the deflocculation effect is already accounted for by the increase of the cement virtual packing density (see section 2.1.2), the decrease of the multiplicative coefficient of K_c' is viewed as a mechanical effect of lubrication between particles, related to the adsorption of admixture on the surface of cement particles.

Therefore, for predicting the yield stress of a mixture containing a superplasticizer at the saturation dosage, the a_i parameters dealing with aggregate can still be taken following equation (2.13), while for the cement a lower value must be used.

Mixtures containing an intermediate dose of superplasticizer

We have already studied the effect of an intermediate dose of super-plasticizer on the packing ability of cement. It appears that the virtual packing density of cement in the presence of superplasticizer can be taken as a parabolic function of the superplasticizer dosage (equation (2.4)). As for the yield stress, an equation of the type of equations (2.12) and (2.15) has to be developed. Here the aggregate terms are not likely to change, as they are the same with no superplasticizer, and with superplasticizer at the saturation dosage. Only the a_i term of the cement must be interpolated, between the value corresponding to no superplasticizer (1.134) and that of highly superplasticized mixes (0.244). The few experimental data acquired for mixtures with intermediate dosages of superplasticizer (Ferraris and de Larrard, 1998) suggest that the lubrication effect is even more rapidly obtained than the deflocculation one when the superplasticizer dosage increases from 0 to the saturation dose. As a preliminary step, we propose the following model for the yield stress, based upon a cubic interpolation:

$$\tau_0 = \exp\{2.537 + 0.540K_g' + 0.854K_s' + [0.224 + 0.910(1 - P/P^*)^3]K_c'\} \tag{2.16}$$

which gives, in the general case,

$$\tau_0 = \exp \left\{ 2.537 + \sum_{\text{aggregate}} [0.736 - 0.216 \log(d_i)] K_i' \right.$$

$$\left. + [0.224 + 0.910(1 - P/P^*)^3] K_c' \right\} \quad (2.17)$$

The trend observed in the experiments is to a certain extent reproduced by this model (Fig. 2.19). However, more experimental data would be needed to check its validity.

2.1.4 Abrams cone slump

Slump vs. yield stress and specific gravity

As the Abrams cone remains the most widely used test for characterizing concrete consistency, it is important to establish a model linking the slump to the concrete mixture proportions. There have been several attempts to find a relationship between slump and yield stress. A review can be found in Hu (1995). More precisely, it can be shown from a dimensional analysis that the slump is governed by the quantity $\tau_0/\rho g$, where τ_0 is the yield stress, ρ is the specific gravity of fresh concrete, and g is the acceleration due to gravity (Hu, 1995; Pashias *et al.*, 1996). This analysis neglects the role played by concrete heterogeneity at the scale of

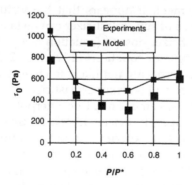

Figure 2.19 Yield stress vs. dosage of superplasticizer. The mixtures containing intermediate dosages were interpolated (in terms of composition) between the central mixture in Fig. 2.12 (without SP) and the corresponding one with SP (Ferraris and de Larrard, 1998). However, the water dosage was not constant, so that this curve should not be taken as a 'saturation curve' (see section 3.4.2).

the test sample. In particular, for high slumps the thickness of the sample is no longer high with regard to the maximum size of aggregate. Also, some segregation phenomena may take place during the slump test (it is common to see a clump of coarse aggregate surrounded by a lake of mortar). Finally, the friction exerted by the substratum also plays a role in the final position of the slump test.

Neglecting these secondary phenomena, and using finite-element calculations, Hu established the following model for slump (Hu, 1995; Hu *et al.*, 1996):

$$s = 300 - 0.27 \frac{\tau_0}{\rho} \tag{2.18}$$

where s is expressed in mm and τ_0 in Pa, while ρ is dimensionless.

On the basis of a large experimental plan, an empirical correction has been proposed (Ferraris and de Larrard, 1998) leading to the following equation, which we shall adopt in the range of slump higher than 100 mm:

$$s = 300 - 0.347 \frac{(\tau_0 - 212)}{\rho} \tag{2.19}$$

The validation of this model is given in Fig. 2.20 (mean error 24 mm).

Therefore the prediction of slump will be performed through the yield stress model (see section 2.1.3), from which the slump will be deduced by using equation (2.19).

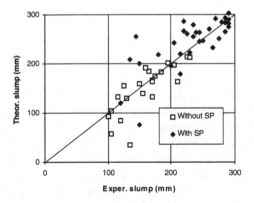

Figure 2.20 Comparison between experimental slumps and predictions given by equation (2.19). Data from Ferraris and de Larrard (1998).

Validation 1: effect of mix-design parameters

In a large unpublished experimental programme performed at LCPC, 31 mixtures were produced with the same set of constituents for investigating the effect of various mix-design parameters (water content, cement content, and coarse/fine aggregate ratio) on slump and air content. The coarse aggregate was a round sea flint from Crotoy having a maximum size of 12.5 mm. The fine aggregate came from the Seine river, and was corrected with a fine sand from Fontainebleau. The cement was an ordinary Portland cement CEM I 52.5 from Cormeilles-en-Parisis. Three series of mixtures were prepared: mortars, concretes without superplasticizer and concretes with a saturation dosage of melamine-type superplasticizer. The mixture proportions and experimental results are presented in Table 2.3.

On the basis of grading and packing density measurements performed on the raw materials – that is, on a purely predictive manner – the theoretical slump was calculated with the model presented above. Comparisons between predictions and measurements are given in Fig. 2.21.

While most experimental slumps are below 100 mm, which is in principle out of the scope of the theory, the agreement of the model with experiments is quite satisfactory. All the trends are reflected, including the effect of superplasticizer, although the chemical nature of the product is not the same as that of the superplasticizer used in the first programme (Ferraris and de Larrard, 1998). Surprisingly, the model systematically underestimates the slump of mixtures without super-plasticizer[8], while the prediction of the slump of mixtures with superplasticizer is excellent. Nevertheless, the mean error is only 45 mm, which is somewhat less[9] than the sum of errors of the mix design/yield stress model (about 150 Pa, or 22 mm of slump) and the yield stress/slump model (27 mm).

Validation 2: mixtures made up with various types of aggregate

Let us examine a study on the effect of aggregate on compressive strength of high-performance concrete (de Larrard and Belloc, 1997). Twelve mixtures were produced with the same cement paste (a blend of Portland cement, silica fume, water and superplasticizer, with a water/binder ratio of 0.26). Five types of aggregate were compared: a round flint from Crotoy, a hard crushed limestone from Le Boulonnais, a soft (semi-hard) limestone from Arlaut, a basalt from Raon l'Etape, and a quartzite from Cherbourg. For each type of aggregate, mortars and concretes were designed and tested. The paste volume was adjusted to produce slump values around 250 mm. This strategy led to free water contents ranging from 143 to 241 l/m^3, depending on the shape and size

Table 2.3 Influence of various factors on slump and air content.

Parameter investigated	Mixture	CA (kg/m³)	FA (kg/m³)	Sand (kg/m³)	OPC (kg/m³)	SP (kg/m³)	Water (kg/m³)	Slump (mm)	Air (%)
Control mortar	MO	0	1092	364	518	0.00	283	95	5.4
Effect of water content	MO − 30	0	1151	384	518	0.00	253	15	7
	MO − 15	0	1121	374	518	0.00	268	45	6.1
	MO + 10	0	1072	357	518	0.00	293	185	3.1
	MO + 20	0	1052	351	518	0.00	303	240	2
Effect of cement content	MO − 200	0	1218	406	319	0.00	283	45	6.1
	MO − 100	0	1155	385	419	0.00	283	85	4.8
	MO + 100	0	1029	343	618	0.00	283	95	3.9
	MO + 200	0	966	322	717	0.00	283	60	4.1
Control NSC	BO	1043	559	186	340	0.00	190	85	1.5
Effect of CA/FA ratio	BO 0.5	602	903	301	344	0.00	192	5	4.8
	BO 1	897	673	224	342	0.00	190	45	2
	BO 2	1188	446	149	339	0.00	189	170	1
	BO 3	1332	333	111	338	0.00	188	100	0.8
Effect of cement content	BO − 200	1136	609	203	146	0.00	189	45	1.5
	BO − 100	1090	584	195	243	0.00	189	75	1.4
	BO + 100	996	534	178	438	0.00	190	60	1.6
	BO + 200	949	509	170	536	0.00	190	20	2
Effect of water content	BO − 30	1087	582	194	340	0.00	160	5	2.9
	BO − 15	1065	571	190	340	0.00	175	40	1.6
	BO + 10	1033	553	184	337	0.00	197	145	1.2
	BO + 20	1014	543	181	341	0.00	209	225	0.9
Control HPC	BHP	1062	569	190	437	12.98	135	200	1.9
Effect of CA/FA ratio	BHP 1	914	685	228	439	13.03	136	25	2.3
	BHP 2	1210	454	151	436	12.94	135	220	0.8
	BHP 3	1356	339	113	434	12.89	134	215	0.9
Effect of cement content	BHP − 100	1112	596	199	340	10.09	135	10	1.3
	BHP + 100	1012	542	181	535	15.88	135	210	2.4
Effect of water content	BHP − 15	1084	581	194	437	12.98	120	5	2.1
	BHP + 5	1055	565	188	437	12.99	140	230	1.5
	BHP + 10	1048	561	187	437	12.99	145	250	1.2

CA, coarse aggregate; FA, fine aggregate; OPC, ordinary Portland cement; SP, superplasticizer. Unpublished tests performed at LCPC.

distribution of the various aggregate combinations. The details of material characteristics and mixture proportions are given in Tables 2.12 and 2.13. For the computation of the theoretical slump an a_i coefficient of 1.25 was taken for the silica fume, as determined in the first experimental programme (Ferraris and de Larrard, 1998). Also, the contribution of the fine part of the aggregate to the yield stress has been modelled by equation (2.13), which means that the lubricating effect of superplastizer on this granular fraction has been neglected. The mean

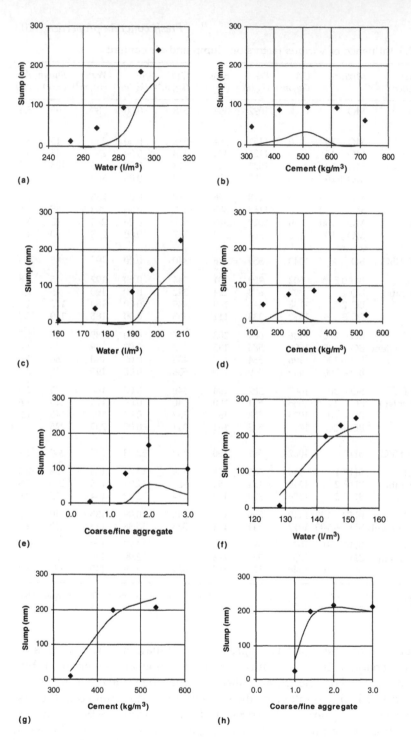

Figure 2.21 Comparison between experimental values of slump and predictions of the model, for the mixtures detailed in Table 2.3: (a),(b) mortars; (c)–(e), concretes; (f)–(h) HPCs.

error provided by the slump model, as compared with the experimental data, is equal to 39 mm (Fig. 2.22). Therefore it can be concluded that the model is able to reflect the contribution of any type of aggregate to the slump, giving a rather universal character to the a_i coefficients.

Validation 3: concretes with other admixtures

Finally it can be of interest to extend the applicability of the model to concretes containing fly ash or entrained air. We shall refer to the experimental programme already used for validating the plastic viscosity model (de Larrard *et al.*, 1996b; Table 2.2). Mixtures 1 and 3 are plain Portland cement concretes, and have been simulated without any fitting process. For mixes 2 and 4 (which contained fly ash), it has been assumed that, owing to their smooth and spherical form, fly ash grains do not contribute directly to the mixture yield stress, which means a value of 0 for the corresponding a_i. In air-entrained mixtures (mixes 10–15) air bubbles have been modelled as a fine aggregate having a uniform distribution between 10 and 200 µm, which corresponds to the usual grading range for entrained air (Pigeon and Pleau, 1995). The contribution of bubbles to the yield stress received a multiplicative parameter of 0.56, obtained by a fitting process. The mean error of the model, for the 17 mixtures, is 46 mm (see Fig. 2.23).

Finally it is realized that, in spite of the scatter encountered in the model linking slump and yield stress, we have built a reasonably predictive model for slump,[10] which may work among mixtures containing a wide range of components. The precision of the model is not sufficient to design a mixture and to obtain at the first batch the required slump, with the usual margin of ±20 mm. However, after a few trial batches the suitable water content will be quickly found (see section 4.2.3). Minor changes in the water content between the first and the last batch will be of little consequence in terms of hardened concrete

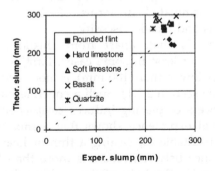

Figure 2.22 Comparison between experimental values of slump and predictions of the model, for the mixtures detailed in Table 2.12. Experimental data from de Larrard and Belloc (1997).

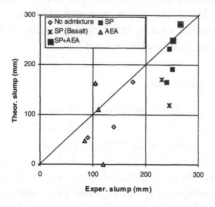

Figure 2.23 Comparison between experimental values of slump and predictions of the model, for the mixtures detailed in Table 2.2. Experimental data from de Larrard *et al.* (1996b).

properties, given the sensitivity of slump to water dosage. Then the selection between products previously made on the basis of simulations is likely to be relevant.

2.1.5 Placeability

Need for a placeability criterion

Placeability of concrete can be defined as the ability of the fresh material to be placed with a given procedure in a given form. By 'placed' is meant that the mixture takes the shape of the piece to be moulded, with a sufficient degree of consolidation. The moulding phase could be considered as a rheological concept, because changing the shape of a given volume essentially requires a *flowing* process. To some extent a placeability criterion could be formulated in terms of rheological properties, for example yield stress and plastic viscosity.

However, the author's opinion is that, apart from rheology, a placeability requirement still holds, at least in some cases. First, a dry or plastic mixture (with a slump lower than 100 mm) is not, strictly speaking, covered by rheology (see section 2.1.1). Such a concrete has a tendency to bulk, so that the process of placing induces *compaction* as much as flowing. Second, practical experience shows that a mix that looks workable in mass can be unsuitable for casting a thin or heavily reinforced piece, even if the distance between rebars is more than the maximum size of aggregates. Then, in the light of the packing theory developed in Chapter 1, we shall consider concrete as an assembly of particles, the aim of the placing process being to compact the mix in a mould, so that the voids content after compaction equals the theoretical water content. Of course, the compaction process is never perfect, but the air content has to remain in a low range (below 2–3% in volume) if an adequate strength and durability is aimed for.

Therefore the compaction index of concrete with no air (called K') will be the indicator of concrete placeability: the higher K', the less placeable the mixture will be. Hence, strictly speaking, this indicator is not a property as it cannot be measured, but only calculated.[11] However, it has a physical meaning: K' expresses the amount of energy necessary to compact the granular mix to the same extent as in the theoretical mixture. To ensure the placeability, the following inequality will have to be verified:

$$K' \leqslant K^* \tag{2.20}$$

where K^* is the compaction index dealing with the placing process. The value of this parameter is discussed in section 4.1.1.

Calculation of the placeability of a mix

Calculation of the placeability is carried out by using equation (1.47) (in section 1.2.1). Here, Φ is the water/(water + solid) ratio (in volume). As for the residual packing density of grain fractions, it must be corrected to account for the wall effect exerted by the boundaries of the piece. These boundaries can be either the form walls or the reinforcement. The formula is

$$\bar{\beta}_i = [1 - (1 - k_w)V_p]\beta_i \tag{2.21}$$

(see section 1.3.1 for the definition of symbols).

From the large experimental programme already used in the previous sections (Ferraris and de Larrard, 1998), Fig. 2.24 has been extracted, giving the relationship between slump and placeability K'. Within the group of concretes without SP there is a rough correlation between slump and placeability, which explains the success of the slump test in designating the ease of placement of fresh concrete. However, the correlation is lost when the whole set of mixtures is considered. This finding explains why slump is generally not used for mortar (in addition to the fact that the quantity of mixture to be tested in the slump test is of course excessive with regard to the maximum size of aggregate). More interestingly, it also provides an explanation for the fact that higher slumps (close to 200 mm and more) are generally required for high-performance concrete (see sections 4.1.1 and 5.5).

However, this diagram also suggests that the concrete compaction index, as defined hereafter, cannot be considered as an absolute index for

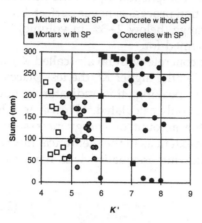

Figure 2.24 Relationship between slump and placeability, for 66 mixtures produced with the same materials (Ferraris and de Larrard, 1998).

the placeability. If the system considered is the solid skeleton (aggregate + binders), it appears that the superplasticizer decreases the amount of energy necessary to obtain a full compaction. As an example, self-compacting mixtures (which are highly superplasticized and display a slump in the range 260–290 mm) are clearly easier to compact than 50 mm slump ordinary concretes, even if they have a higher compaction index value. This inconsistency has to be corrected by specifying a maximum compaction index K^*, which depends not only on the packing process but also on the presence of superplasticizer.

Effect of confinement on placeability

As an application, let us calculate the effect of confinement on placeability in a typical case: a high-performance concrete to be cast in a large, highly congested piece. We shall assume that the structural element contains 500 kg/m³ of reinforcement, mostly in the form of 20 mm dia. rebars. Here the perturbed volume is essentially located around the reinforcement, as the structure is assumed to be of large dimensions (Fig. 2.25). Then the perturbed volume in a unit total volume is

$$V_p = 1.6 \times 10^{-4} d_i(40 + d_i) \tag{2.22}$$

where the diameter d_i is expressed in mm. Table 2.4 shows that the confinement effect will be essentially significant for the coarse aggregate fractions.

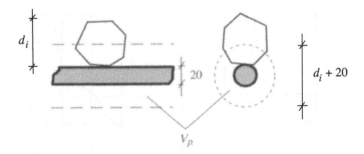

Figure 2.25 Perturbation created by a deformed bar in a concrete piece.

Table 2.4 Perturbed volume (in a unit total volume) as a function of the grain size

d_i *(mm)*	V_P
40	0.512
20	0.192
10	0.080
5	0.036
1	0.007

In the following simulation the HPC mixture is formulated with three fractions of aggregate (12.5/20, 5/12.5 and 0/4 mm), a Portland cement, silica fume, and superplasticizer. The cement dosage is taken equal to 425 kg/m³, with 30 kg/m³ of silica fume, for a free water content of 155 l/m³. For proportioning aggregate fractions the placeability is maximized (which means that the minimum value of K' is sought). Without taking account of the confinement a coarse/fine aggregate ratio of 1.68 is found, while in the opposite case the ratio drops to 1.55 (Fig. 2.26). As expected, the fact of considering the confinement tends to diminish the amount of coarse aggregate in the final mixture (see section 1.4.5). However, this effect is not very significant.

2.1.6 Entrapped air

As already stated, concrete is basically a dry mixture of granular materials, mixed with a volume of water greater than its original porosity. However, although there is in any workable concrete enough water to fill the voids of the solid skeleton, a small volume of air remains in the concrete after consolidation. This *entrapped* air must not be confused with a volume of air intentionally placed in the material to provide freeze–thaw protection (*entrained* air).

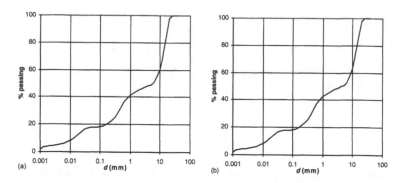

Figure 2.26 Effect of confinement on the optimum grading curve, for an HPC mixture: (a) without confinement; (b) with confinement.

A quantitative evaluation of entrapped air is necessary for two purposes: first, to know the exact composition of a unit volume of in-situ concrete; and, second, to predict the mechanical properties of hardened concrete. In particular, the entrapped air affects the compressive strength to the same extent as the same volume of water (see section 2.3). Various authors have described the air entrainment mechanism (Powers, 1968), but no research work has been found providing a quantitative model of entrapped air in fresh concrete. We shall use a set of original data to calibrate a model, which we shall validate with the help of four other sets.

Experiments and modelling

The data have been partially presented in section 2.1.1 (see Table 2.3). Supplementary results are displayed in Table 2.5. In the whole programme, 47 mixtures were produced with the same materials. The mortars were made up with a siliceous river sand, corrected by a fine sand, a type I Portland cement, and tapwater. Normal-strength concretes (NSCs) were obtained by adding to the constituents a rounded flint coarse aggregate having a maximum size of 12.5 mm. Another series dealt with plasticized concretes, in which a lignosulphonate plasticizer was used. High-performance concretes (HPCs) were produced with a superplasticizer of the melamine type. Finally, silica fume HPCs contained an amount of densified silica fume corresponding to 10% of the cement weight. Small batches (of about 15 litres) were prepared, and the air content was measured by the pressure method, on 81 samples consolidated by vibration. Obtained values, which ranged between 0.8 and 6.1%, are given in Tables 2.3 and 2.5.

Table 2.5 Supplementary data from the experimental programme presented in Table 2.3.

Parameter investigated	Mixture	CA (kg/m³)	FA (kg/m³)	Sand (kg/m³)	OPC (kg/m³)	SF (kg/m³)	P (kg/m³)	SP (kg/m³)	Water (kg/m³)	Slump (mm)	Air (%)
Control plasticized mix	BP	1032	553	184	389		1.95		180	85	3.0
Effect of cement content	BP − 200	1127	603	201	194		0.97		180	55	2.1
	BP − 100	1079	578	193	292		1.46		180	165	1.9
	BP + 100	985	527	176	487		2.43		180	45	2.4
	BP + 200	937	502	167	585		2.92		180	10	2.6
Effect of water content	BP − 15	1054	565	188	389		1.94		165	30	3.7
	BP + 10	1017	545	182	389		1.95		190	165	2.2
	BP + 20	1003	537	179	389		1.95		200	225	1.4
Control SF mix	BTHP	1053	564	188	437	44		16.23	119	210	2.0
Effect of cement content	BTHP − 300	1225	656	219	145	15		5.39	119	5	1.4
	BTHP − 100	1110	595	198	340	34		12.61	119	190	1.7
	BTHP + 100	995	533	178	535	54		19.87	119	135	2.2
	BTHP + 200	937	502	167	633	63		23.50	119	25	2.6
Effect of water content	BTHP − 15	1074	576	192	437	44		16.22	104	30	2.3
	BTHP + 5	1045	560	187	437	44		16.23	124	250	1.6
	BTHP + 10	1038	556	185	437	44		16.23	129	270	1.3

CA, coarse aggregate; FA, fine aggregate; OPC, ordinary Portland cement; SF, silica fume; P, plasticizer; SP, superplasticizer. Tests performed at LCPC in 1993.

It appears in the experimental results that the most important factor controlling the volume of entrapped air is the sand content, as all mortars have more air than the concretes (Fig. 2.28). The slump appears to be a secondary factor, as is the presence of admixtures. The explanation could be the following:

- At the end of the mixing process, a certain air volume remains in the interstices of the granular system; the size distribution of the voids is believed to be in the sand range (say from 0.05 to some millimetres).
- This volume increases with the concrete yield stress, since the yield stress tends to overcome the consolidation action of gravity.
- Organic admixtures, especially if they are tensio-active, also tend to increase this air volume.
- When fresh concrete is placed in a mould and submitted to vibration, the yield stress is greatly reduced, and each individual bubble tends to rise up to the surface of the sample. However, there is a certain probability that this void may be trapped under an aggregate. More precisely, the 'dangerous' fractions for the air bubble are the ones that have simultaneously a sufficient particle size and a high specific surface. This is why the sand fractions are the most efficient at retaining air, while the cement cannot fix a significant amount of entrapped air (Fig. 2.27). Hence fluid grouts are known to incorporate low amounts of air.

Therefore a model has been calibrated, as the product of two terms: the first illustrates the appearance of interstitial air during mixing, while the second deals with the 'trapping capability' of the skeleton:

$$a = (1 + 0.882pl + 0.0683sp - 0.00222SL)(-0.000988CA + 0.00368\,FA)$$

$$(2.23)$$

where a is the volume of entrapped air, as a percentage of the whole volume; pl and sp are the masses of dry extract of plasticizer and superplasticizer (in kg/m^3) respectively; SL is the slump, in mm; and CA and FA are the masses of coarse and fine aggregate in kg/m^3 respectively. The mean error[12] provided by this model is equal to 0.31% (Fig. 2.28).

Cement   Fine aggregate   Coarse aggregate

Figure 2.27 Mechanism of air entrapment during consolidation.

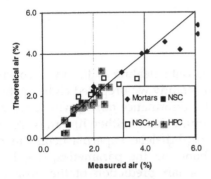

Figure 2.28 Fitting of the entrapped air model, for the data presented in Tables 2.1 and 2.5.

This equation suggests several comments:

- the plasticizer used (an 'old-fashioned' lignosulphonate) has a much higher air entrapment effect than the superplasticizer; however, nowadays water reducers often contain more efficient molecules, so that the corresponding coefficient is likely to be conservative for modern plasticizers.
- The superplasticizer, which is used in higher quantities, still has some effect on the air volume, which partially counteracts the slump effect.
- The slump has been taken as a control parameter instead of the yield stress (which was not measured in these experiments). As expected, the higher the slump (or the lower the yield stress), the lower the air. However, a maximum value of the slump (300 mm) still gives a positive volume of air (which is fortunate for extrapolation purposes).
- As expected, the parameter dealing with fine aggregate is higher than that for coarse aggregate.

This last parameter is even negative, which may be considered as a non-physical feature of the model. However, it is believed that coarse aggregates act as internal vibrators in fresh concrete submitted to vibration (Aïtcin and Albinger, 1989). Then the presence of coarse aggregate probably reduces the sand ability to trap air bubbles. Nevertheless, the use of this model (equation (2.23)) for a high coarse/fine aggregate ratio may provide negative values for the predicted air content. Thus it is proposed to adopt the following model for air content:

$$0.5 \leqslant a = (1 + 0.882pl + 0.0683sp - 0.00222SL)$$
$$\times (-0.000988CA + 0.00368FA) \quad (2.24)$$

as air content values of less than 0.5% are virtually non-existent (in the author's experience).

Further validation of the model

Let us first investigate two sets of data dealing with small concrete batches (less than 100 l). In a previous study we prepared eight mixtures (four normal strength and four high-performance mixes) with a crushed limestone aggregate having a maximum size of either 10 mm or 20 mm (de Larrard and Belloc, 1992). A naphthalene superplasticizer in dry powder was used. Without any change in the parameters, the model provided by equation (2.24) gives a fair prediction of the measured entrapped air (mean error equal to 0.25%; Fig. 2.29). In the data published by Rollet *et al.* (1992) eight non-air-entrained mixes are also found, where the paste composition is changed, at constant slump. Following our model, only changes in the superplasticizer content may create significant variations of air. Here again the agreement between predictions and measurements is excellent (mean error 0.19%; Fig. 2.29).

It seems however that entrapped air is affected by a *size effect*. First, plant mixers generally provide a more efficient mixing as the mixing time is systematically lower (often less than 1 minute) than for laboratory mixers. Second, in the bottom of a large mixer the pressure due to gravity is high, so that air voids are more strongly expelled from fresh concrete than in a small batch. This is why the predictions of the model seem conservative as soon as concrete is produced in higher volumes, say more than 200 l.

In Le Roy's data (Le Roy, 1996) 10 mixes, mainly HPCs, were prepared to cast large creep and shrinkage specimens (see section 2.5). When a reduction coefficient of 0.65 is applied to the calculated air content, the error given by the model remains comparable to that of previous data (0.27%). In other data (de Larrard *et al.*, 1996b) dealing with 11 mixtures

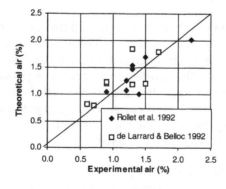

Figure 2.29 Direct validation of the model for two data sets: small batches.

with or without superplasticizer, with a wide range of compressive strengths, the reduction coefficient is equal to 0.49 (mean error 0.15%). These data are illustrated in Fig. 2.30.

As our aim in this chapter is to predict concrete properties as measured *in the laboratory*, we shall keep unchanged the model given by equation (2.24). We shall only keep in mind that if concrete is to be produced in larger quantities, a relative reduction in air content of up to 50% can be obtained. This statement does not apply for entrained air, which is more stable because of the small size of air bubbles.

The most popular American mix-design method (ACI 211) states that the air content is a decreasing function of the maximum size of aggregate (MSA). This statement is consistent with our model, since the higher the MSA, the lower the sand content in an optimized mixture. Let us finally address the case of non-workable concretes, where the model should be applied with caution. In the stiff consistency range the concrete may lack water to fill the voids of granular materials. Therefore the equation provided should work when concrete has a slump of some millimetres. When the water content is lower, as in roller-compacted concretes (see section 5.4.1), the total void content (water + air) should be predicted with the CPM model (see Chapter 1), with a suitable compaction index.

2.1.7 Stability (prevention of bleeding and segregation)

The filling diagram of the de-aired concrete

In section 1.5 a theory was proposed for analysing granular mixtures from the viewpoint of proneness to segregation. The filling diagram was defined. It gives the size distribution of the filling proportions of each clustered class. A clustered class is the reunion of grains, the size of

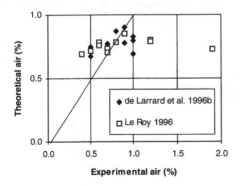

Figure 2.30 Validation of the model for two data sets dealing with large batches (about 200 l). Reduction coefficients have been applied to the model predictions.

which has a maximum mutual ratio of about 2.6 (coarse/fine grain ratio). The filling proportion is the ratio between the actual volume of the class in the mixture and the maximum volume that could be 'virtually' placed, the rest of the partial volumes being constant. From the filling diagram, the segregation index has been defined, expressing the complement of the minimum filling proportion.

Let us now apply these concepts to fresh concrete. Because the air content is generally low, and not known a priori, we shall consider the de-aired concrete (as we did for modelling the rheological properties). The width of each clustered class will be fixed at a value of 2.5, since this ratio is covered in Renard's series[13] of sieve sizes (in the French standard NFP 18-560). When the size distributions are measured following other standards, it is still possible to adopt such a series by means of interpolation. In the next development the series given in Table 2.6 is used. An example of a filling diagram is given in Fig. 2.35a.

It can already be seen that the filling diagram of a real concrete is rather different, as compared with that for the ideal mixtures dealt with in section 1.5.3. Here, although this concrete comprised three fractions of aggregate, including a fine correcting sand, a certain gap appears between the cement and the sand ranges, and a minor one between sand and gravel. Otherwise, because of the natural distribution of granular fractions, the filling diagram looks uneven.

Interpretation of the filling diagram

In the large experimental programme already presented (Ferraris and de Larrard, 1998), an examination of the top surface of mixtures in the rheometer was carried out. Two types of feature were noted: the degree of bleeding at the end of test, and the amount of coarse aggregate rising

Table 2.6 Sieve sizes for the clustered granular classes, considered in the concrete filling diagrams

Mean size (mm)	Minimum size (mm)	Maximum size (mm)
0.0016	0.001	0.0025
0.004	0.0025	0.0063
0.01	0.0063	0.016
0.025	0.016	0.04
0.063	0.04	0.1
0.16	0.1	0.25
0.4	0.25	0.63
1.0	0.63	1.6
2.5	1.6	4
6.3	4	10
16	10	25
40	25	50

up from the mortar surface (due to the dilatancy of the granular skeleton), taken as an indication of segregation proneness. Only visual, subjective assessments were performed, in order to establish in which area of the experimental plan (presented in Fig. 2.12) these phenomena were the most likely to occur. Bleeding and coarse aggregate segregation were ranked from 0 to 3. These numbers were compared with the K_c' and K_g' coefficients (see section 2.1.3). Following the definition of compaction index, K_c' is the sum of the K_i' in the range of cement grading span, while K_g' is the sum of the K_i' in the range of coarse aggregate grading span. Thus K_c' and K_g' are linked with the integral of the filling diagram in the cement and the coarse aggregate ranges respectively.

The relationship between K_c' and bleeding is shown in Fig. 2.31. It turns out that K_c' closely controls the bleeding of mixtures without super-plasticizer. The trend is still observed in the presence of superplasticizer, although the data are more scattered. Moreover, the admixture itself has a dominant effect on the amount of bleeding. A value of 1.6–1.7 for K_c' seems to guarantee a lack of bleeding without superplasticizer, while a higher value (about 4–4.5) is necessary in the presence of a large amount of superplasticizer.

Bleeding is the result of percolation of water through the porous phase of fresh concrete (Powers, 1968). The total amount of bleeding is driven by the amount of initial compaction (the lower the compaction, the higher the bleeding), while the kinetics is controlled more by the specific surface of the mixture. Therefore it is rather logical that, when the cement is closely packed in the voids of the aggregate skeleton, both final amplitude and kinetics of bleeding decrease. Let us note that the data corresponding to low slump concretes were removed from Fig. 2.31,

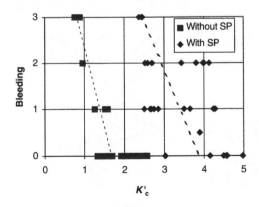

Figure 2.31 Relationship between K_c' (contribution of cement to the compaction index of de-aired concrete) and degree of bleeding. Slump > 80 mm. Data extracted from Ferraris and de Larrard (1998).

since for these mixtures the high compaction (that is, the low placeability) limited the amount of bleeding, even in the case of low cement dosage. In conclusion it is seen that for workable mixtures the filling proportion in the fine range of cement largely controls the bleeding phenomenon.

The tendency of coarse aggregate to be expelled out of the concrete during shear is well correlated with the K'_g coefficient (Fig. 2.32). In contrast to the previous property investigated, there is no significant difference between mixtures with or without superplasticizer. Things happen as if there was a critical value of K'_g (equal to 2 in the present case) above which the coarse aggregate segregates during shear. This phenomenon corresponds to the pattern given in Fig. 2.33b, while a well-proportioned, non-segregating mixture corresponds to Fig. 2.33a.

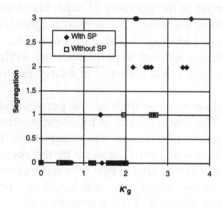

Figure 2.32 Relationship between K'_g (contribution of coarse aggregate to the compaction index of de-aired concrete) and degree of segregation. Data extracted from Ferraris and de Larrard (1998).

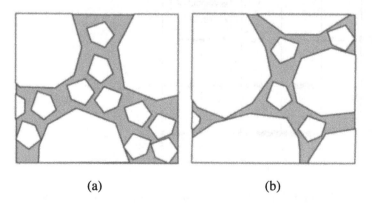

(a)   (b)

Figure 2.33 The granular structure of concrete, with different values of K'_g: (a) $K'_g < 2$; (b) $K'_g > 2$.

A summary of the above-mentioned observations is given in Fig. 2.34, which provides a guide to interpretation of the filling diagram. Note that at this point of the research no quantitative bounds are given. In the author's view the filling diagram is a tool to interpret the result of a trial mix (in terms of bleeding and/or segregation). It may provide help for correcting the initial mix, rather than allowing one to *predict* the behaviour. In fact bleeding and segregation are strongly affected by size effects, and a theory enabling engineers to fully understand and predict these effects has still to be developed.

At this point one could argue that the grading curve of concrete also permits this type of diagnosis. However, the filling ratio has two advantages:

- It takes the packing density (or water demand) of the grain fractions into account. For instance, the volume of crushed coarse aggregate leading to segregation is lower than if a rounded aggregate of the same size is used.
- It reflects the interactions between the different fractions. This means that the minimum amount of cement needed to prevent excessive bleeding will be higher for a coarse sand (having a high fineness modulus) than for a well-graded sand.

Examples of filling diagram

Some typical filling diagrams have been extracted from the study referred to. The mixture proportions and degrees of bleeding/segregation are given in Table 2.7.

The first mixture (BO1B') is a well-proportioned ordinary mixture, having good stability properties (Fig. 2.35a). The BO2C mixture exhibited

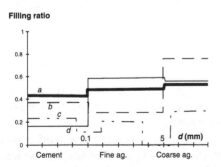

Figure 2.34 Summary of the interpretation of the filling diagram: (a) stable mixture; (b) risk of segregation due to an excess of coarse aggregate; (c) risk of segregation by excess of fluidity (low K' value) and lack of continuity in the grading curve; (d) risk of bleeding by lack of fine elements.

Table 2.7 Mixture proportions and properties of a selection of mixtures.

Mixtures	CA (kg/m³)	FA (kg/m³)	Fine sand (kg/m³)	OPC (kg/m³)	SP (kg/m³)	w/c	Slump (mm)	Bleeding	Segregation
BO1B'	952	614	190	360	–	0.567	100	0	0
BO2C	986	635	197	235	–	0.936	170	3	0
BO4A	1207	405	126	356	–	0.595	80	0	1
BHP8C	904	563	161	634	15.86	0.262	265	0	0
BHP4B	1270	416	119	424	10.59	0.388	120	3	2

Source: Ferraris and de Larrard (1998).

a high bleeding due to its low cement content, together with its high slump value. This behaviour corresponds to low cement peaks in the filling diagram (Fig. 2.35b). The BO4A mixture contains an excess of coarse aggregate, leading to a separation of coarse aggregate from the rest of the mixture. Hence the coarse aggregate peak is almost twice as high as the other ones in the filling diagram (Fig. 2.35c).

The BHP8C is a self-compacting concrete. In spite of its very high slump, it displays neither bleeding nor segregation, which could be anticipated from its quite even filling diagram (Fig. 2.35d). Finally the BHP4B, which has a lower cement and sand content and consequently an excess of coarse aggregate, produces a high degree of bleeding and segregation (Fig. 2.35e).

Segregation potential

In section 1.5, we have defined for dry granular mixtures a number called segregation potential, S, with the view of providing a numerical optimization tool. The equation defining S was:

$$S = 1 - \underset{1 \leqslant i \leqslant n}{\text{Min}} \left(\frac{K_i}{1 + K_i} \right) \tag{2.25}$$

This index has been used to optimize grading curves, in order to obtain even filling diagrams. Such diagrams are believed to produce mixtures having a minimal tendency to separation.

Application of this concept to fresh concrete is tempting, and could be especially useful for mixtures that are more exposed to segregation, such as self-compacting concretes. However, one should consider the following problems:

- The partial compaction index K_i' must be calculated for clustered grain fractions, with a width of 2.5–2.6.

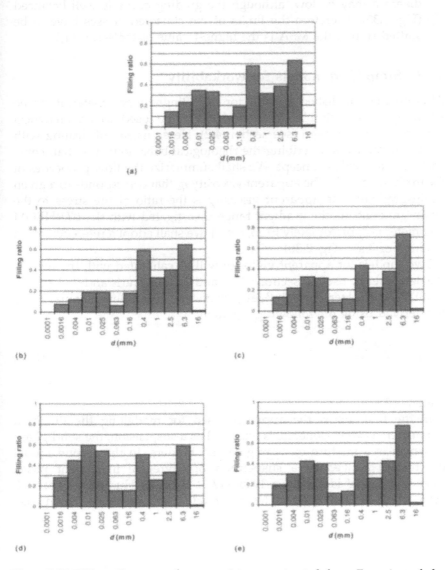

Figure 2.35 Filling diagrams of some mixtures extracted from Ferraris and de Larrard (1998): (a) BO1B′; (b) BO2C; (c) BO4A; (d) BHP8C; (e)BHP4B.

- In the ultrafine range of the grading span, the colloidal nature of the grains makes segregation impossible. Then the values of i, for $d_i < 1$ μm, are not to be considered in equation (2.25).
- In the coarsest range, if the maximum size of aggregate does not match the limits of the cluster series, the right-hand peak in the filling

diagram may be low, although the grading curve is well balanced (Fig. 2.35). Therefore the limits of the clustered classes have to be shifted so that the MSA is the highest value of the series (d_1).

2.1.8 Simplified models for workability

After having established a model for each Bingham parameter, it can be useful to derive simpler models that can be used in an analytical approach to mix design (see section 4.2). Here, instead of dealing with several parameters describing the rheological behaviour, we shall come back to a *workability* concept. We shall summarize the flow properties of the fresh material by the apparent viscosity η_a that corresponds to a given strain gradient. The apparent viscosity is the ratio of the stress to the strain gradient. As the gradient range investigated with the BTRHEOM rheometer is approximately [0.25, 6 s^{-1}], we shall adopt a reference strain gradient of 3 s^{-1} (Fig. 2.36).

Two models for apparent viscosity will be calibrated with the help of the data already used in sections 2.1.1 and 2.1.2 (Ferraris and de Larrard, 1998). From the Herschel–Bulkley parameters, experimental values for η_a are derived using the following equation:

$$\eta_a = \frac{\tau_0 + m\dot{\gamma}^n}{\dot{\gamma}} \tag{2.26}$$

with $\dot{\gamma} = 3$ s^{-1}.

In the first model the approach used for modelling the Bingham parameters is kept. Concrete is analysed as a suspension, the solid phase of which is a mixture of aggregate and binders. In the second model concrete is viewed as a suspension of aggregate in cement paste, following Farris' idea (Farris 1968; Hu *et al.*, 1995; Hu and de Larrard, 1996; Mørtsell *et al.*, 1996). The two models are necessary for an easy

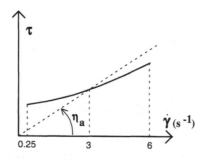

Figure 2.36 Definition of the apparent viscosity.

demonstration of various mix design rules, which will be presented in Chapter 4.

Solid concentration model and Lyse's rule

When plotting the apparent viscosity against the ratio of solid volume to the packing density of the dry materials (normalized solid concentration), two master curves appear, one for ordinary mixes and one for superplasticized mixes (Fig. 2.37). As we did for the modelling of plastic viscosity, the air content was not considered in these calculations. In other words, the Φ term is the ratio between the solid volume on the one hand, and the sum of the solid volume and the water volume on the other.

A simple hyperbolic model gives a reasonable prediction of the apparent viscosity. The same function of the normalized solid concentration is kept for the two concrete families (without or with superplasticizer). Only a multiplicative parameter changes. We have then

$$\eta_a = \frac{41.1}{1 - \Phi/\Phi^*} \tag{2.27}$$

for mixtures without superplasticizer, and

$$\eta_a = \frac{15.9}{1 - \Phi/\Phi^*} \tag{2.28}$$

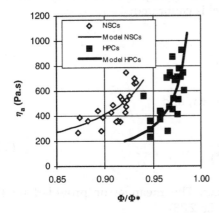

Figure 2.37 Apparent viscosity vs. normalized solid concentration (solid volume/packing density), for the mixtures produced in Ferraris and de Larrard (1998).

for mixtures with superplasticizer. The mean error provided by this model, for the whole concrete series, is equal to 96.2 Pa.s in absolute value.

Note that this model allows the demonstration of the famous Lyse's rule (Lyse, 1932; cited by Powers, 1968). This rule states that in the range of usual concrete mixtures, produced with a given set of constituents, the water content essentially controls the consistency. Within the frame of the present model, the water content is equal to $1 - \Phi$. Hence if the aggregate is properly optimized (see section 4.2) and the cement content is close to the optimal dosage (in terms of packing density), the quantity Φ^* will experience little variation between the various concretes. Therefore a constant water content will imply a constant Φ value, which will give a nearly constant apparent viscosity according to equation (2.27) or (2.28).

Paste/aggregate model

As for this second model, the two parameters are

- the normalized concentration of the *aggregate* in the concrete, g/g^*;
- the normalized concentration of the cement in the matrix, v_c/α_c (where v_c is the concentration of the cement in the matrix, and α_c is the packing density of the cement).

The g^* term has been calculated for each mix with the help of the CPM, with a compaction index of 9. The same procedure has been applied to evaluate α_c, which takes different values with and without admixture (0.6311 and 0.5886 respectively). The overall viscosity will then be the product of a term describing the paste viscosity and a second term dealing with the contribution of the aggregate skeleton. By numerical optimization, the following model is calibrated:

$$\eta_a = \frac{36.2}{(1 - v_c/\alpha_c)^{1.08}} \cdot \frac{1}{(1 - g/g^*)} \tag{2.29}$$

for mixtures without superplasticizer, and

$$\eta_a = \frac{19.8}{(1 - v_c/\alpha_c)^{1.08}} \cdot \frac{1}{(1 - g/g^*)} \tag{2.30}$$

for mixtures with superplasticizer. The mean error provided by this second model is only 37.2 Pa.s (Fig. 2.38).

Therefore it appears that the paste/aggregate model gives a better description of the experimental results than the solid concentration

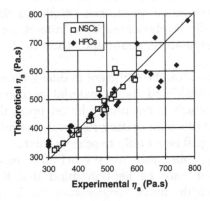

Figure 2.38 Comparison between experimental values of apparent viscosity
and predictions of the paste/aggregate model.

model. These two simplified models will be appropriate tools for an
analytical treatment of the mixture-proportioning problem.

2.2   ADIABATIC TEMPERATURE RISE

In this section the heat generation of concrete during hardening is
addressed. In massive structures such a phenomenon creates a
temperature rise, which can reach more than 60 °C for a high cement
content. When temperature drops towards the ambient one, a thermal
shrinkage appears. When restrained, this shrinkage, which can be
superimposed on autogenous shrinkage (see section 2.5.5), provokes
cracking. This cracking should be avoided as much as possible, because it
is a factor for poor durability, especially in a severe environment.
Moreover, a high temperature during hardening tends to diminish the
final strength of in-place concrete (Laplante, 1993). Even when the
maximum temperature is higher than about 60 °C a delayed ettringite
formation may occur (Neville, 1995). Thus the temperature rise during
concrete hardening must be controlled in a global approach to durability.

Temperature rise depends not only on the concrete used but also on
the dimensions of the structure. Knowledge of the full heat development
of concrete over time is necessary for calculating the maximum
temperature in a concrete piece, because there is a competition between
the heat production due to the binders, and the heat diffusion towards
the cold zones (environment). Nowadays there are numerical techniques
for predicting the heat development of Portland cement concrete from its
composition (Van Breugel, 1991; Bentz *et al.*, 1998). Moreover, if the heat
development test of the cement is known (Langavant's bottle test NF P

15-436), one may compute the heat development of any concrete containing this cement plus an amount of pozzolans (Waller, 1998). Such an adiabatic curve can be finally implemented in a finite-element code for the computation of the temperature, strain and stress fields in the structure (Torrenti *et al.*, 1994).

At the mixture-proportioning stage it can be sufficient to deal with the *final* adiabatic temperature rise, considered as a material property. Therefore the aim of this section, which is essentially based upon the work of Waller (1998), is to evaluate this temperature rise from the mix-design parameters. However, if the goal is not only to select a mixture of limited heat production capacity, but also to predict the exact temperature rise in the structure, it must be kept in mind that this temperature rise depends slightly on the initial temperature: the lower the initial temperature, the higher the temperature rise (Ohta *et al.*, 1996).

We shall deal first with the heat capacity of concrete. Then, for evaluating the heat produced during hydration, an estimate of the final degree of consumption of binders will be necessary, as binder hydration is the only heat source. From these parameters, and from a knowledge of the heat released in the consumption of one unit mass of binder, the total heat released by a unit volume of concrete will be calculated. Finally, with the help of these preliminary calculations, the adiabatic temperature rise will be modelled.

2.2.1 Heat capacity

The heat capacity of fresh concrete, C^{th}, is simply evaluated by adding the heat capacity of the various constituents:

$$C^{th} = \sum_{\text{constituents}} m_i c_i^{th} \tag{2.31}$$

where m_i is the mass of the ith constituent per unit volume of concrete, and c_i^{th} is its heat capacity. Typical values of c_i^{th} are given in Table 2.8 for the most usual concrete components.

When concrete hardens in adiabatic conditions, two effects take place that may change the heat capacity. First, the transformation of cement and pozzolans in hydration products leads to a decrease of the heat capacity, essentially because of the decrease of free water content. Second, the temperature rise itself creates a significant increase of heat capacity (up to 10% between 10 and 80 °C). On the whole, the heat capacity experiences very little change when the two effects are added. Therefore it turns out that the original value (fresh concrete heat capacity) is a good basis for modelling the adiabatic temperature rise (Waller, 1998).

Table 2.8 Heat capacity of concrete constituents at 20 °C.

Constituent	c_i^{th} (kJ/K/kg)
Siliceous aggregate	0.73
Calcareous aggregate	0.84
Dolomitic limestone aggregate	0.89
Anhydrous Portland cement	0.76
Silica fume	0.73
Class F fly ash	0.73
Water	4.19

Note in Table 2.8 that water is the most significant constituent, as far as heat capacity is concerned. Thus concrete with a lower water content will exhibit a higher temperature rise for a given heat production. For example, high-performance concretes have not only a low water content but also a high amount of binders. As for roller-compacted concretes, although their binder dosage is generally low, they are often used for very massive structures. Therefore it can be deduced that an evaluation of adiabatic temperature rise will be critical for these two categories of concrete.

2.2.2 Degree of consumption of binders

In this section the aim is to predict the final degrees of transformation of binders into hydrates when concrete is placed in a sealed container in adiabatic conditions (that is, without thermal exchanges with the environment). It is believed that these degrees are essentially governed by the paste mixture proportions (water/cement ratio, pozzolan/cement ratio), as hydration is a local phenomenon occurring at the cement grain scale. This assumption leads us to neglect any role of the inert phases (paste/aggregate interface, inert filler) in the hydration process. Actually, as seen in section 2.3.7, chemically inert phases are believed to influence the kinetics of hydration significantly, but the final extent of the phenomenon only marginally.

Degree of hydration of Portland cement in absence of pozzolans

Based upon the previous assumptions, the aim is therefore to evaluate the final degree of hydration of Portland cement in a mature cement paste. It is well known that the coarsest grains of cement never reach a full state of transformation, even in the very long term (Neville, 1995). This is because at a given stage of the process, hydrates that have already formed have precipitated around the clinker grains, forming a separation between the water and the anhydrous phase. Thus water has to percolate through this hydrate layer to form further hydrates. The coarser the cement, the more important is this phenomenon. Moreover, at a low water/cement ratio there is a lack of water for transforming the total volume of initial cement. A stoichiometric calculation shows that an amount of water equal to about 0.24 g is necessary to complete the hydration of 1 g of cement. However, as hydration goes on, not only are the kinetics lowered by diffusion phenomena through the hydrate layers, but there is also a self-desiccation phenomenon: that is, a decrease of internal humidity in the porosity of the system (see section 2.5.5). This lower humidity tends to diminish the hydration rate. Powers stated that if the unevaporable water and the adsorbed water[14] were added, the total amount would be about 42% of the mass of cement (Powers and Brownyard, 1946–47).

Waller found various experimental results for final degree of hydration in the literature. He selected those dealing with test periods longer than 400 days, and plotted the values obtained against water/cement ratio, together with his own data[15] (Fig. 2.39). Although cement pastes, mortars and concretes made up with different cements were involved, the various points fitted well with an empirical model, the equation of which is

$$h_c = 1 - \exp(3.38w/c) \tag{2.32}$$

where h_c is the final degree of hydration of the cement. From these data it appears that the decrease of cement hydration when the water/cement ratio decreases is a continuous and gradual phenomenon.

Degree of consumption of binders in cement/pozzolan systems

When the system contains both Portland cement and pozzolans the prediction of degrees of consumption becomes more complex. A stoichiometric calculation shows that 1 g of pozzolan needs about 1.3 or 1.1 g of lime to be fully consumed, for silica fume or fly ash respectively. If one considers that 1 g of cement has released 0.3 g of lime after hydration, at the end of the process either the pozzolan is fully hydrated (and a

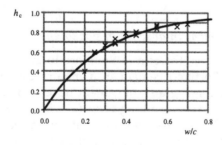

Figure 2.39 Empirical model for the final degree of cement hydration (Waller, 1998).

certain amount of lime remains in the system), or the part of the pozzolan that has reacted has exactly consumed the lime released by the cement. The balance equation leads to the following formulae:

$$h_{SF} = \min\left(1; \frac{0.23h_c}{sf/c}\right) \tag{2.33}$$

for silica fume, and

$$h_{FA} = \min\left(1; \frac{0.27h_c}{fa/c}\right) \tag{2.34}$$

for fly ash, where h_{SF} and h_{FA} are the degree of consumption of silica fume and fly ash respectively.

It is believed that the pozzolanic reaction

lime + silica $\rightarrow$ CSH

does not involve any additional source of water. However, as for cement, the hydrate phase has a high specific area and adsorbs a significant amount of water, which is no longer available for further hydration. Then the cement degree of hydration decreases when a pozzolan is added to the system at constant water/cement ratio, as shown by Justnes for silica fume (Justnes *et al.*, 1992a, b). Equation (2.32) needs to be corrected by subtracting from the total water the part of the water adsorbed by the pozzolanic CSH. Another correction has to be performed, as this reduction effect seems to increase with the water/cement ratio. By numerical optimization, Waller (1998) calibrated the following formulae for the cement degree of hydration in the presence of pozzolans:

$$h_c = 1 - \exp[3.38(w/c - \delta)] \tag{2.35}$$

with

$$\delta = \exp(1.63w/c)\,\frac{0.6h_{SF}sf}{c} \tag{2.36}$$

in the presence of silica fume, and

$$\delta = \exp(1.63w/c)\,\frac{0.42h_{FA}fa}{c} \tag{2.37}$$

in the presence of fly ash.

Whatever the nature of the pozzolan, it turns out that the cement degree of hydration is given by an implicit equation, which is easily solvable by iteration.[16] Figure 2.40 shows the effect of various percentages of silica fume on the final degree of hydration of Portland cement. Here, two remarks can be made:

- The decrease in the final degree of hydration of cement, at a 10% silica fume rate (a fairly usual dosage in high-performance concrete) is significant. The supplementary heat release due to the pozzolanic reaction is then partially overcome by a lower release due to the cement hydration. This could be why some researchers have claimed that the pozzolanic reaction does not produce any heat.
- This decrease is stabilized when the silica fume dosage is above that for which all the lime is consumed (that is, for $sf/c > 0.2$).

For a direct validation of this model experimental degrees of consumption in cementitious systems are needed. However, the measurement of these data becomes difficult when several binders are used together. It can be performed by X-ray diffraction or nuclear magnetic resonance techniques, but the precision is rather poor (Waller, 1998). This is why we prefer to validate the final temperature rise model, which is the main goal of this part of the work.

Ternary system (cement/fly ash/silica fume)

For a ternary mixture it is necessary to generalize the previous models. For this purpose, a first hypothesis is made: silica fume is assumed to react before fly ash. Fly ash will react only if some lime is available after the maximum consumption of silica fume.

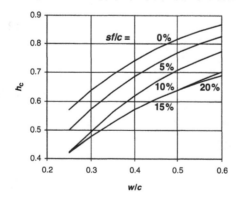

Figure 2.40 Effect of silica fume on the cement degree of hydration (simulations with Waller's model).

For solving the system, it is first assumed that silica fume is the only pozzolan to react. Then h_c and h_{SF} are calculated from equations (2.33), (2.35) and (2.36). If $h_{SF} < 1$ then the second hypothesis is correct, and we have $h_{FA} = 0$. If $h_{SF} = 1$ then a part of the fly ash has reacted. This amount is quantified by accounting for the rest of lime available:

$$0.3h_c c - 1.3sf = 1.1h_{FA}fa \tag{2.38}$$

Finally, whatever the value of h_{SF}, the problem is summarized by the following system of equations:

$$h_{SF} = \text{Min}\left(1; \ \frac{0.23h_c}{sf/c}\right) \tag{2.39}$$

$$h_{FA} = \text{Max}\left[0; \ \text{Min}\left(1; \ \frac{0.27h_C}{fa/c} - \frac{1.18}{fa/sf}\right)\right] \tag{2.40}$$

and

$$h_c = 1 - \exp[3.38(w/c - \delta)] \tag{2.41}$$

with

$$\delta = \exp(1.63w/c)\ \frac{0.60h_{SF}sf + 0.42h_{FA}fa}{c} \tag{2.42}$$

As in the binary case, the system is solved by iteration. We then have a model for the final degree of consumption of the three binders, in any mixture containing Portland cement, fly ash and silica fume. For calculating the total heat released, it is now necessary to evaluate the heat released by one unit mass of each particular binder.

2.2.3   Heat of hydration

Specific heat of hydration of binders

The quantity dealt with in this section is the amount of heat released by a unit mass of fully hydrated (or transformed) binder. There are some tests for the measurement of cement heat of hydration, such as the Langavant's bottle test (NF P 15-436), which is a semi-adiabatic test performed on a standard mortar. However, it appears that semi-adiabatic tests are not reliable as far as total heat is concerned, because the evaluation of heat losses becomes very inaccurate at the end of the tests

(Waller, 1998). Therefore the heat of hydration will be calculated from the Bogue composition. The reactive part of a Portland cement (clinker) is made of four phases: C_3S, C_2S, C_3A and C_4AF (see section 3.2.4). Each phase has a specific heat of hydration, which is given in Table 2.9.

The cement heat of hydration is then a linear combination of the phase heats, weighted by the phase percentages. From Table 2.9 it appears that the two phases that are known for giving early strength (C_3S and C_3A) also provide a high heat of hydration. This is why cements that provide rapid strength development are also known to promote increased thermal cracking.

Concerning the pozzolans, Waller measured their heat of hydration by performing adiabatic tests on mixtures of pozzolan, lime and water, with an excess of lime to attain a maximum transformation of the pozzolans into hydrates. The values obtained are given in Table 2.10. It appears that, given the uncertainty on these measurements, the various products do not exhibit significantly different values, within the two categories (namely silica fume and class F fly ash). This is a surprising, but

Table 2.9 Heat of hydration of the clinker phases (Waller, 1998).

Clinker phases	Heat of hydration (kJ/kg)
C_3S	510
C_2S	260
C_3A	1100
C_4AF	410

Table 2.10 Heat of hydration of some pozzolans (silica fume and Class F fly ashes).

Product	Heat released by unit mass of reacted pozzolan (kJ/kg)
Silica fume A	810 ± 60
Silica fume C	910 ± 70
Silica fume D	890 ± 70
Mean value	**870 ± 70**
Fly ash A	500 ± 90
Fly ash B	560 ± 100
Fly ash C	610 ± 70
Fly ash D	560 ± 60
Fly ash E	630 ± 70
Mean value	**570 ± 80**

Source: Waller (1998).

simplifying result: no special identification test will be necessary to evaluate the contribution of such a mineral admixture to the total heat released by concrete.

Concrete heat of hydration

From the previous developments, the calculation of the concrete heat of hydration, Q, is straightforward: it is the sum of the heats released by the various binders, weighted by their degrees of hydration/transformation:

$$Q = (510t_{C3S} + 260t_{C2S} + 1{,}100t_{C3A} + 410t_{C4AF}) \frac{h_c\,c}{100}$$

$$+ 870h_{SF}sf + 570h_{FA}fa \quad (2.43)$$

where Q is in kJ/m^3, and c, sf and fa are in kg/m^3. t_{C3S} stands for the percentage of tricalcium silicate in the cement. The degrees of consumption are evaluated by using the model given in section 2.2.2.

2.2.4   Adiabatic temperature rise

Calculation of adiabatic temperature rise

The final adiabatic temperature rise of concrete $\Delta\theta$, in K,[17] is just the ratio of the heat released (from equation (2.43)) to the heat capacity (equation (2.31)):

$$\Delta\theta = \frac{Q}{C^{th}} \quad (2.44)$$

The model will now be evaluated by using the extensive data set published by Waller (1998).

Waller performed a large series of adiabatic tests on both mortars and concretes made up with limestone aggregates. Two Portland cements were used, pure or in combination with either a fly ash or a silica fume. Wide ranges of water/cement ratio (0.30–0.65), silica fume/cement ratio (0–0.30) and fly ash/cement ratio (0–0.50) were investigated. Several types of pozzolan were compared, and some mixtures also included a limestone filler.

The 60 tests performed were used first to calibrate the model of cement degree of hydration in the presence of pozzolan (as given by equations (2.36) and (2.37)). Then the theoretical temperature rises as predicted by the model were compared with the experimental data (Fig. 2.41). The mean error of the model is less than 2 K. Ninety-five per cent of the

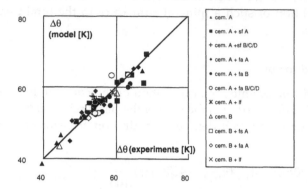

Figure 2.41 Validation of the adiabatic temperature rise model (Waller, 1998). *sf* = silica fume; *fa* = fly ash; *lf* = limestone filler.

values are even inferior to 5 K. In the prediction of temperature rise relative to mixtures containing limestone filler, this product has been considered inert. It will be shown later that limestone filler tends to accelerate the cement hydration, and also contributes directly to the compressive strength through the formation of carbo-aluminates (see section 2.3.7). In theory this second effect could produce some heat, but the amount seems at first sight negligible compared with the other heat sources.

Interest of pozzolans for the limitation of temperature rise

A consequence of this model is rather surprising, and follows from the large amount of heat released by the pozzolanic reaction. As far as the final adiabatic temperature rise is concerned, there is a limited benefit in using a pozzolan (silica fume or fly ash) to replace the cement at constant workability and strength (Waller *et al.*, 1996). For decreasing $\Delta\theta$, the most efficient strategy is to decrease the water content as much as possible by optimizing the proportions of fine particles and aggregate fractions, by using a superplasticizer with a possible addition of silica fume, or by specifying a mixture having a dry consistency. This finding seems to violate the common observation dealing with the interest in using fly ash in mass concretes to decrease the maximum temperature rise and induced cracking.

This apparent paradox lies in the difference between adiabatic and semi-adiabatic temperature rise. In the former, all the heat produced is stored throughout the material, and participates directly in the temperature rise. But in the latter (the only case in practice), heat diffusion takes place simultaneously with heat production. This phenomenon decreases the temperature, which in turn decreases the rate of hydration. Here, the influence of activation energy takes place. As pozzolans have a much greater activation energy than Portland cement, the lowering of hydration rate is much more significant. As a result, the thinner the piece, the higher the benefit of using pozzolans in terms of maximum temperature rise.

Therefore the limitation of temperature rise in a real structure will be performed at the material level by decreasing as much as possible the adiabatic temperature rise predicted by the model, *and* by using a maximum of pozzolans in replacement of cement. However, attention must be paid in parallel to durability. A sufficient amount of free lime should be kept in the hardened concrete porosity to avoid too rapid a carbonation.

2.3   COMPRESSIVE STRENGTH

The compressive strength dealt with in this section is the result of laboratory tests performed on standard cylinder specimens cured in water. For strength higher than 60 MPa it is assumed that special precautions have been taken with regard to the capping process. The models presented herein are assumed to apply equally to cube test results, provided that a suitable conversion coefficient is used. Such a coefficient is not universal, but depends on such factors as the aggregate type, the strength level, and the maximum size of aggregate (de Larrard *et al.*, 1994d).

To analyse the relationship between composition and compressive strength we shall consider concrete as a two-phase material: a hard, stiff, inert phase (the aggregate) dispersed in a matrix (the cement paste) that can be considered as homogeneous at the meso-scale. The paste, being the weakest part of the composite, will be investigated first.

2.3.1   Mature paste of Portland cement

A microstructural model

Hardened cement paste is a porous material. Its solid volume fraction (1 minus the porosity) is directly related to the compressive strength, as pointed out by various authors (Powers and Brownyard, 1946–47). Here, we briefly present a simple model, which is not expected to give quantitative agreement with experience, but which explains the power-law relationship between the two quantities (de Larrard *et al.*, 1988).

Let us assume that the microstructure of hardened cement paste is comparable to a lattice structure. The CSH gel between two pores may form either rods or plates (Fig. 2.42).

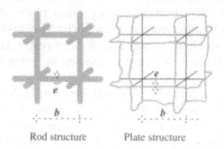

Rod structure   Plate structure

Figure 2.42 A microstructural model for hardened cement paste.

Let b be the dimension of the elementary cell, and e the transverse dimension of the solid parts. For the rod structure, the solid volume fraction is

$$\Phi = \frac{3e^2 b}{b^3} = 3 \left(\frac{e}{b}\right)^2 \tag{2.45}$$

If a compressive stress, σ, parallel to one of the lattice axes, is applied on the structure, the stress in the rods of this direction is

$$\sigma_s = \sigma \cdot \left(\frac{b}{e}\right)^2 \tag{2.46}$$

Now, let us assume that the mechanism of failure is the buckling of the rod. The failure stress in a loaded rod is attained when the force equals the critical Euler's load:

$$\sigma_s \cdot e^2 = \pi^2 \frac{E_s}{b^2} \frac{e^4}{12} \tag{2.47}$$

where E_s is the intrinsic elastic modulus of the solid phase. Replacing the dimension ratio e/b by its expression as a function of Φ (from equation (2.45)), we obtain:

$$\sigma = fc_p = \frac{\pi^2}{108} E_s \Phi^2 \tag{2.48}$$

where fc_p is the compressive strength of the cement paste. The solid content of a fresh cement paste is

$$\Phi_0 = \frac{v_c}{v_c + v_w + v_a} \tag{2.49}$$

where v_c, v_w and v_a are the volume of cement, water and entrapped air respectively. After full hydration, a unit volume of cement produces about 1.6 volume of hydrates (while the paste volume remains approximately the same). Therefore there is a direct relationship between Φ and Φ_0:

$$\Phi \approx 1.6\Phi_0 \tag{2.50}$$

so that the following relation is found between the strength and the cement concentration in the fresh paste:

$$fc_p \propto \left(\frac{v_c}{v_c + v_w + v_a} \right)^2 \tag{2.51}$$

in which we recognize the classical Féret's equation (Féret, 1892); see section 2.3.2.

With the plate structure, the same type of calculation can be done. We have:

$$\Phi = \frac{3eb^2}{b^3} = 3\left(\frac{e}{b} \right) \tag{2.52}$$

$$\sigma_s = \sigma \cdot \left(\frac{b}{e} \right) \tag{2.53}$$

$$\sigma_s \cdot eb = \pi^2 \frac{E_s}{b^2} \frac{e^3b}{12} \tag{2.54}$$

$$\sigma = fc_p = \frac{\pi^2}{324} E_s \Phi^3 \tag{2.55}$$

$$fc_p \propto \left(\frac{v_c}{v_c + v_w + v_a} \right)^3 \tag{2.56}$$

Thus a similar equation is found as compared with Féret's, but with an exponent of 3 instead of 2. This equation is similar to the classical Powers' gel-space ratio equation (Powers and Brownyard, 1946–47). The gel space ratio equals the concentration of hydrates in the paste when the cement is fully hydrated. Powers' expression, in addition to the cement concentration, accounts for the degree of hydration, α. The expression is then more general, but requires the evaluation of α, which is not easy in practice.

Experiments and fitting

In this series of experiments (Marchand, 1992) an attempt has been made to produce a set of cement pastes of various cement concentrations, accounting for the numerous difficulties of paste testing. For the high water/cement ratio mixtures, which were prone to bleeding, a two-step mixing was carried out (Mehta, 1975; Markestad, 1976). A preliminary dilute paste with a w/c of 5 was first produced and continuously mixed for 24 hours. The obtained suspension, already containing hydrated cement particles, was used as mixing water for preparing the samples. The CSH particles acted as a thickening agent, limiting the amount of bleeding while not changing the final composition of the hardened paste. Then the cylindrical specimens were cast, demoulded after 2 days, carefully ground on a lapidary mill, and seal-cured under an aluminium adhesive sheet. The densities were checked for correcting the water/cement ratio, for pastes where bleeding was not completely avoided. The cement used was a high-strength one (CEM I 52.5 PM following the French standard NFP 18 301), with a Blaine SS of 3466 cm^2/g and an ISO strength of 64 MPa; its composition is given in de Larrard and Le Roy (1992). For the low w/c mixtures moderate amounts of a superplasticizer of the melamine type were used to facilitate the placement. Obtained results are given in Table 2.11. Each strength measurement is the mean value obtained on four 38 × 76 mm specimens.

Table 2.11 Results of compressive strength measurements at 28 days performed by Marchand (1992).

w/c	fc_p exp	Model
0.23	153.8	153.8
0.28	114.3	119.8
0.38	81.7	77.0
0.43	63.2	63.2
0.53	41.9	44.2
0.63	33.7	32.2

The experimental values have been regressed by a power-law model (Fig. 2.43) of the following form:

$$fc_{\mathrm{p}} = 730.4 \left(\frac{v_{\mathrm{c}}}{v_{\mathrm{c}} + v_{\mathrm{w}}} \right)^{2.85} \qquad (2.57)$$

Because of the fluid consistency of the pastes, the entrapped air volume was neglected, so that the term in parentheses may be considered as the cement concentration. The mean error of the model is equal to 2.3 MPa in absolute value, a value comparable to the standard deviation of the measurements. This confirms the soundness of the hypothesis of local buckling as the mechanism in which the compressive failure originates. The fact that the exponent is not an integer value is probably due to incomplete final hydration in low water/cement ratio mixtures (see section 2.2.2). However, it falls between 2 and 3, as predicted by the simple models presented above.

Finally a more general model, applicable to the 28-day compressive strength of sealed-cured cement pastes, can be written as follows:

$$fc_{\mathrm{p}} = 11.4 Rc_{28} \left(\frac{v_{\mathrm{c}}}{v_{\mathrm{c}} + v_{\mathrm{w}} + v_{\mathrm{a}}} \right)^{2.85} \qquad (2.58)$$

where Rc_{28} is the ISO strength of the cement at 28 days (see section 3.2.5), and assuming that the value of the exponent remains the same with another cement strength.

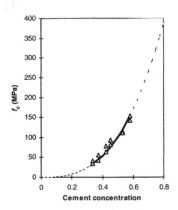

Figure 2.43 Relationship between compressive strength and cement concentration, after tests from Marchand (1992).

2.3.2 Effect of cement concentration on concrete compressive strength

When a cement paste is mixed with aggregate the relationship between strength and composition is no longer as simple. As a first step it is generally accepted that the cement concentration in the paste is still the major factor. Depending on countries and educational background, various models are referred to. Most populars are Féret's (Féret, 1892), Bolomey's (Bolomey, 1935) and Abrams' (Abrams, 1919) equations. Here we compare their effectiveness on a set of data (Kim *et al.*, 1992) from which the seven pure Portland cement concretes were taken. These mixtures have the same slump at the fresh state, the same aggregate nature and proportions, and w/c ratios ranging from 0.31 to 0.60.

The basic parameter in Féret's equation is the cement concentration in the fresh paste, as an explicit term (see previous section). With the adjustment of a single parameter, we obtain the following formula, which matches the data with a mean error as low as 1.2 MPa (Fig. 2.44):

$$fc = 290 \left(\frac{v_c}{v_c + v_w + v_a} \right)^2 \tag{2.59}$$

It is noteworthy that this empirical formula was originally developed for mortars in the 10–20 MPa range. In the old days, when electronic calculators were not available, Bolomey had the idea of proposing a linearized form of Féret's formula. Here, such an equation also gives a good fit to the data (mean error of 1.4 MPa; Fig. 2.45):

$$fc = 24.6 \left(\frac{c}{w} - 0.5 \right) \tag{2.60}$$

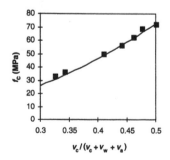

Figure 2.44 Fitting of the data with Féret's formula.

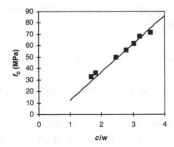

Figure 2.45 Fitting of the data with Bolomey's formula.

where c and w are the masses of cement and water per unit volume of concrete. However, the reference to air content has disappeared, which can impair the precision of the equation for dry concretes, and when air is intentionally entrained in the mix.

Also, note that the slope of the experimental points is lower than that of the model in Fig. 2.44. This defect could be easily corrected by changing the 0.5 coefficient and the multiplicative coefficient in the formula, although the 0.5 constant is generally not adjusted when the formula is applied in concrete mix design (Dreux, 1970).

Finally, independently of the 'European school', Abrams proposed an exponential equation with two adjustable parameters, which is still popular in North America (Popovics and Popovics, 1995). Such a formula:

$$fc = 147 \times 0.0779^{w/c} \tag{2.61}$$

again provides a good approximation to the present data (mean error 2.1 MPa; see Fig. 2.46).

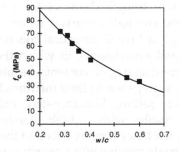

Figure 2.46 Fitting of the data with Abrams' formula.

To summarize, in the range of usual w/c ratios (or cement concentration), the three models give close predictions. In the author's opinion, Féret's formula has four advantages:

- It refers to the cement concentration, which is a physical parameter.[18]
- It takes the air content into account.
- It has a mathematical form that can be physically justified (see section 2.3.1).
- It gives reasonable trends when the cement concentration is extrapolated. This is also true for Abrams' formula, but not for Bolomey's. The latter predicts a negative strength for water/cement ratios higher than 2; and, more importantly, the strength tends towards infinity for ultra-low w/c ratios.

Coming back to Féret's equation, however, it no longer gives such outstanding agreement when other data are investigated. This is simply because the strength of concrete is not controlled *only* by the strength of the paste. Moreover, it is not clear at this stage why the exponent is equal to 2 and not 2.85, as for the cement paste. Therefore, in the following sections, we shall retain from Féret only the power law of the cement concentration, trying to complete the model to account for the aggregate related parameters.

2.3.3   Granular inclusion: effect of the topology

By topology we mean the geometrical parameters of the granular inclusions (grading, shape of the grains, concentration in the composite), regardless of the material that constitutes the grains. In a set of concrete mixtures produced with the same constituents (aggregate, cement and water), some systematic trends are to be expected.

Effect of aggregate volume

Stock *et al.* (1979) published a comprehensive review of the effect of aggregate volume on concrete compressive strength. Most of the cited references tend to support the fact that the strength decreases when the paste content increases, at least in the range of 'usual' structural concrete. These authors also performed some original experiments, in which they produced a series of concretes having different aggregate contents, keeping the same grading and matrix and avoiding segregation in fluid mixtures by continuously rotating the specimens before setting. Unfortunately, cube tests were performed, giving strength results with a high standard deviation. However, the above-mentioned tendency is reproduced in these experiments (Fig. 2.47). Note that the aggregate volume effect, which is not monotonic, can be masked by the increase of entrapped air when

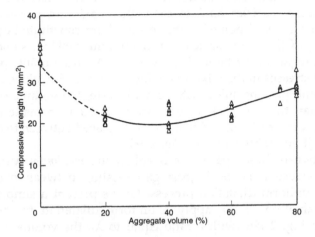

Figure 2.47 Effect of aggregate volume on the compressive strength of concrete (Stock *et al.*, 1979).

workability decreases (see section 2.1.6). According to Féret's equation, the air volume plays a role similar to that of water as regards strength.

Effect of the maximum size of aggregate

In the early 1960s Walker and Bloem (1960) published a paper in the ACI journal that generated many discussions (Alexander *et al.*, 1961). The major finding was the negative effect of the maximum size of aggregate (MSA) on the compressive strength of concrete, at constant w/c. Indeed, this statement astonished the scientific community, since it was – and still is – generally assumed that the higher the MSA, the more compact is the concrete, inducing higher strength and better durability. Here again, two causes exert contradictory effects: for a given amount of cement, increasing the MSA tends to reduce the water demand (see Chapter 1 and section 2.1), while the strength obtained for a given w/c is lower. Therefore, when the cement content and the slump are kept constant, there is an effect of MSA on compressive strength that is not monotonic. An optimal maximum size of aggregate exists, which decreases when the amount of cement increases (Cordon and Gillespie, 1963).

A unifying concept: the maximum paste thickness (MPT) (de Larrard and Tondat, 1993)

Let us show that the two previous effects are two facets of a single physical parameter, dealing with the topology of aggregate, considered as an inclusion in the cement paste matrix.

In a dry packing of particles loaded in compression it has been recognized that coarse particles tend to carry the maximum stresses (Oger, 1987): they act as 'hard points' in the soft medium constituted by the porous packing of the finer particles. In a random packing, some adjacent coarse aggregates may be in direct contact. At these points of contact, the highest stresses in the mixture are found (Fig. 2.48a). Now, if we consider the same packing filled with cement paste, the volume of which is greater than the dry packing porosity in order to provide a certain workability, it can be imagined that the paste placed between two close aggregates will be highly stressed (Fig. 2.48b).

The distance between these aggregates is called the *maximum paste thickness* (MPT), because it is the highest gap existing between two particles in such a uniform dilatation process. Let us present a simple method for its evaluation. If we apply a uniform dilatation to the aggregate grains in Fig. 2.48b (with a ratio equal to λ), the volume of aggregate will become

$$g^* = \lambda^3 g \tag{2.62}$$

where g is the aggregate volume in a unit volume of concrete. When grains are in contact with their neighbours, g^* is equal to the packing density of the aggregate, considered as a granular mix. But this particular packing density is also that of the original aggregate (prior to the dilatation), since a similarity – multiplying all distances by the same coefficient – does not change the porosity of a packing of particles.

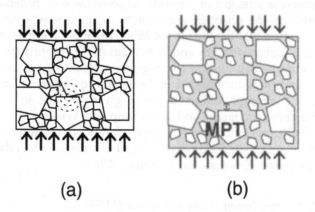

(a)　　　　　　　　(b)

Figure 2.48 (a) Dry packing of particles submitted to a uniaxial loading. (b) Concrete as a dry packing injected with cement paste. It is assumed that the network of particle centres of gravity is uniformly dilated, so that the largest gap is formed between two particles of aggregate that were previously in contact.

On the other hand, we have

$$\lambda = \frac{MPT + D}{D} \tag{2.63}$$

where D is the MSA. Therefore we deduce the equation for calculating the MPT:

$$MPT = D\left(\sqrt[3]{\frac{g^*}{g}} - 1\right) \tag{2.64}$$

D and g can be easily derived from a mixture composition. As for g^*, it can be measured, or accurately calculated with the compressible packing model (see Chapter 1).

To highlight the effect of the MPT on the compressive strength, we shall first use a particular set of data, where 16 concretes have been produced with cement content ranging from 300 to 575 kg/m³ and MSA between 5 and 19 mm (Hobbs, 1972). When plotting the compressive strength against the cement concentration in the fresh paste, a certain correlation is found, but with a significant scatter (Fig. 2.49a). Then the compressive strength is corrected by a multiplicative term, which is the MPT to the power 0.13. The improvement of the correlation is clear in Fig. 2.49b.

A similar treatment was applied to four sets of data (de Larrard and Tondat, 1993). The experimental strength, divided by the square of the cement concentration (as in Féret's equation), has been plotted against the MPT (Fig. 2.50). Quantitatively, the data support the following

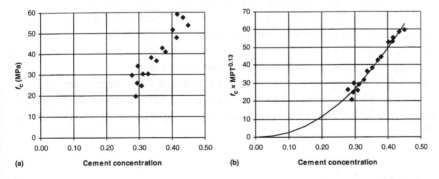

Figure 2.49 Relationship between strength and cement concentration in Hobb's data: (a) experimental strength; (b) strength multiplied by $MPT^{0.13}$.

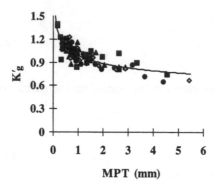

Figure 2.50 Relationship between compressive strength (after the effect of water/cement ratio has been eliminated) and maximum paste thickness, for four independent sets of data (de Larrard and Tondat, 1993). Non-dimensional parameter vs. MPT in mm.

empirical equation:

$$fc \propto MPT^{-r} \tag{2.65}$$

where r is close to 0.13–0.16. The mean error made by Féret's formula in predicting the compressive strength of concrete is halved by the introduction of this MPT term. Note that the influence of MPT is far too great to be negligible. According to Fig. 2.50, the compressive strength of concretes having the same water/cement ratio but different paste content and/or different MSA may range from 1 to 2.

Physically, the MPT could be related to the interfacial transition zone (ITZ). However, it is surprising to realize that when MPT decreases, the compressive strength increases, while the proportion of interstitial space (between two coarse aggregates) occupied by the ITZ also increases. Another explanation would deal with the orientation of CSH crystals: the thinner the MPT, the more oriented the hydrates, which would enhance their bearing capacity for compressive loadings. More research is needed to clarify the physical origin of this phenomenon.

The MPT concept incorporates the effects of both aggregate volume (term g) and MSA (term D) in equation (2.64). An intermediate form of a general model for the compressive strength of mature concrete is then the following:

$$fc \propto MPT^{-r} \cdot \left(\frac{v_c}{v_c + v_w + v_a} \right)^2 \tag{2.66}$$

But a systematic effect of the nature of the rock from which the aggregate originates remains, irrespective of its volume and grading. This is not described by the MPT concept, and will be addressed in the next section.

2.3.4 Granular inclusion: effect of the rock type (de Larrard and Belloc, 1997)

Let us first consider some experiments carried out for displaying the contribution to the strength of the nature of the aggregate.

Materials

Aggregates from five sources were used: a rounded flint from Crotoy (marine aggregate), a crushed hard limestone from the Boulonnais, a crushed semi-hard limestone from Arlaut, a crushed basalt from Raon l'Etape, and a quite heterogeneous crushed quartzite from Cherbourg. Characterization of these aggregates can be found in Table 2.12.

A high-strength Portland cement, in combination with 10% densified silica fume and a melamine-type superplasticizer, was used in all mixtures. Mortars and concretes were designed in order to have the same paste composition (w/c[19] = 0.285) and a fluid consistency, say slump value higher than 200 mm (4 in.). For each type of aggregate and each MSA the grading of the mixture was optimized to obtain the minimum paste content, using RENÉ-LCPC software (Sedran and de Larrard, 1994). Mixture proportions are given in Table 2.13.

Tests

Paste, mortar and concrete 110×220 mm cylinders were cast and consolidated with a vibrating table. No segregation or bleeding was experienced with any mixtures, including the pure paste, owing to the low w/c and the presence of silica fume. The samples were demoulded at 20 hours, and test faces of the specimens were carefully ground with a lapidary mill. Compressive strength at 1, 3 and 28 days was measured for all mixtures (with three specimens). Tensile splitting strength and elastic modulus were measured at 28 days. For tests after 1 day, the cylinders were sealed with an adhesive aluminium sheet to avoid any moisture exchange. The curing temperature was 20 ± 2 °C. More details of the experiments are given in the original publication. The results of the mechanical tests are given in Table 2.13.

Table 2.12 Characteristics of the aggregates used in the present experiments.

	Seine sand	Crotoy	Crotoy	Boul.	Boul.	Boul.	Arlaut	Arlaut	Arlaut	Raon	Raon	Raon	Raon	Cherb.	Cherb.	Cherb.
Aggreg. fractions	–	5/12.5	12.5/25	0/5	5/12.5	12.5/20	0/5	5/12.5	12.5/25	0/2	1/4	6/10	10/20	0/3.15	4/10	10/20
f_c (MPa)		285	285	160	160	160	111	111	111	250	250	250	250	–	–	–
Packing density	0.711	0.653	0.634	0.778	0.635	0.597	0.708	0.6022	0.584	0.742	0.615	0.601	0.600	0.758	0.543	0.570
Sieves (mm)														*Percentage passing*		
0.005														0		
0.0063										1				1		
0.008										1				1		
0.01										2				2		
0.0125										3				3		
0.016										4				3		
0.02										4				4		
0.025										5				5		
0.0315										6				5		
0.04										7				6		
0.05										7				7		
0.063				0						8				7		
0.08	0			8			0			9				8		
0.1	1			11			1			12				11		
0.125	3			14			2			14				14		

Sieve (mm)	1	2	3	4	5	6	7	8	9	10	11	12	13	14	15	16
0.16	7			19			4			16				17		
0.2	12			24			8			19				21		
0.25	17			30			13			22				26		
0.315	24			34			18			26				31		
0.4	35			40			26			30				37		
0.5	48			46			34			35				44		
0.63	60			51			44			41				53		
0.8	70			57			52			46	0			58		
1.00	76			63			61			54	1			68		
1.25	80			68			72			64	2			77		
1.60	84			74			80			74	2			85		
2.00	88			79			88			86	3			92		
2.50	92			83			95			96	4			97	0	
3.15	95			88			97	0		100	7			100	0.5	
4.00	99			95	0		98	0.5			12				1	
5.00	99	0		100	1		99	2			22	0			2	
6.30	100	7			11		100	10			40	1			4	
8		20			33			31			72	2	0		17	0
10		41			57	0		64	0		95	9	1		48	1
12.5		89	0		93	1		100	1		100	39	4		89	2
16.0		100	4		100	10			14			84	30		100	12
20.0			48			53			38			100	78			41
25.0			95			92.5			92				96			82
31.5			100			100			100				100			100

Table 2.13 Mixture proportions and mechanical properties of paste, mortars and concretes.

Number	1	2	3	4	5	6	7	8	9	10	11	12	13
Aggregate	—	Crotoy	Crotoy	Crotoy	Boulonnais	Boulonnais	Boulonnais	Arlaut	Arlaut	Raon l'Etape	Raon l'Etape	Cherbourg	Cherbourg
Mixture	Paste	Mortar	μ-concrete	Concrete	Mortar	μ-concrete	Concrete	Mortar	Concrete	Mortar	Concrete	Mortar	Concrete
G (kg/m³)			1051	653		876	661		638	0	725		627
G' (kg/m³)				446			285		345	0	210		163
S (kg/m³)		1528	771	753	1407	844	789	1401	683	674	187	1193	774
S' (kg/m³)										854	701		
OPC (kg/m³)	1543	644	444	431	717	539	525	670	500	673	513	826	612
SF (kg/m³)	154.3	64.4	44.4	43.1	72.56	53.9	52.5	67	50	67	51.3	82.6	61.2
SP (kg/m³)	30.9	16.1	11.1	10.8	17.9	13.5	13.1	16.8	12.4	16.9	12.8	20.6	15.2
Water (kg/m³)	417	190	136	134	204	156	153	215	173	201	158	231	174
g^*		0.767	0.864	0.871	0.829	0.885	0.89	0.755	0.855	0.789	0.852	0.813	0.869
g		0.579	0.711	0.719	0.527	0.645	0.649	0.543	0.651	0.548	0.651	0.457	0.598
D (mm)		2.5	12.5	20	3.15	12.5	20	2	25	5	20	2	20
MPT (mm)		0.246	0.839	1.32	0.513	1.39	2.22	0.232	2.378	0.646	1.8768	0.423	2.654
Air (%)		3.5	0.8	0.7	2.5	1.1	0.8	3.8	0.8	1.7	0.5	3.1	1.4
Slump (mm)	(80)	255	240	235	260	250	250	250	220	260	230	220	215
f_{c1} (exp.) (MPa)	109.2	56.9	45.9	47.5	72.0	72.4	61.8	56.3	53.4	61.4	46.3	71.8	61.9
f_{c1} (mod.) (MPa)		56.0	47.7	45.0	75.3	68	64.8	67	53.4	55.1	48.0	72.2	60.6
f_{c3} (exp.) (MPa)		76.4	64.9	62.7	94.9	90.7	82.6	79.2	71	76.9	64.2	87	76.7
f_{c3} (mod.) (MPa)		76.4	65.1	61.4	95.0	86.5	82.6	83.0	67.4	75.2	65.5	88.9	75.9
f_{c28} (exp.) (MPa)	152.3	106.3	90.5	86.8	112.1	109.5	107.7	101.9	89	104.9	90.9	108.4	101.8
f_{c28} (mod.) (MPa)		106.6	90.8	85.6	119.3	109.5	105	101.9	84.7	104.9	91.3	108.4	94.3
f_{t28} (MPa)	2.57	6.34	5.91	5.48	5.45	4.71	4.86	5.23	4.59	6.53	5.62	5.60	5.70
Ei_{28} (MPa)	29.3	45.4	55.4	55.5	44.7	52.6	53.8	37.6	42	45.5	52.1	38	42.6

G, G', S, S' stand for the different sizes of aggregate used, from the coarsest to the finest.

Confirmation of the effect of MPT

A direct comparison between the compressive strength developments of the paste and the composites (Table 2.13) leads to the following remarks:

- The strength of the paste is always higher than that of the composite.
- At a given age, and for a given type of aggregate, the strength decreases when the MSA (and the MPT) increases.

In order to account for the MPT effect, the composite strength is plotted against the matrix strength in Fig. 2.51. Here the matrix strength fc_m is defined as the paste strength fc_p multiplied by a term describing the effect of MPT:

$$fc_m = fc_p \cdot MPT^{-0.13} \qquad (2.67)$$

The MPT is calculated from equation (2.64). It is noticeable that all points dealing with a given type of rock can be plotted on a single master curve, a characteristic of the rock (Fig. 2.51). This is a further confirmation of the soundness of the MPT concept.

Ceiling effect

For some aggregates (Crotoy, Raon) the development of composite strength is strictly proportional to that of the matrix (at least in the range investigated), while for the others (Boulonnais, Arlaut, Cherbourg) a non-linearity is found, owing to a limiting effect of the aggregate. The master curves have been smoothed with an empirical hyperbolic model of the following type:

$$fc = \frac{p \cdot fc_m}{q \cdot fc_m + 1} \qquad (2.68)$$

where fc is the composite (concrete or mortar) strength (in MPa), and p and q are two empirical constants depending on the type of aggregate (the values of which are given in Table 2.14).

The mean accuracy of this model, given by equation (2.68), is 2.2 MPa. For very high matrix strengths the composite strength tends towards p/q, so that this ratio is expected to be controlled by the intrinsic strength of the rock. Hence from Table 2.14, the ranking between the aggregate sources given by p/q is:

Arlaut $\approx$ Cherbourg $<$ Boulonnais $<$ Raon l'Etape $\approx$ Crotoy,

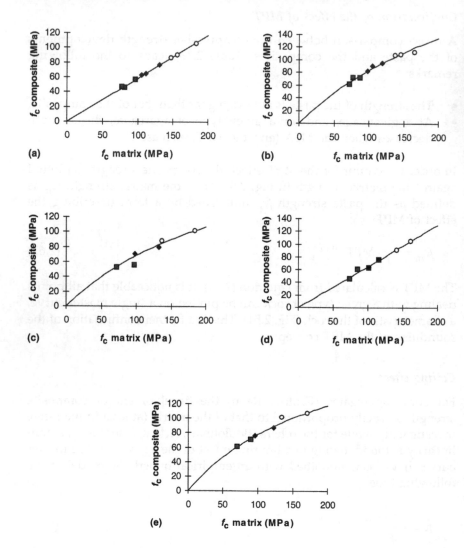

Figure 2.51 Relationship between the strength of the matrix (defined by equation (2.67)) and the strength of the composite, for various aggregates of various origin: (a) Crotoy; (b) Boulonnais; (c) Arlaut, (d) Raon l'Etape; (e) Cherbourg. ■, 1 day; ◆, 3 days; ○, 28 days.

while the ranking corresponding to the compressive strength of the rock is

Arlaut < Boulonnais < Raon l'Etape < Crotoy.

The heterogeneous nature of the Cherbourg aggregate makes it impossible to obtain any significant compressive strength.

Table 2.14 Values of the parameters of the master curves, for each aggregate nature.

	p	q (MPa^{-1})	p/q (MPa)	% of debonded aggregate	fc (MPa)	E_g (GPa)
Crotoy	0.583	0	∞	13.0	285	77
Boulonnais	1.111	0.0033	337	0	160	78
Arlaut	0.960	0.0040	241	0	111	62
Raon	0.651	0	∞	13.6	250	90
Cherbourg	1.145	0.0047	244	0	–	86

For low matrix strengths equation (2.68) is equivalent to

$$fc \approx p \cdot fc_m \qquad (2.69)$$

In this case, where the stress supported by the aggregate is small compared with its own strength, the parameter p would essentially describe the bond between paste and aggregate. It has long been recognized that limestone aggregates provide an excellent bond with the cement matrix, which is confirmed by the high p-values of the Boulonnais and Arlaut aggregates. It is not surprising that the lowest p-value is obtained with the round flint aggregate from Crotoy. To provide further confirmation of this statement, a systematic quantification of the amount of debonded aggregate was carried out on the ruptured surface of the specimens after the 28 day splitting tests (de Larrard and Belloc, 1997). Only the Crotoy and Raon aggregates exhibited debonded aggregate; this corresponds to the two low p-values.

Let us now return to the question of the exponent in Féret's equation. We have seen in section 2.3.1 that it had theoretically a value of 3 for pure cement paste, but because of incomplete hydration in low water/cement ratio mixtures the experimental value we found was 2.85. Moreover, increasing the cement concentration will have a lower effect on concrete strength because of the increase of paste volume (which leads to an increase of MPT), and because of the ceiling effect. This is why the apparent value on concrete is generally close to 2, as in the original Féret's formula. But it is now clear that this apparent value will differ from one rock type to another, being higher with hard aggregates than with soft ones.

Further comparison between aggregate sources

All master curves obtained with the aggregates tested are drawn together in Fig. 2.52. It must be borne in mind that these curves express the strength potential of a rock, irrespective of the grading and shape of

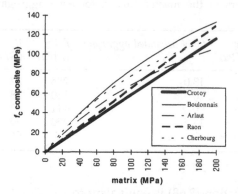

Figure 2.52 Comparison of the master curves of various aggregates.

particles. For instance, the Crotoy flint is not as bad as shown in this figure as, because of its rounded shape, it is possible to obtain smaller MPT values for a given cement content, workability and MSA.

It is interesting to note that, depending on the strength range, the hierarchy between two rocks may change, as already pointed out by Day (1995). For instance, this is the case for Raon and Arlaut. For normal-strength concrete the latter is preferable, while the former could be preferred for HPC (if comparable grading and particle shape can be obtained with the two aggregate sources).

Theoretical justification of the relation between matrix and concrete strength

Let us consider the following simple model for hardened concrete: two phases are put in parallel, one of which is made of matrix and aggregate (called the series phase), while the other is made of matrix only (Fig. 2.53).

As far as the strength of the right-hand phase is concerned, it is obviously limited by the strength of the aggregate. When the matrix strength is lower than that of the aggregate, the phase strength will be governed not only by the matrix strength but also by the bond: with a perfect bond the stress state in the matrix will be of a triaxial nature, so the strength will be higher than the uniaxial strength. Conversely, in the case of a poor bond between aggregate and matrix the strength of the phase will be closer to the uniaxial matrix strength. Then the strength of the series phase (right-hand phase in Fig. 2.53) is qualitatively as given in Fig. 2.54.

The strength of concrete is obtained by adding the strength of the two phases (if we assume that each phase has a plastic behaviour). The left-hand phase is made of matrix only, so that its intrinsic strength grows

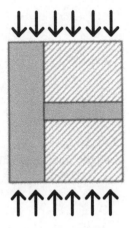

Figure 2.53 A composite model for hardened concrete crushed in compression.

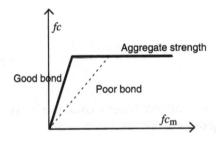

Figure 2.54 Relationship between the series phase strength and the matrix strength.

linearly with that of the matrix. A broken line is finally obtained, but without a horizontal plateau (Fig. 2.55). Therefore this simplified model gives a 'cubist' picture of the experimental curves shown in Fig. 2.52. The pure matrix phase is necessary to reflect the fact that the concrete strength may in some cases be higher than that of the aggregate (Baalbaki, 1990).

Summary

We have now built a comprehensive model for the compressive strength of mature Portland cement concrete (at 28 days), accounting for the cement strength, the cement concentration in the fresh paste, the MPT, the bond between paste and aggregate, and the intrinsic strength of the rock. The matrix strength is first calculated,[20] and then the concrete

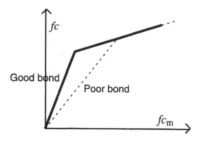

Figure 2.55 Relationship between the concrete strength and the matrix strength.

strength is deduced:

$$fc_m = 13.4Rc_{28} \left(\frac{v_c}{v_c + v_w + v_a} \right)^{2.85} \cdot MPT^{-0.13} \tag{2.70}$$

$$fc = \frac{p \cdot fc_m}{q \cdot fc_m + 1} \tag{2.71}$$

A simplified version, which is a generalized Féret's equation, may be used when the properties of the aggregate are unknown:

$$fc = K_g \cdot Rc_{28} \left(\frac{v_c}{v_c + v_w + v_a} \right)^{2} \cdot MPT^{-0.13} \tag{2.72}$$

K_g is a constant that must be calibrated on some available results dealing with the aggregate used. But a lower precision is to be expected. Evaluation of the predictive capabilities of these models will be carried out in section 2.3.6.

2.3.5   Strength development vs. time

The question addressed in this section is the modelling of the compressive strength at any age (different from 28 days). In particular, the aim is to develop formulae for the strength after 1 day. Before this age, the strength development is very dependent on the curing conditions of the concrete, and it is strongly affected by minor changes in the raw materials. Therefore a predictive model based upon the general characteristics of the constituents is not likely to give sufficiently accurate results for this period. Fortunately, early-age strengths are by definition quickly obtained after concrete batching, so that a trial-and-error approach is less time consuming than if later strength is dealt with.

We shall refer to the same data already used in section 2.3.2. For these seven concretes, the compressive strength has been measured at 3, 7, 28, 90, 180 and 365 days,[21] providing a comprehensive set of data with a wide range of water/cement ratios. When we plot the compressive strength development against time, we find that all curves are roughly parallel (Fig. 2.56) (de Larrard, 1995). Clearly, this observation cannot hold at very early age, where all curves should depart from 0. But this period is out of the scope of our study, as already explained.

Therefore the extension of the model to different ages is simple. A kinetics term, independent of the cement concentration, should be added. In the previous section, it was shown that the effects of aggregate – bond effect and ceiling effect – were the same at all ages (at least between 1 and 28 days). So the kinetics term should be inserted in the matrix strength equation, which becomes

$$fc_m(t) = 13.4Rc_{28}\left[d(t) + \left(\frac{v_c}{v_c + v_w + v_a}\right)^{2.85}\right]MPT^{-0.13} \qquad (2.73)$$

where $d(t)$ is the kinetics parameter at age t. It is supposed to be a characteristic of the cement. This equation, together with equation (2.68) for the concrete strength, has been calibrated on the data from Kim *et al.* The cement strength has been arbitrarily taken to be equal to 60 MPa,[22] and values of 1.15 for p and 0.0070 for q have been found by optimization on the strengths at 28 days. They are reasonably close to the values found for similar aggregates in section 2.3.4 (Kim *et al.* used a single source of limestone for producing their concretes). Then the $d(t)$ values have been calibrated on the rest of the data. The obtained kinetics parameter development is given in Fig. 2.57.

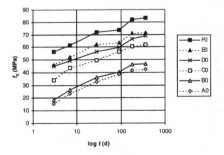

Figure 2.56 Strength development of Portland cement concretes at different water/cement ratios (Kim *et al.*, 1992).

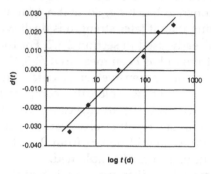

Figure 2.57 Evolution of the kinetics parameter $d(t)$ with time.

It can be seen that this evolution is reasonably linear in a semi-logarithmic scale, at least after 7 days. This could be a general feature of the strength development of Portland cement concretes (Baron *et al.*, 1993), and thus of most cements. Therefore knowledge of the strength development of a cement between 7 and 28 days allow us to predict the concrete compressive strength at any age after 7 days. As for the present data, the mean error of the model is equal to 2.20 MPa (Fig. 2.58).

2.3.6   Contribution of pozzolanic admixtures

Action of pozzolanic admixtures in hardening concrete

Let us first replace the volume of cement v_c by its mass c in the equation for the matrix strength, as the latter is a more usual parameter. If ρ_c is the specific gravity of the cement, and w and a are the volumes of water and

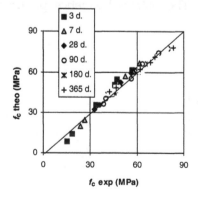

Figure 2.58 Comparison between model and experience for Portland cement concretes. Data from Kim *et al.* (1992).

air respectively, we have

$$fc_m(t) = 13.4Rc_{28} \left[d(t) + \left(1 + \rho_c \frac{w+a}{c} \right)^{-2.85} \right] MPT^{-0.13} \qquad (2.74)$$

Now, let us consider the contribution of pozzolanic admixtures to the compressive strength. The most common products of this category are fly ash (ASTM class C) and silica fume, while natural pozzolans, metakaolin and rice husk ash can also be considered as other types covered by this section.

The main mechanism from which pozzolans contribute to concrete strength is the well-known pozzolanic reaction. Each unit mass of cement releases during hydration a certain mass of free lime, which is more or less constant for any cement (as it comes from the silicate phases of cement, which are predominant). This lime may in turn be converted into additional CSH when combining with available silica (from the pozzolan) and water. If c_{eq} is the mass of equivalent cement as regards strength (at a given age), we then have

$$c_{eq} = c[1 + \Psi(pz/c, ...)] \qquad (2.75)$$

where pz is the mass of pozzolan per unit volume of concrete, and Ψ is a function. When pz is fixed and c decreases, c_{eq} should tend towards 0 (as a concrete containing only pozzolanic admixtures but no Portland cement never hardens). It is therefore reasonable to think that the Ψ function would tend towards a finite limit for high pz/c ratios. But one may wonder whether Ψ would also depend on the water/cement ratio. At low water/cement ratio the cement hydration is not complete, so that less lime is available for the pozzolanic reaction. The appearance of additional CSH is also governed by the solubility of the silica, which is influenced by the concentration of alkalis in the interstitial solution.

General form of the Ψ function

Replacing c by c_{eq} in equation (2.74) allows us to determine the Ψ function for different values of w/c and pz/c. This has been done for two sets of data. The first set (de Larrard and Le Roy, 1992) deals with silica fume concretes. Four mixes were prepared, with a constant w/c ratio of 0.33 and a silica fume addition of 0, 5, 10 and 15% of the cement weight. The corresponding values of Ψ (at 28 days) appear in Fig. 2.59. It can be seen that Ψ increases with the silica fume/cement ratio in a non-linear way.

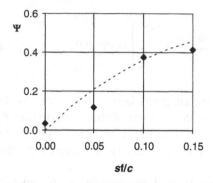

Figure 2.59 Contribution of silica fume to the concrete compressive strength at 28 days. Data from de Larrard and Le Roy (1992).

The same treatment has been applied to Kim's data. Here, in addition to the seven Portland cement concretes already dealt with, 19 fly ash concretes were produced, with percentage replacements ranging from 0 to 41% (of the remaining cement weight). The Ψ values have been calculated for all fly ash mixtures at the age of 1 year (when most of the pozzolanic reaction has already taken place), and are plotted in Fig. 2.60. The same observations are made as for silica fume concretes. Moreover, while the water/cement ratio ranges from 0.31 to 0.86, the Ψ coefficients are mainly governed by the fly ash/cement ratio.

We shall assume that Ψ depends only on the pz/c ratio, for a given pozzolan and a given age. For the data considered, it turns out that the

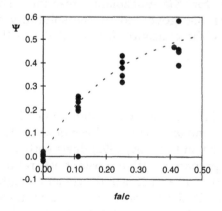

Figure 2.60 Contribution of fly ash to the compressive strength of concrete at 1 year. Data from Kim *et al.* (1992).

following equation gives a good fit:

$$\Psi = \Psi_{max} \left[1 - \exp\left(-K_p \frac{pz}{c} \right) \right]$$ (2.76)

K_p is an activity coefficient describing the effect of the pozzolan on the compressive strength. Clearly, it depends on the pozzolan, and on the age of the concrete. We shall assume that it is rather independent of the cement considered. In other words, the kinetics of the pozzolanic reaction are supposed to be controlled by the reactivity of the pozzolan (which is slow), and not by the rate of lime release due to the cement hydration. In fact we found that the most effective pozzolan (silica fume) gives similar contributions to strength with very different cements (de Larrard *et al.*, 1992).

As far as the Ψ_{max} coefficient is concerned, as it is related to the maximum amount of lime released, we would expect it to increase with time. However, as strength at very early age is out of our scope, and to avoid too complicated a model, we shall try to find a unique value giving a reasonable picture of the available data.

Calibration and validation of the model

An optimization of the model (Féret-type model given by equation (2.72) with a kinetics term, enriched by equations (2.75) and (2.76)) has been performed (Waller *et al.*, 1996). Seven sets of data have been used, resulting in 142 different mixes and 552 compressive strength results. Twenty-one mixtures contained only Portland cement, 57 contained fly ash, 51 contained silica fume, and 5 contained a mixture of both products. After calibration, the mean error given by the model (in absolute value) is 2.7 MPa (see Table 2.15 and Fig. 2.61), a value that is hardly higher than that obtained for pure Portland cement concretes.

The optimal value for Ψ_{max} is 1.10. This means that a very high dosage of an active pozzolan may *double* the cementing efficiency of a Portland cement. Hence this remark supports the hypothesis that the pozzolanic reaction is not the only mechanism involved in strength enhancement. For instance, it has been found that silica fume may in some cases combine directly with already formed CSH, increasing the volume of hydrates and decreasing the Ca/Si ratio.

The activity coefficients found in the calibration process are displayed in Fig. 2.62. Note that, as expected, the fly ash values (thin lines) are lower than the silica fume values (thick lines). Within the silica fume set, lower K_p values could be due to incomplete deflocculation of the product, especially when the silica fume is densified (as was the product used by Rollet *et al.*). As for the fly ash coefficients, Naproux's data are of

Table 2.15 Results of the calibration of the compressive strength model.

Data	Pozzolan used	Water/ cement ratios	Pozzolan/ cement ratio (%)	Ages considered (d)	Mean error (MPa)
Rollet *et al.* (1992)	SF	0.30–0.88	0–10	1–90	2.4
de Larrard and Le Roy (1992)	SF	0.28–0.5	0–15	1–28	3.7
Malhotra (1986)	SF	0.4–0.71	0–43	3–28	2.9
Yamato *et al.* (1986)	SF	0.25–0.79	0–43	7–90	5.1
Yamato *et al.* (1989)	SF	0.55–0.79	0–43	7–90	1.7
Naproux (1994)	FA/SF	0.28–0.76	0–50	2–365	2.3
Kim *et al.* (1992)	FA	0.31–0.86	0–41	3–365	2.2

SF, silica fume; FA, class C fly ash.
Source: Waller *et al.* (1997).

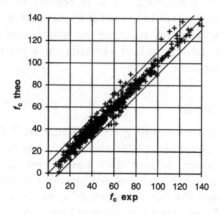

Figure 2.61 Comparison between experimental and theoretical strengths of pozzolan concretes (Waller *et al.*, 1997).

interest, since many fly ashes from the same origin but with different fineness were tested. In this case, the specific surface is the major parameter governing the activity, especially at later ages. While fine selected fly ash may approach the performances of silica fume (see B8), coarse ashes (G65) may have a slightly negative effect on strength. This could be due to the hollow spheres contained in the pozzolan, which would act as voids in the system.

Let us look at the Ψ function for a typical silica fume at 28 days ($K_p = 4$) and for a typical fly ash ($K_p = 0.5$, see Fig. 2.63). To have 80% of the pozzolanic cementing effect ($\Psi/\Psi_{max} = 0.8$), it requires a pozzolan/ cement ratio of 0.4 for silica fume, and 3.2 for fly ash. Hence in this case

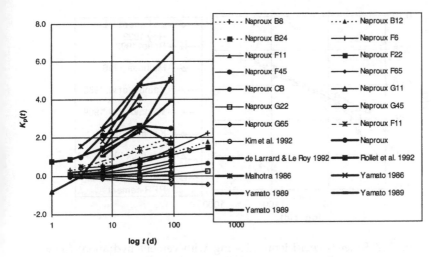

Figure 2.62 Activity coefficients found in the calibration process (Waller *et al.*, 1997). G22 means fly ash G with a maximum size of 22 μm.

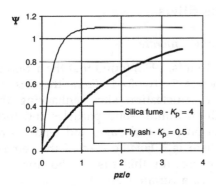

Figure 2.63 Ψ function (defined in equation (2.76)) for a typical silica fume and a typical class C fly ash at 28 days.

maintenance of sufficient alkalinity in the pore solution could be a problem. In practice much lower dosages are used,[23] so that the risk of steel corrosion due to lack of free portlandite is no longer real (Wiens *et al.*, 1995), except in the case of severe carbonation.

This validation also allowed us to collect more data about the kinetics function $d(t)$ introduced in section 2.3.5 (Fig. 2.64). As already noted, a linear kinetics in a semi-logarithmic diagram seems to be a general feature of cements after 7 days (Baron *et al.*, 1993).

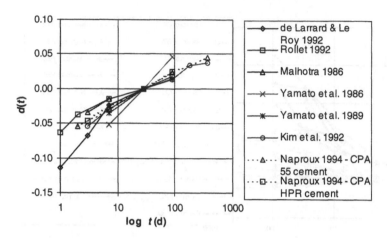

Figure 2.64 Kinetics coefficients (dealing with cement hydration) found in the calibration process (Waller *et al.*, 1997).

2.3.7   Contribution of limestone fillers

Action of limestone filler in hardening concrete

Several investigators (Detwiler and Tennis, 1996) found that the action of limestone filler in hardening concrete is twofold. First, there is an *acceleration effect*. The grains act as nucleation sites, enhancing the probability that dissolved CSH will encounter solid particles and then precipitate. This effect is significant only at early age, and tends to be negligible after 28 days. Second, there is a *binding effect* if the cement contains a significant aluminate phase. In this case carboaluminate compounds are produced, which have a certain cementing capability.

For the first effect, we shall assume that the incorporation of a given surface of limestone filler will accelerate the cement hydration kinetics by a constant time shift Δt, whatever the age of concrete. We have seen in section 2.3.5 that a given concrete often has a logarithmic strength development after 7 days. We may then write

$$d(t) = A \log(t/28) \qquad (2.77)$$

where A is a constant depending on the cement considered.

From equation (2.74) the hardening rate (matrix strength) is then

$$\Delta fc_m = 13.4 Rc_{28} \frac{A}{t} \Delta t MPT^{-0.13} \qquad (2.78)$$

Therefore the strength increment resulting from the acceleration effect will be inversely proportional to time. We may safely[24] extend this hypothesis to the strength between 1 and 7 days, as in this range the hydration kinetics, expressed in a semi-logarithmic diagram, tend to be faster than at later ages (Fig. 2.64). To build an expression applying to any cement, we may replace A by another quantity proportional to a $d(7)$ coefficient (which expresses the hydration rate between 7 and 28 days).

The matrix strength taking account of the acceleration effect of the limestone filler finally becomes

$$fc_m(t) = 13.4Rc_{28}\left[d(t) - B\,\frac{S_{FI} \cdot d(7)\,fi}{t}\,\frac{1}{c} + \left(1 + \rho_c\,\frac{w+a}{c}\right)^{-2.85}\right]MPT^{-0.13}$$

$$(2.79)$$

where S_{FI} is the Blaine specific surface of the filler, fi is the mass of filler per unit volume of concrete, and B is a coefficient to be adjusted on experimental results. Note that the filler contribution is always positive, since $d(7)$ is negative.

Let us now deal with the second effect. In an approach similar to that used for the contribution of pozzolans, we may state that there is an equivalent cement content in the presence of limestone filler, the difference with the actual cement content being proportional to the amount of tricalcium aluminate[25] (C_3A) in the cement. We then write

$$c_{eq} = c\left[1 + t_{C_3A}\Psi'\left(\frac{fi}{t_{C_3A} \cdot c}\right)\right]$$

$$(2.80)$$

where t_{C_3A} is the percentage of tricalcium aluminate in the Bogue composition of the cement, and Ψ' is a function. As for the pozzolans, we shall state that

$$\Psi' = \Psi'_{max}\left[1 - \exp\left(-K_{FI}\,\frac{fi}{t_{C_3A} \cdot c}\right)\right]$$

$$(2.81)$$

where Ψ'_{max} is the maximum value for Ψ', and K_{FI} is an activity coefficient. Unlike the pozzolanic reaction, we shall assume that the carboaluminate formation is rapid, so that these two coefficients are constant regardless of the age.

Replacing c by c_{eq} in equation (2.79), we then have a general model for limestone filler concretes, with three adjustable parameters. As limestone fillers are rather pure products (unlike fly ash, which contains a variety of subcompounds that may influence their activity), we shall assume that these coefficients do not depend on the filler. In other words, the only parameter of the filler used in the final model would be its specific surface.

Calibration on UNPG mortar data

A comprehensive study has been carried out to investigate the strength contribution of limestone filler (Giordano and Guillelmet, 1993). The strengths of mortars were measured at 7, 28 and 90 days. All possible combinations of 15 limestone fillers (the fineness of which ranged between 200 and 1000 m^2/kg) and four Portland cements (with C_3A content between 4 and 8%) were investigated. Besides the pure Portland cement mortars (ISO composition), filler mortars have been formulated by replacing 25% of the cement, at an equal water/fines ratio of 0.5. The obtained compressive strengths (192 strength results) are given in Table 2.16.

Table 2.16 Results of the mortar tests (Giordano and Guillelmet, 1993). The data dealing with fillers J and Q have been excluded from the calibration because, owing to their extreme fineness, excessive air entrainment is believed to have taken place. Compressive strengths are given in MPa.

	Cements											
	Le Teil			*La Malle*			*Montalieu*			*Beffes*		
C_3A (%)	4.50			6.20			8.36			10.14		
Blaine SS (m^2/kg)	325			320			352			293.5		
Age (d)	**7**	**28**	**90**	**7**	**28**	**90**	**7**	**28**	**90**	**7**	**28**	**90**
Control	41.7	63.9	74.9	43.4	54	(57.4)	44.3	60.2	70.3	43	56.3	64.3
Type of filler (Blaine SS)												
A (580)	31.9	47.4	57.2	33.9	45.5	48.9	34.5	46.1	52.8	37.9	47.3	48.5
B (504)	29.4	45.3	54.3	31.9	43.2	48.3	35.2	48.2	56.5	36.1	49.0	52.5
C (270)	25.9	43.1	56.1	30.5	43.3	50.0	34.1	44.4	51.1	35.1	44.9	46.6
D (203)	25.4	41.1	50.9	32.4	42.2	50.4	32.2	43.2	51.8	32.2	43.0	50.7
E (412)	29.8	45.2	53.5	32.3	43.5	50.6	33.2	45.0	55.0	35.5	46.7	50.2
F (295)	29.1	44.3	51.4	32.9	42.8	47.4	32.2	43.9	51.1	36.2	45.4	48.2
G (307)	26.3	42.8	50.3	31.4	41.6	46.3	33.4	47.3	54.1	35.3	43.0	50.3
H (491)	30.3	46.7	57.4	33.4	43.8	50.2	33.6	44.1	54.1	40.0	46.5	54.0
J (1013)	*35.6*	*50.2*	*53.0*	*34.4*	*44.3*	*48.3*	*37.0*	*46.8*	*49.6*	*37.2*	*49.0*	*49.9*
K (601)	34.4	50.6	53.0	36.5	48.0	48.0	34.8	45.8	52.0	40.2	48.4	51.1
L (523)	33.5	48.0	52.1	33.4	42.3	46.2	37.3	50.3	53.0	38.4	50.1	50.2
M (350)	30.9	46.3	54.5	30.4	40.6	46.7	33.1	44.8	52.9	36.7	46.3	50.0
N (445)	28.0	44.9	55.2	31.7	42.1	48.0	34.0	44.3	51.6	37.2	46.5	51.3
P (440)	27.5	44.6	53.0	32.5	46.0	51.5	34.0	44.1	52.4	37.0	47.7	53.8
Q (900)	*32.5*	*48.2*	*53.9*	*33.9*	*44.0*	*46.3*	*33.8*	*45.2*	*51.6*	*39.5*	*48.8*	*50.6*

Before we calibrate the model, it is interesting to display separately the two effects of the limestone filler on the compressive strength. A calibration of the Portland cement model (equations (2.73) and (2.71)) has been performed, allowing us to calculate the strength of filler mortar with the hypothesis of inert filler. Then, by difference with the experimental strength of these mortars, the absolute contribution of the filler to the strength has been evaluated. For the cement having the lowest C_3A content, the contribution of filler at 7 days essentially comes from the acceleration effect (Fig. 2.65a). It can be seen that it is reasonably correlated with the specific surface of the filler. Conversely, if we select the cement that has the highest C_3A content, the contribution of filler at 90 days originates in the binding effect (Fig. 2.65b). Here, no correlation appears with the fineness of the filler. Moreover, the maximum contributions that can come from both effects are of a similar order of magnitude. The effect of limestone filler in hardening concrete cannot be reduced to either an acceleration or a binding effect.

In Table 2.17 the model parameters are given, as found after calibration on the results given in Table 2.16 (except strength results dealing with fillers J and Q). The mean error of the model is only 1.91 MPa (see a comparison between theoretical and actual strengths in Fig. 2.66).

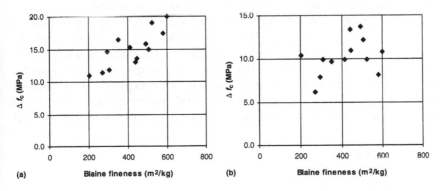

Figure 2.65 Net contribution of limestone filler to the compressive strength of mortar: (a) acceleration effect (Le Teil cement, 7 days); (b) binding effect (Beffes cement, 90 days).

Table 2.17 Values of the parameters describing the effect of limestone filler on strength.

Acceleration effect	Binding effect	
B	Ψ'_{max}	K_{FI}
0.0023	0.017	79

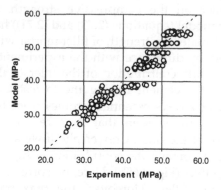

Figure 2.66 Comparison between theoretical and experimental compressive strengths of mortars.

The variation of the $\Psi' \cdot t_{C_3A}$ term, which represents the relative cementing contribution of the filler, with the filler/cement ratio is displayed in Fig. 2.67, for cements having different C_3A contents. For very low filler contents, the C_3A content of the cement does not matter, as all the filler finds available aluminate to react with. For higher dosages, a saturation appears: the lower the C_3A content, the sooner this occurs. When comparing the cementing capability of limestone filler with that of pozzolans (shown in Fig. 2.63), two remarks can be made. First, as expected, the maximum cementing capability of limestone is much lower than that of pozzolans. Second, the limestone dosage necessary to obtain the cementing effect is also much lower. This explains why cement manufacturers like to introduce some small percentages of limestone filler in their Portland cement (as authorized by French and European

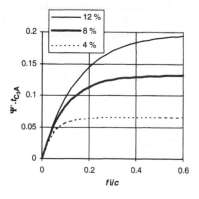

Figure 2.67 Relative cementing contribution of limestone filler, as a function of the filler proportion (for cements having different C_3A contents).

regulations).[26] A further advantage of the limestone filler is that the effect appears at early age. It is also completed by the acceleration effect, already analysed.

Further validation on some concrete data

A series of 14 concretes were produced to investigate the effect of fillers (either calcareous or siliceous) on the durability-related properties of concrete (Caré *et al.*, 1998). Eight standard mortar tests for measuring the activity index of the fillers were carried out in parallel. Two cements (of low and high C_3A contents) and four fillers (two limestone, two quartz) were used in these experiments. All strength results have been simulated with the model developed herein. Only three parameters had to be fitted: the aggregate parameter K_g (a Féret-type model was used for the concretes), and the kinetics parameters $d(7)$ of the two cements, which have not been measured. As for the siliceous fillers, only the acceleration effect has been accounted. Their pozzolanic effect, which is probably minor because of their crystalline nature, has been neglected. A comparison between theory and experiment is given in Fig. 2.68. The mean error of the model is equal to 2.4 MPa.

2.3.8 Summary: a general model of compressive strength

Final equations of the compressive strength model, and reservations about its use

After a step-by-step approach, by which a comprehensive model for the compressive strength of water-cured concrete or mortars has been built, it may be useful to summarize it. The aim is then to establish equations that lead to prediction of the compressive strength at an age t between

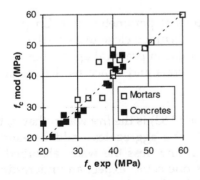

Figure 2.68 Comparison between simulated and measured strengths for 8 mortars and 14 concretes containing various amounts of filler (Caré *et al.*, 1998).

1 day and 1 year. The concrete contains normal-weight[27] aggregate of various fractions, Portland cement, water, air and, possibly, pozzolanic admixtures and limestone filler.

The first step is to calculate the equivalent cement content, by adding the cementing contributions of the pozzolans (pz_i) and fillers (fi_j):

$$c_{eq}(t) = c\left\{1 + 1.1\left[1 - \exp\left(-\frac{\sum K_{p,i}(t) \cdot pz_i}{c}\right)\right]\right.$$

$$\left. + 0.017t_{C_3A}\left[1 - \exp\left(-79\frac{\sum fi_j}{t_{C_3A} \cdot c}\right)\right]\right\} \quad (2.82)$$

Then the matrix strength is evaluated. Here, the accumulated surface of the limestone filler exerts an acceleration effect:

$$fc_m(t) = 13.4Rc_{28}\left[d(t) - 0.0023\frac{d(7)}{t} \cdot \frac{\sum S_{FI,j}fi_j}{c}\right.$$

$$\left. + \left(1 + \rho_c\frac{w+a}{c_{eq}}\right)^{-2.85}\right]MPT^{-0.13} \quad (2.83)$$

Finally the concrete strength is deduced from the matrix strength, by the equation already given in section 2.3.4:

$$fc(t) = \frac{p \cdot fc_m(t)}{q \cdot fc_m(t) + 1} \quad (2.84)$$

With proper calibration of the component parameters (see Chapter 3), this model is expected to give a mean accuracy close to 2–3 MPa, or 5% in relative value. Before promoting its use in practical applications, note the following list of precisions, limitations and concerns that must be kept in mind:

- The water dosage used in equation (2.83) is the *free* water: that is, the total water (originally present in the aggregates + added water + water coming from admixtures, if any) less the water absorbed by aggregate. For porous aggregates one could expect an underestimation of low-w/c concrete strength by the model, since some water is likely to migrate from aggregate to matrix, when self-desiccation will take place (see section 2.5.3). Nevertheless, if the q parameter (describing the ceiling effect) is properly calibrated, this effect will also be present in the calibration experiment, so that it will not significantly impair the precision of the model.

- The model has not been validated for pure pastes containing pozzolanic admixtures. The author would expect an overestimation of strength in many cases, because the activity coefficients of pozzolanic admixtures are determined on mortar (or concrete), where a part of the enhancement originates in the bond between aggregate and matrix.

- Often, the model underestimates compressive strength for water/cement ratios higher than 0.65, especially at early age: hence the paste model has been calibrated on a set of data where all water/cement ratios were lower than 0.63 (see section 2.3.1). However, contemporary good practice rules (as laid down in some standards, including European ENV 206) do not suggest the use of such high water/cement ratio concrete, which will frequently lead to durability problems.

- The model does not account for the presence of organic admixtures. Superplasticizers are not believed to have any *direct* effect on strength except at early age, where they act as a retarding agent. Therefore the model may sometimes overestimate the strength of high-performance concretes in the very first days after mixing. A practical means of correcting this error is to account for a retardation of 0.3–1 day: that is, for predicting the compressive strength at 1 day, to use the model with an age of 1.3–2 days.

- Similarly, the model will predict values that are too high at early age for concretes containing retarders; however, final values will be often underestimated, because retarders generally promote a more complete hydration of cement in the long run.

- In the calculation of the maximum paste thickness (MPT), it is implicitly assumed that the granular skeleton is close to the optimal distribution (see section 1.4); deviations from the model can be expected if the coarse aggregate volume is far from the optimal value, or if the concrete is subject to segregation.

- The model tends to overestimate the compressive strength of mixtures with low fines content. Here the amount of bleeding that takes place after consolidation creates an accumulation of water below the coarse aggregates. This process greatly impairs the aggregate/matrix bond. Conversely, this type of concrete will be greatly enhanced by an addition of filler, which will provide not only a decrease of water demand (see section 4.2.1, rule 12), but also a stabilizing effect, improving the relationship between strength and water/binder ratio (Kronlöf, 1994). Introduction of entrained-air displays the same type of beneficial effect (see Table 4.4). However, the fact that the model becomes less precise in this type of concrete does not matter from a practical viewpoint, as a predictive model for compressive strength is normally used after the size distribution of the mixture has been optimized. In particular, structural concretes with a very low fines content should never be produced, since they are both pathological and uneconomical.

• Finally, one may wonder whether the mineral admixtures (pozzolanic admixtures or limestone filler) should be included in the paste or in the aggregate skeleton when calculating the aggregate volume. The question arises especially for limestone fillers, which can sometimes be considered as inert particles (see section 2.3.7). In fact the effect of accounting fines particles with the aggregate or with the cement paste on the MPT is of little consequence, as it is mainly the ratio g^*/g that matters. Therefore we shall conventionally put any particle bigger than 80 μm in the aggregate skeleton, the rest being considered as part of the matrix.

The case of air-entrained concretes

So far only mortars or concretes without entrained air have been simulated. Let us refer to a last data set of interest in this matter. In this study (de Larrard *et al.*, 1996b), we have produced 17 concretes with a wide strength range (from 25 to 130 MPa), with the following additional variables investigated: the nature of the aggregate (limestone or basalt), the presence of pozzolans (fly ash or silica fume), and the presence of entrained air (between 1 and 7%). The model in its most complete form has been used to simulate the strength results, excluding concretes with water/cement ratio in excess of 0.65, and strengths before 7 days for concretes containing retarder and superplasticizer (see Fig. 2.69). The mean error given by the model is equal to 4.4 MPa (which is, in relative values, similar to the usual precision, if the wide strength range is

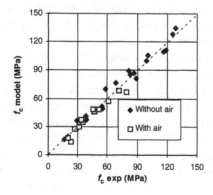

Figure 2.69 Simulation of strength results on air-entrained and non-air-entrained concretes. Data from de Larrard *et al.* (1996b).

considered). The value dealing with air-entrained concretes is 3.6 MPa. Therefore the model is qualified to illustrate the effect of entrained air on compressive strength.

2.4 TENSILE STRENGTH

In this section an attempt is made to propose a suitable model for predicting the *splitting* strength (as measured on cylinders) from the concrete composition. This value gives a conventional index of the tensile strength of concrete. However, it must be kept in mind that this is slightly different from the *direct* tensile strength, and even more from the *flexural* strength (also called the modulus of rupture). Conversion coefficients have been proposed, but they generally depend on the type of aggregate and the curing regime (Neville, 1995). Moreover, the tensile strength of concrete displays a significant scale effect: the larger the specimen, the lower the mean strength, and the lower the coefficient of variations (Rossi *et al.*, 1994). This is also true for compressive strength, but to a lesser extent.

2.4.1 Power-law type relationship between tensile and compressive strengths

There have been several attempts to calculate the tensile strength of concrete by using fracture mechanics theories (Lange-Kornbak and Karihaloo, 1996). However, such theories require a knowledge of fracture mechanics parameters, for the prediction of which no model is available (to the author's knowledge). Moreover, these parameters could be strongly affected by size effects, and their physical significance is still a matter of discussion. Therefore we shall keep to the traditional empirical approach, which is to link the tensile with the compressive strength. There is some logic in this idea, as both strengths are controlled by the nature of the paste (water/cement ratio, presence of mineral admixtures), the spatial arrangement of the grains, the bond of the cement paste with the aggregate, and the intrinsic strength of the aggregate (see section 2.3). With high-performance concrete, it is common to see fracture facies where most coarse aggregates are split by the main crack that provoked the failure of the specimen.

Oluokun (1991) reviewed the existing relationships between tensile and compressive strengths. It appears that most of them are of the power-law type. By compiling 566 data points, Oluokun proposed the following relationship:

$$ft = 0.214fc^{0.69} \tag{2.85}$$

which gave the best fit to the data considered (Fig. 2.70). Hence *ft* is the splitting tensile strength of concrete, and *fc* is the compressive strength at the same age. By statistical analysis, this author could confirm that the age of concrete does not affect significantly the (*ft*,*fc*) relationship (Oluokun *et al.*, 1991).

However, the scatter of the data points suggests that parameters other than compressive strength could play a role in governing the concrete tensile strength. Also, a wide number of parameter sets (namely the multiplicative coefficient and the exponent in equation (2.85)) can be found, which give similar agreements with the data. This is why, among the works reviewed by Oluokun, the exponent ranges from 0.5 to 0.79. Finally, note that the maximum compressive strength in Oluokun's data is about 60 MPa, while much higher strengths are now attainable (especially in the laboratory).

In a set of data already used for various models (de Larrard *et al.*, 1996b), 15 concretes were produced with the same aggregate (a limestone crushed material), with a wide strength range (from 25 to 130 MPa), the presence or absence of pozzolans (fly ash or silica fume), and the presence or absence of entrained air (between 1 and 7%). The 28 days splitting strength of water-cured cylinders appears closely related to the

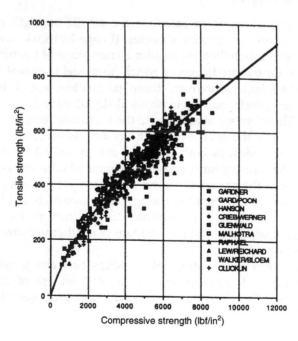

Figure 2.70 Tensile strength vs. compressive strength (Oluokun, 1991). 1 lbf/in$^2 \simeq 6.9$ kPa.

compressive strength by the following relationship:

$$ft = 0.413fc^{0.57} \tag{2.86}$$

The mean error in absolute value is only 0.17 MPa. This fact suggests that the parameters investigated in this study (pozzolans and air) have the same effect on tensile strength as on compressive strength, so that the relationship between both strengths remains the same whatever the type of paste.

In a previous work (Rollet *et al.*, 1992), eight concretes were tested with or without silica fume, made up with another type of limestone aggregate, a different cement, and with a similar strength range (as regards our data). The samples were cured in water over 28 days. The same type of relationship between tensile and compressive strength was found, with a mean error of 0.39 MPa:

$$ft = 0.459fc^{0.57} \tag{2.87}$$

We have previously measured the tensile strength development of a high-performance concrete cured in water, between 1 and 28 days (de Larrard, 1988). Again, we find a similar equation linking the two mechanical characteristics:

$$ft = 0.468fc^{0.57} \tag{2.88}$$

(mean error: 0.3 MPa).

All these data are summarized in Fig. 2.71. It seems therefore that the exponent 0.69 would tend to overestimate the tensile strength of

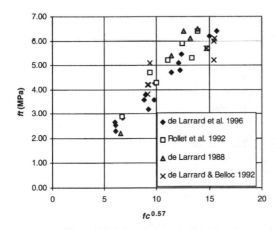

Figure 2.71 Tensile vs. compressive strength (to the power 0.57), in data comprising normal- and high-strength concretes.

high-performance concretes. On the other hand, the value of 0.57, which is intermediate between Oluokum's and the ACI one (0.5), is certainly reasonable for normal-strength concretes.

2.4.2   Effect of aggregate type

From the previous analysis, the following model comes naturally:

$$ft = k_t fc^{0.57} \tag{2.89}$$

where the k_t coefficient essentially depends on the aggregate. But is it only the *nature* of the aggregate (mechanical characteristics of the rock) that matters, or is there also an influence of the spatial arrangement of the particles? To answer to this question, let us refer to the data presented in section 2.3.4 (de Larrard and Belloc, 1997). Here, five sources of aggregate were taken, from which mortars and concretes were prepared, all with the same paste. The specimens were sealed-cured (under aluminium foils). In Fig. 2.72, the k_t values calculated for each data point are plotted against the maximum size of aggregate (MSA).

From this figure it can be deduced that the MSA has no systematic effect on k_t, nor the maximum paste thickness, which is a more physical parameter (see section 2.3.3). By calculating the mean k_t value for each aggregate origin (see Table. 2.18), the comparison of equation (2.89) with the experimental data gives a mean error of 0.17 MPa, which is comparable with the intrinsic error of the test. Thus we may assume that the k_t parameter is governed mainly by the nature of the rock (mechanical and petrographical characteristics), irrespective of the size distribution and of the aggregate concentration in the concrete. From

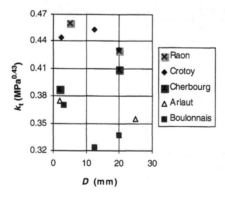

Figure 2.72 k_t coefficients (from equation 2.89) calculated for a series of mortars and concretes having the same paste and different aggregates (de Larrard and Belloc, 1997).

Table 2.18 Mean k_t coefficients (from equation 2.89) by nature of rock, in the data investigated. These values are not directly comparable with those appearing in equations (2.86)–(2.88), because of differences in curing regimes (sealed vs. water curing).

Quarry	Type	k_t
Crotoy	Marine flint	0.442
Boulonnais	Hard limestone	0.344
Arlaut	Semi-hard limestone	0.365
Raon l'Etape	Basalt	0.445
Cherbourg	Quartzite	0.398

Source: de Larrard and Belloc (1997).

Table 2.18 it seems that siliceous aggregates lead to higher k_t coefficients than limestone aggregates.

In conclusion, the model given by equation (2.89) appears reasonable for predicting the tensile splitting strength of concrete at any age. The compressive strength may be either measured or predicted with a suitable model, such as the one developed in section 2.3. The k_t coefficient is a characteristic of the rock type, and may also depend on the curing regime.

2.5 DEFORMABILITY OF HARDENED CONCRETE

General modelling approach

In this section different properties related to the deformability of hardening (or hardened) concrete are dealt with. The triple-sphere model is first introduced, which will give a unified approach for all these properties. The basic cell of concrete as a solid material, describing the arrangement between matrix and aggregate, is taken as a simple geometrical model. In this model a distinction is made between the minimum paste content that would fill the voids of the fully packed aggregate, and the 'overfilling' paste that gives fresh concrete its workability. Submitting the model to a hydrostatic pressure, one may calculate different properties of concrete, such as bulk modulus, shrinkage and creep.

As for the aggregate phase, the main variable influencing concrete deformability is the elastic modulus, which can be measured (at least for aggregates crushed from a massive rock). However, as already seen for compressive strength (section 2.3), it is shown that the properties of the matrix phase suffer certain modifications as compared with those of the pure paste. These modifications are most often quantified by reference to the concrete compressive strength. This property is not only the most commonly measured one; it is also a direct illustration of the matrix compactness, taking account of various effects coming from the aggregate. Finally, quantitative models are built up and validated with the help of various sets of data.

Following the triple-sphere model, it turns out that the packing density of the aggregate phase plays a major role in the deformability of hardened concrete. Unfortunately, this characteristic is lacking in most papers of the international literature presenting experimental data about the elastic modulus, creep or shrinkage. This is why the validation of the models developed will be essentially made with the help of data coming from LCPC. However, some data sets of different origins will be referred to, when the number of mixtures presented allows one to calibrate one or two adjustable parameters, without impairing the significance of the model evaluation.

Different types of deformation

The first characteristic of interest is the elastic modulus, which expresses the ability of concrete to deform when rapidly loaded (say, with a loading duration lasting some minutes or less). The natural order would lead us then to deal with shrinkage deformations, which we separate into autogenous shrinkage (appearing without humidity exchange) and drying shrinkage (the additional shrinkage due to water loss). However, we shall look at creep first, because matrix creep (and relaxation) is of importance in the deduction of concrete shrinkage from matrix shrinkage. Creep strains also are separated into deformations appearing without drying (basic creep) and additional creep due to drying (drying creep). The total creep is the sum of both basic and drying creep. Similarly, the total shrinkage is the sum of autogenous and drying shrinkage. Figure 2.73 summarizes the definitions of instantaneous and delayed deformations.

Codes for concrete structural design generally include some simplified laws to predict creep and shrinkage development, in which the mixture proportions are accounted for only through the compressive strength. A more refined and comprehensive empirical model has been proposed, using some mix design ratios such as cement content or water/cement ratio (Bazant *et al.*, 1991–92). Here an attempt is made to establish links between mixture proportions and delayed deformations, through a unique description of the structure of the material.

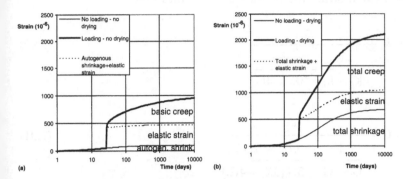

Figure 2.73 Separation between the various types of delayed deformation of concrete. In this example, a concrete sample is either (a)sealed cured or (b) dries from an age of 7 days. A 15 MPa compressive stress is applied at the age of 28 days. Calculations were carried out following the French BPEL 97 code, for a 50 MPa non-silica-fume concrete.

2.5.1 The two-phase nature of hardened concrete: the triple-sphere model

When looking at pictures representing sections of hardened concrete, the following pattern appears: a wide range of granular material (the aggregate particles) dispersed in a continuous matrix. While the size of aggregate ranges from 0.1 mm to some centimetres, the matrix is a porous material with a maximum size of pores close to some micrometres. Even when accounting for the interfacial transition zone (the width of which is 10–30 μm; Maso, 1980), there remains a gap between the characteristic dimensions of aggregate and matrix. It is therefore natural to try to relate the deformability of the composite to one of the phases. It is worth noting that this gap does not exist between coarse aggregate and mortar, except in some particular concretes such as gap-graded ones. Therefore it can be questionable to model hardened concrete as a composite in which coarse aggregate and mortar stand for inclusion and matrix, respectively.

The Hashin–Shtrikman bounds

Le Roy (1996) recently reviewed various composite models used for two-phase materials. If the investigation is restricted to statistically homogeneous, isotropic composite materials, the phases of which have a linear elastic behaviour and are perfectly bonded, it can be shown that the elastic properties always lie between two bounds, commonly called the Hashin–Shtrikman bounds (Hashin and Shtrikman, 1963). For the bulk and shear moduli, the bounds are the following:

$$Kh_i = K_i + \cfrac{c_j}{\cfrac{1}{K_j - K_i} + \cfrac{3c_i}{3K_i + 4G_i}} \qquad (2.90)$$

$$Gh_i = G_i + \cfrac{c_j}{\cfrac{1}{G_j - G_i} + \cfrac{6(K_i + 2G_i)c_i}{5G_i(3K_i + 4G_i)}} \qquad (2.91)$$

where K_i, G_i and c_i are the bulk modulus, the shear modulus and the concentration of phase i respectively. Kh_i and Kh_j (Gh_i and Gh_j) are the two bounds of the composite bulk (shear) modulus; i and j can take the values of 1 or 2, with $i \neq j$. Of course, $c_i + c_j = 1$.

From the bulk and shear moduli, the elastic modulus and Poisson's coefficient can be easily computed. Therefore any model aiming to evaluate deformability properties of two-phase elastic composites should produce values that are bracketed by the Hashin–Shtrikman bounds. The greater the contrast between the phases (in terms of elastic properties), the wider is the range defined by the bounds.

Hashin's two-sphere basic cell for concrete homogenization

One of these models has been proposed by the same author (Hashin, 1962) for calculating the elastic properties of two-phase materials, a phase of which is an assembly of spherical particles dispersed in a continuous matrix. The idea of the model is to consider that the mixture is a packing of composite spheres, which completely fills the space. The composite spheres are made of a spherical aggregate, surrounded by a 'crust' of matrix; the ratio between the aggregate diameter and the crust thickness is constant (Fig. 2.74).

When the composite is submitted to a hydrostatic stress state, it can be shown that each basic cell is in the same state (as both solutions are compatible in strains and stresses). Therefore it becomes easy to calculate the bulk modulus or the free shrinkage of the composite, simply by solving the basic equations of linear elasticity in the body constituted by the double sphere. To deduce the elastic modulus some simple assumptions are generally made, such as considering that Poisson's coefficients of both phases and composite are equal to 0.2. By using more complex models (Christensen and Lo, 1979; Mori and Tanaka 1973), Le Roy (1996) has shown that the quantitative effect of this assumption on the value of the elastic modulus is minor. Finally it can be shown that,

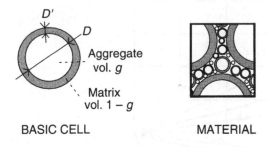

BASIC CELL   MATERIAL

Figure 2.74 The double-sphere cell in Hashin's model.

when the inclusion is stiffer than the matrix, the elastic modulus obtained coincides with the lower Hashin–Shtrikman bound. Using the same model, but replacing the matrix by the inclusion (and vice versa), the upper Hashin–Shtrikman bound is found. Thus it is concluded that these bounds are the most narrow bounds that embrace the elastic properties of all possible composite materials complying with the original hypotheses.

Replacing the bulk and shear moduli in equations (2.90) and (2.91) by their expressions in terms of elastic moduli and Poisson's coefficients, and adopting a value of 0.2 for the Poisson's coefficient of both phases, gives

$$E = \frac{(1+g)E_g + (1-g)E_m}{(1-g)E_g + (1+g)E_m} \, E_m \qquad (2.92)$$

where E_g, E_m and E are the elastic moduli of aggregate, matrix and composite respectively, and g is the concentration of aggregate. A nomograph representing the predictions of this model is given in Fig. 2.75.

In spite of its elegance, this model is seldom used by concrete practitioners, probably because its precision is limited. One reason is that an implicit hypothesis of the model never applies for concrete (nor for other materials with granular inclusions), as will be shown in the next section.

The triple-sphere model (de Larrard and Le Roy, 1992; Le Roy, 1996)

In Hashin's original model the granular concentration may range from 0 to 1. Furthermore, the grading span of the composite spheres has no

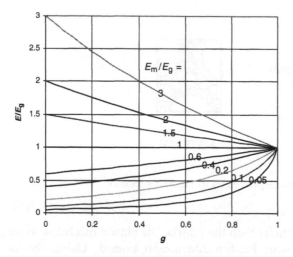

Figure 2.75 Nomograph giving the composite elastic modulus predicted by Hashin's model, as a function of phase modulus ratio and aggregate concentration.

inferior limit, as the packing of these spheres is assumed to fill the space completely. However, we have shown that the attainment of a zero porosity with an actual random packing of particles is physically impossible (see section 1.4). Any aggregate skeleton can fill only a limited part of its container, the relative volume of which is called g^*, its packing density. Therefore, to apply Hashin's concept to concrete, one must complete the actual aggregate phase by some 'matrix particles' added at the fine side. The virtual aggregate so constituted has no inferior size limit, and its packing density is by definition equal to 1.

For calculating the deformability of the virtual aggregate it can be noted that, in this material, real aggregates are in close contact. The small amount of matrix remaining in the interstitial space of the packing can be considered as a dispersed phase (Fig. 2.76). Thus the basic cell of the virtual aggregate may be taken as a Hashin's double sphere in which the outer layer is made of aggregate, and the nucleus of matrix. In other words, when aggregates are densely packed the two phases combine with each other so that the upper Hashin–Shtrikman bound is approached.

In the real material one must account for the dilution of the virtual aggregate. In order to obtain the same aggregate content g as in the original material, the concentration of the virtual aggregate in the composite should be g/g^*. Finally, the pattern obtained for illustrating the matrix-aggregate arrangement is displayed in Fig. 2.77.

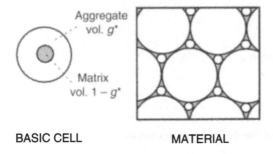

Figure 2.76 The basic cell of the virtual aggregate (which is a dense packing of particles filled up with matrix)

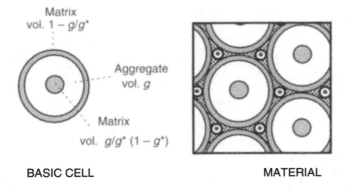

Figure 2.77 The basic cell of concrete according to the triple-sphere model.

Note that the outer layer considered here has nothing to do with the interfacial transition zone considered in other models (Nilsen and Monteiro, 1993). It is actually a way of accounting for the fact that not all the matrix plays the same role with regard to the deformability of concrete. But the matrix is still considered as a homegeneous phase. Moreover, the matrix put in the external layer in the triple-sphere model can be seen as a 'workability matrix': suppressing this layer would lead us to design a concrete with minimum paste volume, without any additional amount that could lubricate the aggregates during the flowing of fresh concrete.

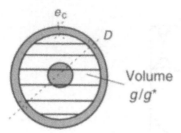

Figure 2.78 Calculation of the outer layer thickness.

To establish a link between the triple-sphere model and the maximum paste thickness concept (see section 2.3.3), let us calculate the thickness e_c of the external crust in the pattern defined in Fig. 2.78.

The volume of the aggregate, including the inner paste nucleus, is equal to g/g^*. We then may write

$$\frac{\pi}{6} D^3 = \frac{g}{g^*}$$

$$\frac{\pi}{6} (D + 2e_c)^3 = 1$$

(2.93)

which gives

$$e_c = \frac{D}{2} \left(\sqrt[3]{\frac{g^*}{g}} - 1 \right)$$

(2.94)

and, from equation (2.64) (see section 2.3.3):

$$e_c = \frac{MPT}{2}$$

(2.95)

This shows the consistency between the packing concepts developed in Chapter 1 and the influence of the aggregate skeleton on all the mechanical properties of hardened concrete.

2.5.2 Elastic modulus

This section deals with the tangent elastic modulus of concrete, determined on concrete cylinders (either water cured or sealed-cured), after two or three loading cycles, with the maximum stress remaining in the linear range: that is, less than 50–60% of the concrete strength.

Composite elastic modulus in the triple-sphere model

The virtual aggregate has a modulus, E_{max}, that is the maximum value attainable with the phases considered. It is obtained from equation (2.92) by permuting the elastic moduli and concentrations of the phases:

$$E_{max} = \frac{(2 - g^*)E_m + g^* \cdot E_g}{g^* \cdot E_m + (2 - g^*) E_g} E_g \qquad (2.96)$$

Now, for calculating the composite elastic modulus, we use again the same equation, replacing E_g by E_{max}, and g by g/g^* (as explained in section 2.5.1):

$$E = \left(1 + 2g \frac{E_g^2 - E_m^2}{(g^* - g)E_g^2 + 2(2 - g^*)E_g E_m + (g^* + g)E_m^2} \right) E_m \qquad (2.97)$$

A nomograph illustrating the predictions of this formula is given in Fig. 2.79. By comparison with Fig. 2.75, the effect of the triple-sphere model is displayed. The difference from Hashin's model is essentially

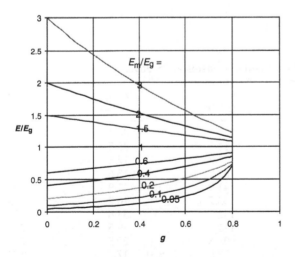

Figure 2.79 Nomograph giving the composite elastic modulus predicted by the triple-sphere model, as a function of phases, moduli and aggregate concentration. Calculations carried out with $g^* = 0.80$. Each curve deals with a constant E_m/E_g value.

significant in the region of low matrix/aggregate modulus ratio, and high aggregate concentration.

To apply this model to concrete one must know the values of the different parameters. The packing density g^* can be either directly measured, or evaluated with the compressible packing model (see Chapter 1). The aggregate volume comes from the concrete composition. The aggregate elastic modulus can be measured, or determined by calibration of the model on concrete data. The only parameter remaining is the matrix modulus.

Elastic modulus of cement paste

In the same series of experiments already referred in section 2.3.1, the elastic moduli of a series of cement pastes were measured (Marchand, 1992). Because of the precautions taken in these experiments – accounting for the risk of segregation by special mixing technique, sealed curing of the specimens, grinding of the tested faces – they constitute a unique data set, which we shall use for modelling purposes. The experimental results are given in Table 2.19, and plotted in Fig. 2.80. Portland cement pastes and silica fume pastes (with a silica fume/cement ratio of 10%) have been prepared. All the results fit consistently with a straight line, so that for this series the elastic modulus, E_p appears to be simply proportional to the compressive strength:

$$E_p = 226fc_p \tag{2.98}$$

(mean error: 1.4 GPa)

Table 2.19 Elastic modulus of cement pastes at 28 days.

w/c	sf/c	fc (MPa)	$E\ exp$ (GPa)	$E\ mod$ (GPa)
0.23	0	153	34.4	34.6
0.28	0	114.3	25.5	25.8
0.38	0	81.7	17.8	18.5
0.43	0	63.2	15.6	14.3
0.53	0	41.9	11.9	9.5
0.63	0	33.7	10.2	7.6
0.23	0.1	142	34.8	32.1
0.28	0.1	111	29.5	25.1
0.38	0.1	93.8	20.5	21.2
0.43	0.1	78.5	17.7	17.7
0.53	0.1	55.9	11.8	12.6
0.63	0.1	40.5	10.4	9.1

Source: Marchand (1992).

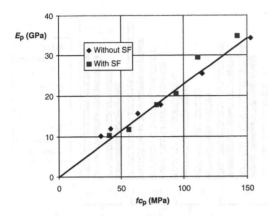

Figure 2.80 Relationship between elastic modulus and compressive strength, in Marchand's data.

Application to concrete

In the modelling of concrete compressive strength, we made the distinction between the pure paste and the matrix characteristics. Here again the paste modulus suffers some modifications when the paste is 'injected' in the aggregate porosity. To highlight this effect, we shall use a set of data (de Larrard and Belloc, 1997) already presented in section 2.3.4 (see Tables 2.13 and 2.14). In these experiments a single paste was mixed with five types of aggregate, for producing either mortars or concretes with various maximum sizes of aggregate. All moduli have been measured at 28 days (pure paste, massive rocks from which aggregates have been produced, and mortars/concrete). As already noted by other researchers for certain mixes (Hirsh, 1962; Nilsen and Monteiro, 1993), it appears that for all these mixtures the composite modulus lies under the Hashin–Shtrikman bounds (Fig. 2.81).

It is clear, then, that the paste modulus cannot be introduced directly into equation (2.97) if the aim is to predict concrete moduli with satisfactory accuracy. Several effects may explain the fact that the equivalent matrix modulus is lower than the pure paste one:

- the presence of an interfacial transition zone – while, in the present case, where only high-performance concrete is dealt with, this ITZ is probably very tiny;
- an incomplete bond between the matrix and the aggregate particles;
- a microcracking pattern due to the restrain of internal shrinkage by aggregates.

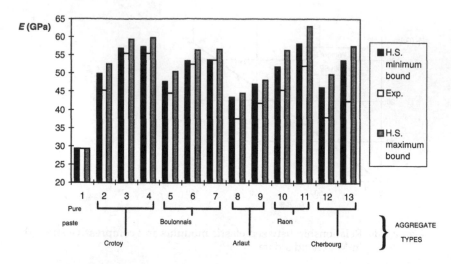

Figure 2.81 Composite moduli from de Larrard and Belloc (1997), compared with the Hashin–Shtrikman (HS) bounds.

Fortunately all these effects also affect the compressive strength, which has already been modelled in section 2.3. Therefore the simplest way to estimate the matrix modulus is to state that the proportionality found in the previous paragraph is still valid, but with the *concrete* compressive strength:

$$E_m = 226fc \qquad\qquad\qquad\qquad (2.99)$$

Validation of the model

In all the following examples the packing density of aggregate skeletons (term g^* in equation) has been calculated within the framework of the compressible packing model (see Chapter 1), with a compaction index of 9.

A first validation is still made by referring to the data used in the previous paragraph (de Larrard and Belloc, 1997), where the aggregate variables are mainly investigated. The mean error of the model (as given by equations (2.97) and (2.99)) is 2.6 GPa (instead of 4.4 GPa if the matrix modulus is taken equal to that of the paste; Fig. 2.82). Note that the cement used in these data (Cormeilles CEM I 52.5 cement) is different from the one used by Marchand in his experiments on cement pastes (Marchand, 1992).

A second set of data is now examined in which the variables investigated are compressive strength (from 25 to 125 MPa), type of

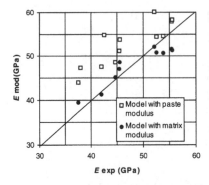

Figure 2.82 Comparison between elastic modulus measurements and model predictions. Data from de Larrard and Belloc (1997).

binder (pure Portland cement[28] or mixtures with fly ash or silica fume), and the presence of entrained air (de Larrard *et al.*, 1996b). The mixture proportions appear in Table 2.2 (see section 2.1.2). The agreement obtained is similar to that dealing with previous data (mean error equal to 2.6 GPa; Fig. 2.83). The model therefore appears equally suitable for normal-strength and high-performance concrete.

Another set of data has been taken from de Larrard and Belloc (1990), in which normal-strength and high-performance concretes were produced. The Boulonnais crushed limestone was used, with or without silica fume, and with two different maximum aggregate sizes (10 and 20 mm). The cement was the Le Teil CEM I 42.5 HTS. Here again the agreement is still fair (mean error 2.9 GPa; Fig. 2.84). Two other HPCs from de Larrard *et al.* (1994c) present the peculiarity of having been intentionally produced for displaying extreme Poisson's coefficients (0.24 and 0.15, respectively) by selection of the aggregates. This does not

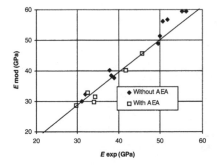

Figure 2.83 Comparison between elastic modulus measurements and model predictions. Data from de Larrard *et al.* (1996b). AEA: air-entraining agent.

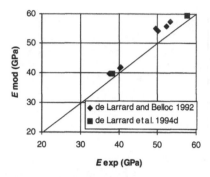

Figure 2.84 Further comparisons between elastic modulus measurements and model predictions.

impair the precision of the model (1.6 GPa), as theoretically demonstrated by Le Roy (1996).

Let us now look at elastic moduli measured at ages other than 28 days. Two important sets of data have been produced at LCPC using sand and gravel from the Seine river (the elastic modulus of which cannot directly be measured). Le Roy produced 11 concretes at constant aggregate content (except for three of them), changing the water/cement ratio (from 0.33 to 0.5) and the silica-fume/cement ratio (from 0 to 0.15) (de Larrard and Le Roy, 1992; Le Roy, 1996). Elastic moduli were measured at 1, 3, 7 and 28 days. To apply the model, the aggregate elastic modulus has been optimized (to a value of 76.6 GPa, which makes one adjustable parameter for 44 experimental values). The mean error obtained is 1.8 GPa (see Fig. 2.85).

During the same period Laplante (1993) investigated the properties of two of these concretes at very early age. Hence the validation provided

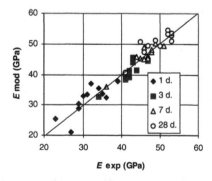

Figure 2.85 Comparison between elastic modulus measurements and model predictions. Data from de Larrard and Le Roy (1992). $g^* = 0.82$.

by his data is one of the most convincing with regard to the introduction of a third layer in Hashin's model (mean error 2.4 GPa; Fig. 2.86); as already pointed out, it is when the contrast in moduli is the highest (that is, at early age) that the two models differ the most significantly. In the same figure a last set of data has been added (dealing with the moduli of a single HPC, measured between 1 day and 4 years; de Larrard, 1988). For this last concrete the mean error is 1.6 GPa. These values are close to those provided by the elastic modulus test.

Finally a last data set will be examined, in which the authors used coarse and fine aggregates of different origins (making concretes with heterogeneous skeletons) (Schrage and Springenschmid, 1996). Eighteen concretes were produced with coarse aggregates of three mineral natures (limestone, basalt or quartz), while the fine aggregate used was always a natural quartz sand. The binders were either pure Portland or blended slag cement, with or without silica fume. The compressive strengths were measured on cubes.

To simulate these results an equivalent aggregate modulus has been calculated for each mix, taking the mean of fine and coarse aggregate moduli[29], weighted by their relative proportions. The packing densities have been estimated from the knowledge of particle shape (rounded or crushed) and maximum size of aggregate. The cylinder compressive strengths have been deduced from the cube strength using a conversion coefficient of 0.85. The comparison between predictions and measurements is given in Fig. 2.87. The mean error of the model is 3.2 GPa, partly because of the lack of precision in the estimate of cylinder compressive strengths.

We shall now go on to use the same approach to deal with the delayed deformations of concrete.

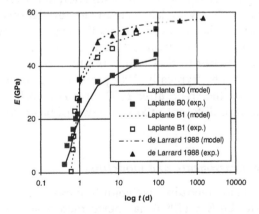

Figure 2.86 Development of elastic modulus vs. time. Experimental values and predictions of the triple-sphere model.

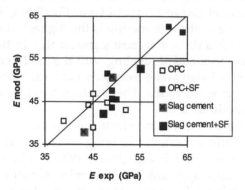

Figure 2.87 Comparison between elastic modulus measurements and model predictions. Data from Schrage and Springenschmid (1996).

2.5.3   Basic creep

The creep strains of concrete are of importance in various applications. They influence the losses of prestress and the deflections of loaded structures, and can sometimes be a critical criterion in concrete optimization. Therefore it is worth investigating creep in a book dealing with mixture proportioning.

Delayed deformations are complex phenomena. Every concrete has its own behaviour, which means that the creep deformation takes place at certain kinetics, and depends strongly on the age at loading, and on humidity exchanges. Moreover, creep deformation can never stabilize, even for loadings lasting several tens of years (Troxell *et al.*, 1958). For the sake of simplicity we shall therefore restrict our investigation to the creep displayed by samples loaded at 28 days. Also, as many extrapolations of creep experiments often appear hazardous, we shall focus on creep deformation occurring after a 1000-day loading duration. This deformation is generally significantly inferior to that which might appear at the end of the expected duration life (typically 100 years for a bridge). However, we shall assume that the creep after 1000 days is a sufficient criterion if the aim is to compare different mixes in terms of basic creep potential.

Basic creep of cement paste: experiments and modelling

As we did for strength and elastic modulus, we shall first examine some data dealing with pure paste. Le Roy (1996) has performed basic creep tests on pure pastes. The samples were continuously rotated during hardening in order to minimize bleeding and segregation, following a technique taken from Stock *et al.* (1979). Also, special precautions were taken to ensure high-quality tests with regard to the application of the load, the strain measurements, and the avoidance of water losses during the tests. Unfortunately, the duration of the tests was only about 90 days, which is sufficient to exhibit the general trends but somewhat short for making reliable extrapolations. Mixture proportions and properties of the pastes are given in Table 2.20, and the results of the creep tests appear in Fig. 2.88.

Table 2.20 Mixtures used in the paste creep tests.

w/c	sf/c	fc (MPa)	E (GPa)	Extrapolated specific basic creep at 1000 d. (10^{-6}/MPa)	Ed (GPa)
0.28	0	114.3	25.5	163	4.95
0.38	0	81.7	17.8	196	3.96
0.5	0	49.3[a]	11.1[b]	288	2.65
0.28	0.1	111	29.5	89	8.14
0.38	0.1	93.8	20.5	87	7.36

[a] Calculated from equation (2.58).
[b] Calculated from equation (2.98).

Source: Le Roy (1996).

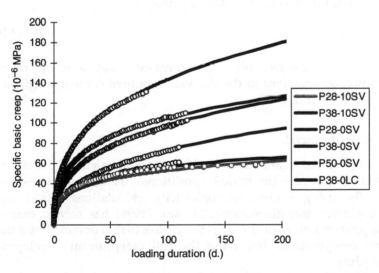

Figure 2.88 Creep curves obtained in the paste tests (experimental data and fitting) (Le Roy, 1996). The samples were loaded at an age of 28 days.

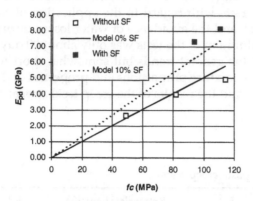

Figure 2.89 Delayed modulus of cement pastes vs. compressive strength at 28 days.

The delayed modulus is defined as the ratio of the applied stress to the total deformation due to this stress (that is, elastic and creep strain). When plotting the delayed moduli vs. compressive strength (Fig. 2.89), the trend looks similar, as compared with elastic moduli (Fig. 2.80), except that the points for silica fume do not match the same line as the points for pure Portland cement pastes. This is consistent with the fact that silica fume greatly reduces the creep of HPC at equal strength (de Larrard *et al.*, 1994d). From this observation, a model is derived for the delayed modulus, having the following shape:

$$E_{pd} = \alpha \left(1 + \beta\Psi\right) \cdot fc_p \tag{2.100}$$

where α and β are two adjustable constants, and Ψ is the relative contribution of pozzolans to the equivalent cement content, as given by equations (2.75) and (2.76).

Basic creep of concrete

As we did for instantaneous modulus, we shall calculate the delayed modulus of concrete (in sealed conditions) using the triple-sphere model. By using refined viscoelasticity calculations based upon Laplace–Carson transformations, Le Roy (1996) has shown that this process provides numerical results very close to that obtained by a more rigorous computation, describing the full creep strain development from loading.

The matrix delayed modulus is taken as the paste modulus, where the paste compressive strength is replaced by the concrete compressive strength, in order to account for the paste/aggregate interaction. We shall then have

$$E_d = \left[1 + 2g \frac{E_g^2 - E_{md}^2}{(g^* - g) E_g^2 + 2(2 - g^*)E_g E_{md} + (g^* + g) E_{md}^2} \right] E_{md}$$

(2.101)

and

$$E_{md} = \alpha(1 + \beta\Psi) \cdot fc \tag{2.102}$$

where E_d and E_{md} are the delayed moduli of concrete and matrix respectively.

The specific basic creep of concrete, ε_{bc}^s, can be derived using the following equation:

$$\varepsilon_{bc}^s = \frac{1}{E_d} - \frac{1}{E} \tag{2.103}$$

As for the creep coefficient, it is equal to the ratio of the creep deformation to the elastic deformation:

$$K_{cr} = \frac{E}{E_d} - 1 \tag{2.104}$$

For the calibration of the two constants in equation (2.102) we shall use some concrete data. The same concretes tested by Le Roy for elastic modulus, made up with the same binders as the pastes, were also subjected to basic creep tests (Le Roy, 1996). The loads were applied at 28 days, and strain measurements were taken over 2–3 years. In spite of a certain scatter of the data with regard to the influence of mix-design parameters, the creep values at 3 years appear more reliable than the previously reported paste data. By numerical optimization, the following two parameters ($\alpha = 50.6$, $\beta = 0.966$) led to a mean error of 2.5 GPa. When applying these parameters to the paste experiments, two straight lines were determined in Fig. 2.89. They provide a reasonable agreement with experimental data (mean error 0.6 GPa), given the low duration of the paste creep tests. Let us note the high efficiency of pozzolans for increasing the delayed modulus (or reducing basic creep): in addition to their effect on compressive strength, the pozzolans can almost double the delayed modulus of cement paste. This dramatic effect could be a consequence of self-desiccation (see section 2.5.5). Also, it seems that pozzolanic hydrates have the ability to overcome local moisture flows, which could be partly the microstructural cause of creep.

For further validation of the basic creep model, 20 more concrete basic creep tests were used. All these tests were performed over a sufficient duration, under a good sealing (aluminium foils[30]) so that extrapolations to 1000 days are considered sound enough. Here no adjustment has been made, except in some cases to the aggregate modulus, which has been determined from the experimental concrete elastic modulus. The g^* parameter (aggregate packing density) has been either calculated from the CPM model (see Chapter 1) or estimated from knowledge of the aggregate origin and maximum size. Details about the mixture proportions of these concretes can be found in Le Roy (1996). Let us only note that, in addition to Le Roy's cement (CEM I 52.5 from St Vigor), seven other cements were used in this concrete series. The mean error of the model on this set is 3.0 GPa (while the type of aggregate and strength level vary over a wide range). A comparison between measured and predicted values is provided in Fig. 2.90. Thus it appears that the accuracy provided by the triple-sphere model is comparable in absolute value, whatever the type of modulus dealt with (instantaneous or delayed). Also, the values of the parameters α and β seem to be reasonably universal, given the variety of the cements.

A last set of data, already used for validation of the elastic modulus model, deserves a special treatment, because of the high number of parameters investigated (Schrage and Springenschmid, 1996). All the concretes of this study have been submitted to basic creep tests. The 'experimental' delayed moduli have been calculated from the actual elastic moduli, and from the extrapolated values of specific creep, which are slightly higher than the values at 1000 days dealt with in our model.

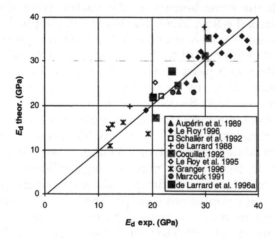

Figure 2.90 Comparison between measured values of delayed modulus (in sealed conditions) and predictions of the model.

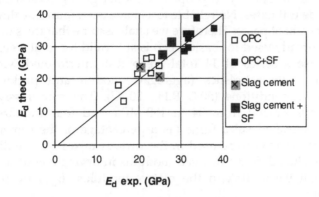

Figure 2.91 Comparison between measured values of delayed modulus (in sealed conditions) and predictions of the model (Schrage and Springenschmid, 1996).

For simulation of the matrix delayed moduli, it has been necessary to estimate the K_p coefficient (activity coefficient; see section 2.3.6) of the silica fume used. A calibration of the strength model (see section 2.3) on experimental strength results dealing with a single type of aggregate (round limestone aggregate) led to the value of 8. This high coefficient could originate in the slurry form of the silica fume (as opposed to the densified form, which could partly impair the pozzolanic activity of silica fume). Moreover, the uncertainty on the cylinder strength still affects the quality of the predictions. Nevertheless the model provides a fair classification of concretes, and the mean error (2.7 GPa) is of the same order as for the tests previously reported (Fig. 2.91). According to these data, it seems that the model could give acceptable predictions for slag concretes as well as for pure Portland cement concretes.

2.5.4 Total creep

Drying creep of concrete is not really a material property: it may have an intrinsic part, but the superficial cracking of drying specimens, which is less when the concrete is loaded, is accounted for in the measured deformation (Acker, 1982). Therefore this deformation can be affected by a significant size effect, not only in terms of kinetics, but also of final amplitude. The amount of superficial cracking, which can be measured by some experimental methods such as the replica technique (Sicard *et al.*, 1992), depends strongly on the drying kinetics and pattern. This is why it can be difficult to build up precise models for quantifying drying, and total creep.

However, especially in the range of ordinary concrete, drying creep is a significant part of the total creep, and this deformation cannot be neglected if the aim is to optimize a concrete having minimum deformations. We shall then try to propose a model giving reasonable orders of magnitude estimates. No reliable tests have been found dealing with total creep of cement paste. Therefore we shall assume that the same type of approach already used for basic creep is still valid for total creep.

We shall first use a series of 14 total creep tests, performed with materials of known characteristics (aggregate moduli and packing densities), in drying conditions (50% R.H., 20 °C). The compressive strengths at 28 days ranged from 34 to 108 MPa, and four mixtures contained 10% of densified silica fume (as a percentage of the cement mass). Six different cements were used in the various mixes. As we did for basic creep, we shall define a delayed modulus in drying conditions, E_d', which we shall try to link to the matrix modulus, E_{md}', by the following equation:

$$E_d' = \left[1 + 2g \frac{E_g^2 - E_{md}'^2}{(g^* - g)E_g^2 + 2(2 - g^*)E_g E_{md}' + (g^* + g)E_{md}'^2} \right] E_{md}'$$

(2.105)

and

$$E_{md}' = \alpha'(1 + \beta'\Psi) \cdot fc$$

(2.106)

The calibration gave the values of 28.5 and 0.935 for the α' and β' coefficients respectively. The plot of theoretical versus experimental delayed moduli is shown in Fig. 2.92. The almost equal values of β and β' are remarkable; it seems therefore that pozzolans have the same reduction effect on basic and on total creep.

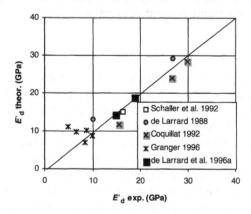

Figure 2.92 Comparison between measured values of delayed modulus (in drying conditions) and predictions of the model after calibration (two parameters calibrated)

We shall refer again to the data by Schrage and Springenschmid (1996) to reach a further validation of the model. These authors published the results of nine drying creep tests, in which the 104 mm diameter specimens were stored at 65% R.H. Experimental delayed moduli have been calculated from the long-term extrapolated specific creep values. In our simulations, the materials parameters, most of them in the original publication, are the same as those used for basic creep (Fig. 2.91). While the conditions of the test are not exactly comparable to ours (smaller specimens, higher humidity and longer loading duration), the agreement between the model and the experimental data is quite fair (Fig. 2.93), especially for concretes without silica fume. For the mixtures containing silica fume, the model underestimates the delayed modulus and thus overestimates the total creep. Hence in some cases the silica fume almost suppresses the drying creep (de Larrard *et al.*, 1994d). Let us remind ourselves that silica fume in the slurry form was used by Schrage and Springenschmid, which ensures a better filler effect than the densified silica fume used in the French studies.

Before dealing with the shrinkage deformations, let us check the consistency of the three types of modulus found for the concrete matrix. We should have in any case

$$E_m \geqslant E_{md} \geqslant E'_{md} \qquad (2.107)$$

or

$$\frac{E_m}{fc} \geqslant \frac{E_{md}}{fc} \geqslant \frac{E'_{md}}{fc} \qquad (2.108)$$

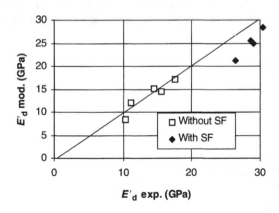

Figure 2.93 Comparison between measured values of delayed modulus (in drying conditions) and predictions of the model (Schrage and Springenschmid, 1996).

which is verified by the present models (from equations (2.99), (2.102) and (2.106)):

$$\frac{E_m}{fc} = 226 \geqslant 99.5 = 50.6(1 + 0.966)$$

$$\geqslant 50.6(1 + 0.966 \ \Psi) = \frac{E_{md}}{fc}$$

$$\geqslant 28.5(1 + 0.935 \ \Psi) = \frac{E'_{md}}{fc} \tag{2.109}$$

The links between compressive strength and the different moduli are summarized in Fig. 2.94.

2.5.5 Autogenous shrinkage

Autogenous shrinkage starts at the concrete set point, or before (but the part of this deformation appearing at the plastic stage has no mechanical consequences at the hardened state). Its development is fast, and the strain stabilizes within a few months. It originates in self-desiccation: that is, the lowering of internal humidity due to water consumption in the hydration process. The amplitude of autogenous shrinkage is essentially significant in the high-performance concrete (HPC) range, where the free water content tends to be low because of the low initial water/binder ratio. The autogenous shrinkage of HPC can be sufficiently high to provoke cracking of a fully restrained member after only some days, even if the surface is protected against evaporation (Paillère *et al.* 1989; Tazawa and Miyazawa, 1996). Therefore an evaluation of autogenous shrinkage is worthwhile for some HPC applications, such as massive structures

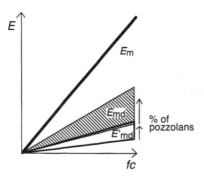

Figure 2.94 Matrix elastic and delayed moduli as a function of the concrete compressive strength.

where successive lifts are cast and restrain each other. As for the other deformability properties of concrete, we shall apply the triple-sphere model to establish a link between paste and concrete property. Then we shall build up a model for paste shrinkage, which we shall calibrate on a series of concrete measurements. We shall try to assess the whole autogenous shrinkage, from the setting to the long term.

Application of the triple-sphere model to shrinkage

The principle of this calculation is similar to that used for elastic modulus. The elasticity equations are solved on the structure comprising the triple-sphere pattern (Fig. 2.76) (Garboczi, 1997). As the stress and strain fields of all triple spheres are compatible with each other, this solution is also applicable to the whole material. With the hypothesis that all Poisson's coefficients are equal to 0.2, the following equation is found (de Larrard and Le Roy 1992; Le Roy, 1996):

$$\varepsilon_{as} = \frac{\left(1 + \dfrac{E_{md}}{E_g}\right)\left(1 - \dfrac{g}{g^*}\right) + \dfrac{4E_{md}/E_g(1 - g^*)g/g^*}{g^* + E_{md}/E_g(2 - g^*)}}{1 + \dfrac{g}{g^*} + \dfrac{E_{md}}{E_g}\left(1 - \dfrac{g}{g^*}\right)} \varepsilon_{as}^m \qquad (2.110)$$

where ε_{as} and ε_{as}^m are the autogenous shrinkage of concrete and matrix respectively. A nomograph displaying the prediction of this model is given in Fig. 2.95. Here we have assumed that the aggregate exhibits no significant shrinkage.

An empirical model for autogenous shrinkage of cement paste

The matrix autogenous shrinkage is the deformation caused by the capillary stresses exerted by water menisci in the cement paste porosity (Hua, 1992). Theses stresses can be summarized by the concept of *hydraulic stress* (de Larrard and Le Roy, 1992; Le Roy, 1996). Referring to the delayed matrix modulus, one may write

$$\varepsilon_{as}^m = \frac{\sigma_H}{E_{pd}} \qquad (2.111)$$

where ε_{as}^m is the matrix autogenous shrinkage, σ_H is the hydraulic stress, and E_{md} is the matrix delayed modulus, already defined in the previous section. σ_H should depend on the matrix mix-design ratios. Moreover, it has been shown that, unlike the other types of delayed strain, the type of

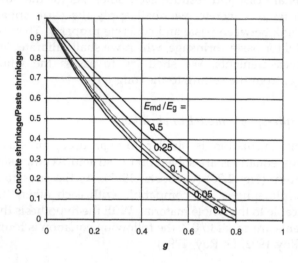

Figure 2.95 Nomograph providing the concrete/matrix shrinkage ratio as a function of the phase delayed modulus ratio and the aggregate content, with $g^* = 0.8$.

cement has a significant influence on autogenous shrinkage (Tazawa and Miyazawa, 1993).

It would be tempting to try to link autogenous shrinkage with the compressive strength of concrete, as we did for the other deformability properties. However, experience has shown that very different autogenous shrinkages can be obtained for similar compressive strength, depending on the proportions of the different binders. In the Civaux II nuclear power plant project an HPC could be produced having an autogenous shrinkage substantially lower than the control normal-strength concrete used in the Civaux I reactor (see section 5.3.3). Then it appears more relevant to refer to the parameter that best illustrates the amount of remaining water after hydration: that is, the water/cement ratio.

The hydraulic stress is expected to increase when w/c decreases, and to be also affected by the presence of pozzolans. The general following form is then proposed:

$$\sigma_{\text{H}} = K_{\text{c}} \cdot ([w/c]_0 - w/c)(1 + \gamma \Psi) \tag{2.112}$$

where $[w/c]_0$ is the maximum water/cement ratio under which autogenous shrinkage appears, K_{c} is a constant describing the influence of the cement on autogenous shrinkage, and γ is another parameter dealing with the effect of pozzolans.

No reliable paste data, with sufficiently long-term measurements, have been found in the literature. Thus the hydraulic stress model will be calibrated directly on concrete.

Calibration and validation on concrete data

Le Roy's 11 mixtures (Le Roy, 1996), together with three of Granger's mixtures made up with the same cement (Granger, 1996) are used to calibrate the model. The shrinkage strains have been measured from demoulding (at 24 hours) to more than one year. In the case of the cement used, Le Roy has shown that the part of shrinkage occurring before demoulding was likely to be negligible (Le Roy, 1996). With $K_c = 13.7$ MPa, $[w/c]_0 = 0.631$ and $\gamma = 3.11$, all experimental values (the range of which is $81 - 190 \times 10^{-6}$) are regressed with a mean error of 22×10^{-6}.

An attempt to validate the model has been made with two sets of data dealing with another cement (Aupérin *et al.*, 1989; Schaller *et al.*, 1992). The six HPCs referred to, one of which is the concrete used in the Pont de Joigny bridge, have a water/cement ratio in the range 0.33–0.41, and silica fume/cement ratio of either 0 or 0.08. Here only the K_c coefficient has been adjusted to a value of 14.4. The mean error is 22×10^{-6}, which is the same as that obtained with the first data set. The two sets are plotted in Fig. 2.96. The precision of this model may seem inferior to that of the previous models, but the low values of measured strains, and the numerous traps of these experiments (possible influence of thermal and drying shrinkage, assessment of shrinkage before demoulding) make it difficult to predict experimental results with a high precision. Also, water migrations between aggregate and matrix could play a role

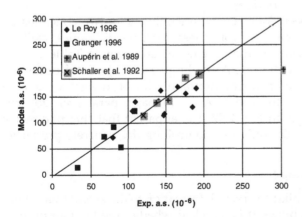

Figure 2.96 Comparison between model predictions (or fitting) and experimental data for autogenous shrinkage.

Table 2.21 K_c values for various cements.

References	Cements	C_3A (%)	K_c (MPa)
Le Roy (1996)	CEM I 52.5 St Vigor	2.73	13.7
Le Roy (1996)	CEM I 52.5 St Pierre la Cour	8.62	19
Granger (1996)	CEM II 42.5 Airvault	7	19.6
Aupérin et al. (1989)	CEM I 52.5 Cormeilles	11.5	14.4
de Larrard et al. (1996a)	CEM I 52.5 Le Teil	9	25
de Larrard et al. (1996a)	CEM I 52.5 Beaucaire	8.2	18
de Larrard (1988)	CEM I 52.5 HTS Le Teil	4.5	17

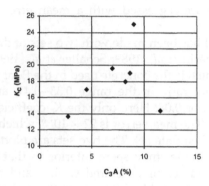

Figure 2.97 K_c parameter plotted against C_3A content of cement. Data from Table 2.21.

in self-desiccation, impairing the precision of a model that treats the matrix separately, as ours does.

Other data have been collected, where only one or two concretes were made with a single cement. From these data, the K_c values, calibrated from autogenous shrinkage measurements, are given in Table 2.21.

The tricalcium aluminate (C_3A) is the phase of Portland cement that consumes the most water during hydration. It is also very reactive. One may therefore expect to find C_3A as a major parameter governing autogenous shrinkage. In Fig. 2.97 it is apparent that this parameter is significant, but is not the only one controlling this concrete property.

2.5.6 Total shrinkage

Total shrinkage is similar to creep deformation to the extent that it takes a very long time to stabilize. It is not clear whether the final amplitude of shrinkage is affected by a size effect. Theories based on diffusion (for the transport phenomena) and capillarity (for the physical phenomenon responsible for the deformation) tend to predict no size effect. Most experimental work supports the opposite thesis (Neville, 1995; Alou, *et al.*, 1987), but it is possible that at least a part of the size effect reflects the fact that for large specimens only a small part of the ultimate shrinkage can be measured within the duration of the test. Moreover, many studies do not refer to early autogenous shrinkage, which is maturity driven (and so develops faster for larger elements).

Another factor that can impair the predictability of shrinkage measurements is the variable amount of skin cracking that can occur during drying. Granger recently proposed a method for assessing the part of shrinkage 'absorbed' by this phenomenon (Granger *et al.*, 1997a,b), by dealing with the curve relating the experimental water loss and the shrinkage deformation. However, such a curve is seldom available in most available shrinkage data. We shall therefore neglect this phenomenon. We shall try to build up a model predicting the total shrinkage of 150/160 mm diameter specimens, from setting to the end of the expected life duration of a common civil engineering structure (about 100 years). The relative humidity is taken to be 50%.

Consistent with the rest of this section, we shall apply the triple-sphere model to predict the total shrinkage from the concrete formulation. It has been long recognized that the aggregate skeleton plays a major role in controlling the shrinkage of concrete, while the nature of the cement, for example, is of minor importance (Neville, 1995). For describing the matrix deformability, it appears after several attempts that the delayed matrix modulus *in sealed conditions*, E_{md}, provides a better quantification than the same parameter in the drying condition, E'_{md}. This apparent paradox comes from the fact that the older the concrete, the less deformable it is, especially with regard to creep and stress relaxation. As the major part of the 'hydraulic loading' comes long after 28 days, accounting for the total creep of concrete when loaded at 28 days would lead one to overestimate the matrix deformability. Also, a part of the apparent drying creep comes from a structural effect (at the specimen level), and therefore does not affect the matrix at the aggregate level.

We shall then have:

$$\varepsilon_{ts} = \frac{\left(1 + \dfrac{E_{md}}{E_g}\right)\left(1 - \dfrac{g}{g^*}\right) + \dfrac{4E_{md}/E_g(1 - g^*)g/g^*}{g^* + E_{md}/E_g\,(2 - g^*)}}{1 + \dfrac{g}{g^*} + \dfrac{E_{md}}{E_g}\left(1 - \dfrac{g}{g^*}\right)}\,\varepsilon_{ts}^m \qquad (2.113)$$

where ε_{ts} and ε_{ts}^m are the total shrinkage of the concrete and the matrix respectively.

The case of shrinking aggregates (which seldom occurs, except in a few parts of the world; Day, 1995; Neville, 1995) can be covered by the model. The concrete shrinkage will be equal to the aggregate shrinkage, ε_{ts}^g, plus an additional shrinkage coming from the difference between the phase-free deformations:

$$\varepsilon_{ts} = \varepsilon_{ts}^g + \frac{\left(1 + \dfrac{E_{md}}{E_g}\right)\left(1 - \dfrac{g}{g^*}\right) + \dfrac{4E_{md}/E_g(1 - g^*)g/g^*}{g^* + E_{md}/E_g(2 - g^*)}}{1 + \dfrac{g}{g^*} + \dfrac{E_{md}}{E_g}\left(1 - \dfrac{g}{g^*}\right)} (\varepsilon_{ts}^m - \varepsilon_{ts}^g)$$

(2.114)

Calibration of a model on mortar data

It is very difficult to perform significant total shrinkage tests on pure cement paste because, owing to the brittleness of this material, the specimens may be destroyed solely by drying-induced skin cracking. To avoid this problem it is necessary to lower the humidity by successive small steps (Ferraris, 1986). But the deformations found seem small compared with what can be expected from measurements dealing with mortars and concretes. This is why we shall calibrate a model on mortar measurements. The benefit is to deal with small samples, which can reach a high proportion of ultimate shrinkage in a reasonable time, with a limited amount of skin cracking.

A series of 10 mortars were prepared by Kheirbek (1994), with a 2 mm maximum size graded sand, a low-C_3A Portland cement, a densified silica fume, and a superplasticizer (melamine type). The water/cement ratio ranged from 0.27 to 0.63, and the percentage of silica fume varied between 0 and 15% of the cement mass. Three supplementary mixes were added: one was made with a second type of superplasticizer (naphthalene type), while two mixes contained a high-C_3A cement instead of the low-C_3A one. All the mixtures had the same paste content, and the workability was adjusted with the help of the superplasticizer. Prisms 20 × 20 × 160 mm in dimensions were cast, and demoulded at 48 hours. After initialization of the length measurements, the prisms were cured under a double layer of polyethylene and adhesive aluminium up to the age of 28 days. This precaution allowed these specimens to reach a satisfactory level of hydration, while too early an exposure to drying would have impaired it. After 28 days the specimens were uncovered and put in drying conditions (at $50 \pm 10\%$ R.H., $20 \pm 2\ °C$). Measurements were collected over 3 years. Based upon diffusion theories, such a drying period would correspond, for prism specimens of 150 × 150 mm cross-section, to a duration of 170 years. While this calculation overestimates the size effect (Alou *et al.*, 1987), the fact remains that these tests are still relevant for estimating long-term shrinkage values, as aimed for in this section. Mixture proportions and shrinkage of the mortars are given in Table 2.22.

The cement used in mixes 1–10 was that used by Le Roy (1996), whose tests served to calibrate the autogenous shrinkage model (see section 2.5.5). Thus the model could be used to correct the data, in order to account for the amount of autogenous shrinkage that had developed before 2 days. According to Le Roy's curves, about 21% of the final autogenous shrinkage appeared before this date, and was therefore added to the measured total shrinkage. Similarly, the cylinder compressive strengths and the corresponding delayed moduli could be evaluated with the models developed in sections 2.3 and 2.5.3 respectively. Finally, by using the triple-sphere model (equation 2.113), the 'experimental' matrix shrinkage could be calculated.

Let us first note that, as already found in the literature, the effect of cement nature is small: no clear tendency appears when comparing mixes 1 and 2 with their companions 12 and 13. Also, the nature of the superplasticizer used does not create any significant difference between mixes 5 and 6. The main parameters affecting the matrix shrinkage are then the water/cement and silica/cement ratios. However, as we did for the elastic modulus and the creep deformations, we shall look first for a relationship between the matrix shrinkage and the mortar compressive strength. Given the unavoidable scattering of shrinkage tests, this relationship appears quite significant (Fig. 2.98).

From Fig. 2.98 it can be concluded that the effects of silica fume on total shrinkage and on compressive strength are similar.

The following empirical model is derived:

$$\varepsilon_{ts}^m = 0.0286 fc^{-0.414} \tag{2.115}$$

where *fc* is the mortar compressive strength (mean error 350×10^{-6}, or 7.2%).

We now have a predictive model for total shrinkage of mortars and concrete, which remains to be validated.

Validation on concrete data

A unique set of data has been published by Sellevold (1992), presenting a systematic investigation of the effect of aggregate volume on shrinkage. Eight types of paste were formulated, with all possible combinations of the three following parameters: nature of the cement (pure Portland cement or blended cement containing fly ash), water/binder ratio (0.4 or 0.6), and presence of silica fume (0 or 10% of the binder content). The 36

Table 2.22 Compositions and properties of mortars subjected to shrinkage measurements.

	Mix no.												
	1	2	3	4	5	6	7	8	9	10	11	12	13
Sand (kg/m³)	1472	1472	1463	1464	1483	1473	1483	1476	1472	1447	1464	1453	1495
Cement (kg/m³)	522	493	598	578	567	562	550	663	616	405	392	516[b]	500[b]
Silica fume (kg/m³)	0	49.3	0	29	56.7	56.2	82.4	0	61.6	0	39.2	0	50
SP (kg/m³)	0	6.87	5.96	8.67	11.86	10.15[a]	13.73	9.95	13.52	0	0	0	22.8
Water (kg/m³)	225	207	193	185	179	180	172	172	157	255	247	222	172
w/c	0.43	0.43	0.33	0.33	0.33	0.33	0.33	0.27	0.27	0.63	0.63	0.43	0.43
sf/c	0	0.1	0	0.05	0.1	0.1	0.15	0	0.1	0	0.1	0	0.1
% air	5.4	5.3	6	5.9	4.6	5.2	4.5	5.2	5.3	7	5.8	6.6	6
LCL flow time (s)	14	7	8	9	8	7	7	7	13	0.5	10	15	9
fc_{28} (MPa)	39.0	64.6	56.6	74.7	95.9	92.6	108.5	78.5	116.5	18.0	34.5	36.3	62.8
E_{md} (GPa)	1.98	4.31	2.86	4.44	6.40	6.18	7.88	3.97	7.78	0.91	2.30	1.84	4.19
Shrinkage from 2 d.	900	793	880	737	793	833	823	637	653	1320	1197	1013	730
Corrected total shrinkage	945	841	930	791	845	887	876	682	708	1320	1197	1083	795
Matrix shrinkage	6104	4785	5607	4410	4469	4657	4374	3980	3483	8621	7456	6776	4757

[a] Naphthalene sulphonate type.
[b] High-C_3A cement.

The numbers in italics are the results of simulations.

Source: Kheirbek (1994).

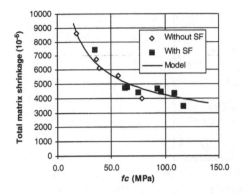

Figure 2.98 Total (corrected) shrinkage vs. theoretical mortar compressive strength, for Kheirbek's mixtures.

mixtures were produced by mixing theses pastes with the same aggregate skeleton, at different concentrations (0, 20, 40 and 60%). A superplasticizer was used in the low water/binder mixes. The pastes were made both with and without superplasticizer, to investigate the effect of this component. Special precautions were taken to avoid segregation and bleeding. The 27 mm diameter specimens were cured in water for about 3 months, and then submitted to drying for 2 years. Owing to the type of curing and the size of specimens, it is believed that no significant self-desiccation took place. Therefore the data deal with *total* shrinkage, and do not need any correction, as the duration of the tests was comparable to that of Kheirbek's tests.

For simulating these data, a number of parameters were lacking, and had to be estimated. The aggregate packing density has been taken to be equal to 0.8, a typical value for a rounded aggregate with an 8 mm maximum size. The compressive strengths have been calculated either with the pure paste model (equation 2.58), in which the cement has been replaced by the equivalent cement content (equation 2.82), or with the concrete model (equations (2.83), (2.84)). No adjustment of any kind has been made. The results are given in Fig. 2.99. Except for some pure pastes,[31] the agreement of the model with the experiments is rather satisfactory. In particular, the prediction of the effect of aggregate concentration is noteworthy (see Fig. 2.100). From this behaviour, it can be assumed that microcracking, which appears at the meso-scale (Yssorche and Ollivier, 1996), does not appear to play a significant role in the restraining effect of aggregate. The mean relative error on concrete shrinkage is equal to 11%. For comparison purposes, most code-type models for shrinkage provide mean errors in the range of 30% of relative values (Le Roy, 1996).

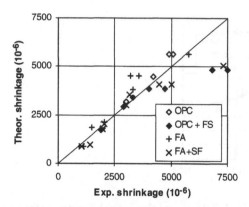

Figure 2.99 Comparison between predictions of the model and measurements. Data from Sellevold 1992). Estimated parameters: $E_g = 55$ GPa (E. Sellevold, private communication, March 1994), $Rc_{28} = 65$ MPa, $p = 1.11$, $q = 0.0033$, $Kp_{fs} = 5$, $Kp_{FA} = 1$. The values of pozzolanic coefficients deal with the age at which the concretes started to dry (about 3 months).

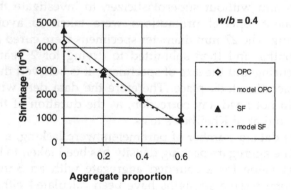

Figure 2.100 Effect of aggregate concentration on total shrinkage. Data from Sellevold (1992).

Another set of shrinkage data will be now investigated (Alou *et al.*, 1987). Thirteen concretes were produced with the same set of constituents (a round aggregate with a maximum size of 32 mm, a Portland cement, and water). The cement content ranged from 250 to 450 kg/m³. The concretes were sealed-cured for 7 days, and then dried in a 65% R.H. environment (which is higher than the other data). The shrinkage tests were performed with great care, and investigations were carried out on the size effect affecting the results. The data considered for validation purposes deal with samples having transverse dimensions equal to 150/160 mm. A correction has been made to account for the autogenous shrinkage appearing before 7 days. The latter deformation has been taken to be equal to 60% of the whole autogenous shrinkage (calculated by the model of section 2.5.5, with $K_c = 14$ MPa).

To use the model of total shrinkage, a g^* value of 0.86 has been taken. The estimate for the calcareous aggregate modulus was 60 GPa. The compressive strength values were given in the reference. A reduction coefficient had to be applied to the calculated strains in order to account for the higher humidity. By optimization, a value of 0.91 has been found. The mean error in relative value is 12% (see Fig. 2.101).

Finally, some concretes from several French studies will be referred to, all the parameters of which have been measured (so that the model will be used in a purely predictive manner). As for the previous data set, corrections have been made by adding an estimate of the early autogenous shrinkage to the total shrinkage measured after demoulding. The agreement is reasonable, but less satisfactory than in the previous sets of data (Fig. 2.102). This could be because the difference between final measurements on 160 mm diameter cylinders and extrapolated long-term shrinkage is generally significant, while smaller specimens tend to reach their ultimate deformation more quickly. Nevertheless, the cloud is rather centred on the line of equality, so that it is not believed that the model would lead to an important and *systematic* error.

Finally, as we did for the different types of moduli, let us compare autogenous and total shrinkage. The comparison is not straightforward, because the two models do not refer to the same parameters. To try to deal with a realistic case, we shall consider the matrix shrinkage of a concrete made up with a siliceous aggregate having a 20 mm aggregate maximum size. The simulations are shown in Fig. 2.103. As expected, it appears that autogenous shrinkage is always lower than total shrinkage. It levels off quite rapidly when the strength increases. Strangely enough, silica fume does not seem to increase the matrix autogenous shrinkage.

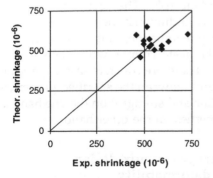

Figure 2.101 Comparison between model predictions and experimental values of shrinkage. Data from Alou *et al.* (1987).

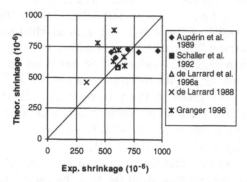

Figure 2.102 Comparison between model predictions and experimental values of shrinkage.

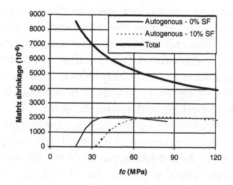

Figure 2.103 Evolution of the different types of matrix shrinkage, for a series of concretes made up with the same aggregate. $w/c = 0.25 - 0.65$; $Rc_{28} = 65$ MPa; $p = 0.6$, $K_p = 4$; $K_c = 20$ MPa.

However, when going from matrix to concrete shrinkage, the matrix delayed modulus plays an important role. The fact that silica fume increases the delayed modulus tends to diminish the restraining effect of aggregate on shrinkage. Another phenomenon, not illustrated in this diagram, is the trend to increase the paste volume when the strength of concrete increases (see section 4.2.1). Therefore concrete shrinkage appears to be the result of complex interactions between the specified properties of concrete (consistency at the fresh state and strength) on the one hand, and between the aggregate/matrix properties on the other hand.

2.5.7   Conclusion: effect of changes in mix-design parameters on concrete deformability

The triple-sphere model provides a unified and fruitful way to analyse the deformations of concrete. In addition to the properties dealt with in this chapter, it could be applied to investigate the coefficient of thermal expansion or the swelling of immersed concrete. It shows the weight of the different mix-design parameters in controlling the deformability.

By a large series of simulations, the following conclusions are drawn (Le Roy, 1996):

- The elastic modulus of aggregate, E_g, has a large influence on that of concrete, and to a lesser extent on creep; conversely, the effect on shrinkage is minor, essentially because the restraining effect of aggregate on shrinkage is controlled by the ratio of aggregate modulus to the delayed matrix modulus, and this ratio is high, whatever the type of aggregate.
- The packing density of the aggregate, g^*, essentially affects the deformations in the case of high contrast between the rigidity of the phases: that is, for elastic modulus at early age, creep and shrinkage. The higher the g^*, the more deformable the concrete, since more paste comes in the external layer in the triple-sphere pattern (Fig. 2.77); however, a more compact aggregate leads to better workability, so that less paste is necessary to maintain the same consistency in the fresh state (see Chapter 4).
- The aggregate concentration is always an important factor. Therefore any strategy aiming to produce stiff concretes should consider the volume of aggregate. This consideration highlights once again the importance of packing concepts in the rational mix design of concrete.

2.6 FACTORS AFFECTING CONCRETE PERMEABILITY

For concrete structures, one of the most common mechanisms of degradation is corrosion of the steel reinforcement. Apart from the environmental and structural factors, which are outside of the scope of this book, the process is governed by some concrete material parameters. Carbonation is a critical phenomenon for corrosion initiation, but is not likely to be a problem if corrosion development cannot take place. This development will be controlled by the concrete transport properties. Any concrete that allows an easy transport of alien species because of its high open porosity will carbonate more or less rapidly. Thus the most important properties for controlling the structural durability of concrete are the transport properties. Among them, permeability is one of the most commonly measured properties, and can be considered as a good indicator for the contribution of concrete to the durability of reinforced concrete structures: 'The permeability of concrete is the most important parameter controlling durability to corrosive environments' (Mehta, 1990).

In theory, permeability of a porous material should be a material property, and should not depend on the type of percolating fluid or on the type of specimen. However, the microstructure of concrete changes with the development of hydration, which is affected by the type of curing. Moreover, concrete is a three-phase material, in which the gaseous phase can never be totally removed. If an attempt is made to remove the liquid phase, microcracking takes place and greatly changes the solid microstructure, affecting subsequent transport phenomena. This is why it is impossible to define, or to measure, a unique index for describing concrete permeability. No universally accepted standard exists in this field, so we shall investigate some permeability data sets, each of them dealing with a different measurement technique.

In a first set of data (Rollet *et al.*, 1992), 16 mixtures were produced with the same aggregate (crushed limestone), with compressive strength ranging from 20 to 120 MPa. The corresponding range of variation of the water/binder ratio was 0.88–0.28, and half of the mixes were air-entrained. Concretes pertaining to the upper strength category contained silica fume, while some intermediate strength levels were attained either with or without silica fume, for comparison purposes. Gas permeability at 28 days was measured for each concrete (with the CEMBUREAU method), after a curing regime comprising a drying period at 50 °C (using a Belgian standard for rocks). The water porosity was also determined, after a similar curing treatment.

A similar set of concretes has already been used for other concrete properties (de Larrard *et al.*, 1996b; see Table 2.2). In this series a large strength range is covered, with mixtures containing either pure Portland cement or additions of fly ash or silica fume. Air permeability was measured after a 28 d. drying period at 80 °C.

A comparable strength range was covered in another study (Yssorche and Ollivier, 1996). Four of the eight mixes produced contained silica fume, and the effect of aggregate type was also investigated. In addition to compressive strength, the gas permeability was measured with an original procedure, at variable pressure.

Finally, the fourth set of data examined deals essentially with fly ash concretes (Al-Amoudi *et al.*, 1996). Five levels of water/binder ratio (between 0.35 and 0.55) and five levels of cement replacement percentage (from 0 to 40%) were investigated. Compressive strength, water permeability and porosity were measured for each mix at 28, 90, 180 and 365 days. The porosity was assessed on oven-dry specimens (at 105 °C), while permeability was evaluated by measuring the volume of water intruded at a constant pressure of 6.9 MPa.

2.6.1 Permeability and porosity

A natural idea is to try to correlate permeability with the final porosity of hardened concrete. Hence a very good link appears between permeability and porosity in two of the data sets referred to (Fig. 2.104), whatever the type and the age of the concrete. A calculation of concrete porosity is achievable from knowledge of the degree of hydration of binders (see section 2.2.2), so that a prediction of permeability range from the concrete composition could be feasible.

However, after various attempts to predict the porosity, it appears that total porosity, evaluated in terms of initial water, air content and non-evaporable water contained in hydration products, exhibits little variation within the range of strength or permeability. It is believed that most water porosity measurements are *directly* affected by the concrete

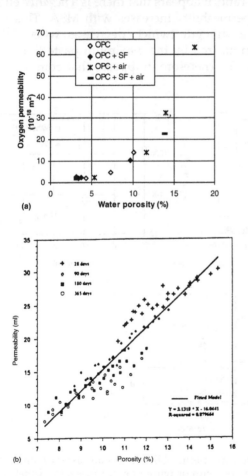

Figure 2.104 Link between permeability and porosity: (a) data from Rollet *et al.* (1992); (b) data from Al-Amoudi *et al.* (1996).

permeability, to the extent to which water penetrates only in the skin of the specimens (especially in the HPC range). Therefore the experimental porosity is no longer a material property, but depends strongly on the size of the specimen. The contradictory needs of statistical homogeneity and limited specimen size (for allowing the penetration of water) make almost impossible the experimental determination of concrete total porosity, which is not the major factor governing the transport properties. The pore size distribution is more probably the dominant factor (Baroghel–Bouny, 1994).

Another reason for not relying upon a porosity–permeability relationship deals with the effect of maximum size of aggregate (MSA). An increase of MSA at fixed water/binder ratio generally leads to a lower paste volume, and then to a lower porosity. But when permeability is plotted against water/binder ratio, it appears that there is a *negative* effect of MSA (Mehta, 1990): the permeability increases with MSA. This fact also stands for compressive strength (which decreases when MSA increases), and has been summarized by the maximum paste thickness concept (MPT; see section 2.3.3). Therefore it appears that compressive

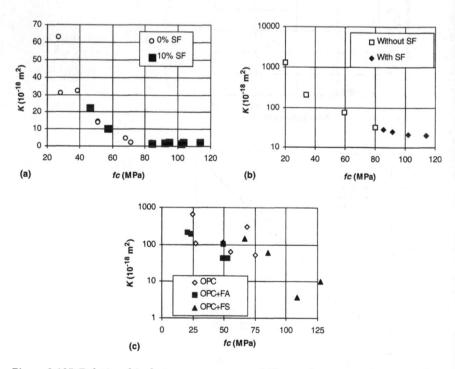

Figure 2.105 Relationship between gas permeability and compressive strengths, for those series of mixes with or without silica fume: (a) Rollet *et al.* (1992); (b) Yssorche and Ollivier (1996); (c) de Larrard *et al.* (1996b) (tests by LMDC Toulouse).

strength could be a better indicator than porosity for predicting concrete permeability.

2.6.2 Permeability and compressive strength

Within the four data sets referred to, there appears to be a direct link between permeability and compressive strength, at least for strengths ranging from 20 to 80–100 MPa (Figs 2.105 and 2.106). It seems that, for higher strengths, internal microcracking counteracts the effect of matrix densification, so that the permeability does not decrease any further. However, this effect remains minor, and does not seem to be harmful for the concrete durability (Yssorche and Ollivier, 1996).

Following the four data sets, the marginal effect of pozzolans on permeability and strength is unclear, as pure Portland cement concretes

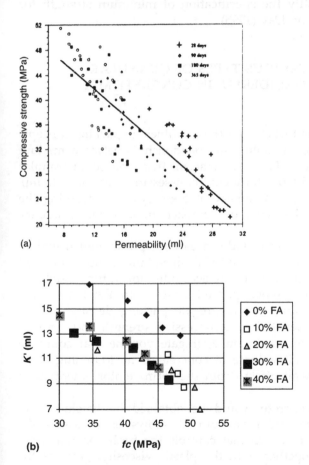

Figure 2.106 Relationship between permeability and compressive strength (Al-Amoudi *et al.*, 1996): (a) all ages; (b) at 28 days.

and concretes containing pozzolans fit with the same master curves. There seem to be some exceptions to this rule in the fly ash concrete data (Fig. 2.106b). However, it is thought that the apparent improvement obtained between 0 and 10% of fly ash is due to an improvement of the workability, leading to a decrease of the air content of the mixes.

It is remarkable that the two natures of aggregate do not lead to separate clouds in Yssorche and Ollivier's data. Then, in the range of concretes containing normal-weight aggregates, Portland cement and pozzolans, it seems that variations in the quality of matrix or matrix/aggregate interface affect both permeability and strength to the same extent. There is not a sufficient amount of available data to claim that a single relationship exists between the two properties, irrespective of the type of materials used. However, from an engineering viewpoint, classification of any mix series based on strength or permeability appears similar. This would justify the specification of minimum strength for durability, as suggested by Day (1995).

2.7 SUMMARY: THE VARIOUS TYPES OF GRANULAR SYSTEM TO BE CONSIDERED IN CONCRETE MIX DESIGN

We have seen throughout this chapter the usefulness of packing concepts in understanding and to modelling the relationships between mixture proportions and mix design. However, it seems that a variety of granular systems have to be considered for a correct assessment of concrete mix design. Figure 2.107 gives a summary of these systems, detailing the nature of the phases, the level of compaction, and the boundary conditions.

As-produced fresh concrete includes a gaseous phase, which will be reduced by the consolidation process. This air volume is not controlled by the formulator, and cannot be easily accounted for in the models.

The aggregate skeleton, when packed apart from the rest of the concrete, gives the g^* parameter (aggregate packing density), which appears in the models devoted to compressive strength and deformability. g^* is actually the maximum aggregate volume that could be physically placed in a unit volume of concrete, if no loosening effect (from the binder) nor wall effect (from the form/reinforcement) took place.

The de-aired fresh concrete would be obtained by batching and compacting the mixture under vacuum. It allows us to assess the structure of the granular system that controls the fresh concrete yield stress. This property, together with the plastic viscosity, governs the flowing behaviour of fresh concrete. As for plastic viscosity, it was modelled by comparing the solid volume of de-aired concrete with the

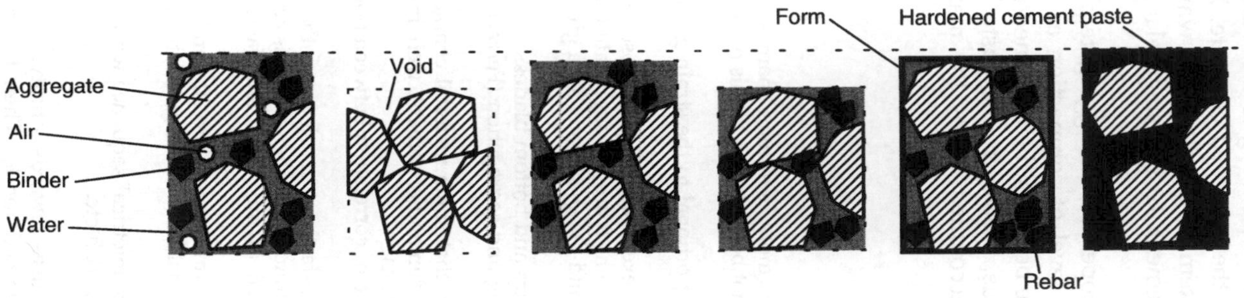

	Granular system					
	1 *Fresh concrete*	2 *Aggregate*	3 *De-aired fresh concrete*	4 *Squeezed fresh concrete*	5 *In situ, de-aired fresh concrete*	6 *Hardened concrete*
Boundary conditions	In bulk	In bulk	In bulk	In bulk	Form + reinforcement	In bulk
Granular phase	Aggr. + binder	Aggregate	Aggr. + binder	Aggr. + binder	Aggr. + binder	Aggregate
Interstitial phase	Water + air	Void	Water	Water	Water	Cement paste
Compaction index	$K < 9$	$K = 9$	$K = K' \leqslant 9$	$K = 9$	$K \leqslant 9$	$K < 9$
Related concrete property	–	Aggregate packing density[31]	Yield stress	Plastic viscosity	Placeability	Strength, deformability

Figure 2.107 The various granular systems to be considered for modelling the concrete system.

maximum possible solid volume (Φ^*) corresponding to a K value of 9. Φ^* corresponds to the solid volume of the squeezed concrete. Finally, when the mix is placed, the granular system is submitted to a wall effect exerted by the form and by the reinforcement. At this stage placeability can be evaluated.

To deal with the hardened concrete properties, the aggregate phase of concrete is emphasized, with the same level of compaction as that of fresh concrete. Here the interstitial phase becomes the hardened cement paste, which is considered homogeneous. However, the paste heterogeneity (interfacial transition zone) is accounted for by the maximum paste thickness concept.

NOTES

1 Rather than m and n, a and b are more common names for the Herschel–Bulkley parameters; they have not been chosen in this book because of other use.
2 Which means that the stress is simply proportional to the strain gradient.
3 Without water.
4 In other words the fine part of the cement exerts a negligible loosening effect (see section 1.1) on the other components (which are the aggregate fractions).
5 That is, the ratio between the shear stress and the shear rate, for a given strain gradient.
6 Which tends to accumulate between the top and bottom blades.
7 These mixtures did exhibit a high bleeding and segregation (Ferraris and de Larrard, 1998).
8 This discrepancy could be due to the fact that some material characteristics used in the calculations, such as the water demand of cement, were measured long after the occurrence of the concrete tests.
9 This is not surprising, owing to the absence of correlation between both types of error.
10 The accuracy obtained is especially satisfactory in view of the limited repeatability and the rather significant operator dependence of this crude test.
11 To this extent the concept is close to that of mix suitability factor (MSF) proposed by Day (1995).
12 In the calibration process the mixes with slump lower than 20 mm have been excluded, as the air measurement is not considered as fully reliable for these stiff mixes.
13 In Renard's series the sieve sizes form a geometrical sequence with a ratio of $10^{0.1}$: 1, 1.25, 1.6, 2, 2.5, 3.15, 4, 5, 6.3, 8, 10, 12.5 etc.
14 Not available for further hydration.
15 Based upon extrapolations of calorimetric tests towards infinity.
16 An equation of the type $x = f(x)$ is obtained for h_c. Then the following sequence is generated: $u_1 = 1, ..., u_n = f(u_{n-1})$... Convergence is obtained after five or six iterations.
17 Or °C.
18 While the w/c ratio is only an index, with no direct physical meaning.
19 Free water/cement alone.

20 The constant in equation (2.70) is different from that of equation (2.58) because we did not use the same size of past cylinders in the referred study, as compared to Marchand's one. Also, the present equations apply for water-cured concretes.
21 No values were given after 28 days for one of the concretes (mix G1).
22 This process does not impair the validity of the calibration, as Rc_{28}, p and q are not independent parameters in equations (2.71) and (2.73).
23 Even in high-volume fly ash (see section 5.5.2).
24 By safely, it is meant that predictions are likely to be conservative as regards experimental results.
25 Here we neglect the possible contribution of C_4AF to the strength, as this compound has an extremely slow reactivity.
26 At very low filler dosage we have $c_{eq} \approx c + \Psi'_{max} \cdot K_{FI} \cdot fi \approx c + 1.3fi$, so that 1 kg of filler has the cementing capability of 1.3 kg of cement.
27 The case of lightweight aggregate concrete is dealt with in section 5.5.1.
28 Same as Marchand's cement: that is, the St Vigor CEM I 52.5 cement.
29 Given in the paper.
30 Other materials, such as polyethylene sheets and resin coatings, do not provide a sufficient barrier to prevent the concrete from drying in long-term tests (Toutlemonde and Le Maou, 1996).
31 The paste data have first to be considered with caution, given the likelihood of skin cracking. Moreover, as already claimed, the strength model developed in section 2.3 has not been validated for cement pastes containing pozzolans. It probably tends to overestimate the strength, as a part of the efficiency of mineral admixtures comes from the improvement of the interfacial transition zone.
32 Preliminary calculation, which is used in the maximum paste thickness concept (for the modelling of compressive strength), and in the triple-sphere model (for the modelling of deformability properties).

3 Concrete constituents: relevant parameters

In this chapter, the material parameters defined in the models developed in Chapter 2 are listed for each type of component. Methods of determination and ranges of typical values are given. Indications about relevant standards (in the European, US and French systems) are also provided, although this type of data may change rapidly. Also, the concrete properties influenced by the material parameters are recalled. In this analysis it is assumed that a concrete of fixed composition by mass is produced, and the properties that are affected by changes in the parameter considered are highlighted.

Other parameters, not listed in this chapter, can be important for durability issues. The reader is invited to refer to general concrete manuals, such as Neville (1995), and to national/international standards.

3.1 AGGREGATE

The materials addressed in this section are normal-weight aggregates, with a size higher than 75–80 μm. They can be either natural (rounded) or artificial (crushed). Lightweight aggregates are covered in section 5.5.1.

3.1.1 Specific gravity

Let us consider a mass M_D of dry aggregate (dried at constant weight in an oven). Let us immerse the sample in water until maximum absorption, and let us dry the particles superficially. The mass becomes M_{SSD} (mass of the saturated surface-dry specimen). Let V_{SSD} be the water volume displaced by immersion of the SSD sample. The specific gravity of the aggregate can be defined in dry conditions:

$$\rho_D = \frac{M_D}{m_W \cdot V_{SSD}} \tag{3.1}$$

or in saturated-dry conditions:

$$\rho_{SSD} = \frac{M_{SSD}}{m_W \cdot V_{SSD}} \tag{3.2}$$

where m_W is the mass per unit volume of water ($m_W = 1000 \text{ kg/m}^3$).

ρ_{SSD} is commonly used in concrete mix design. When the proportions of constituents in the concrete formula are determined by volumes, the mass computation should account for the SSD specific gravity, but also for the degree of saturation of aggregate in the real conditions of utilization. This last item is difficult to measure in a continuous production process. It is the most important source of concrete variability.

3.1.2 Porosity and water absorption

The porosity is the relative volume of voids in an aggregate particle. However, it is essentially the *porosity accessible to water* that matters in concrete technology. This is why the water absorption is used more commonly. With the symbols defined in the previous section, the water absorption, *WA*, and porosity accessible to water, *n*, are (in percent)

$$WA = 100 \left(\frac{\rho_{SSD}}{\rho_D} - 1 \right) \tag{3.3}$$

$$n = \rho_D \cdot WA \tag{3.4}$$

Methods of measurement are given in national standards for water absorption. Conventionally, water absorption at 24 hours is considered.

Table 3.1 Specific gravity in SSD conditions: summary of data.

Parameter	Specific gravity in SSD conditions
Symbol	ρ_{SSD}
Concrete property affected	All properties when concrete is batched by mass
Range	2.4–2.9
Variability for a given source (% relative)	Low (<2%)
European standard	prEN 1097-6
US standard	ASTM C 127, 128
French standard	NFP 18-554, 18-555

Table 3.2 Water absorption: summary of data.

Parameter	Water absorption
Symbol	*WA*
Concrete property affected	All properties if aggregate is used in dry conditions, with a special emphasis on rheology
Range	0–2% and more. Preferably less than 1% for HPC
Variability for a given source (% relative)	Low (except if the aggregate is heterogeneous, e.g. a mixture in variable proportions of a weak, porous component with a hard, non-porous one)[a]
European standard	prEN 1097-6
US standard	ASTM C 70, 127, 128, 566
French standard	NFP 18-555

[a] A common example is provided by some river gravels, which are a mixture of chalk and flint particles.

However, if the aim is to assess the effect of water absorption on fresh concrete rheology, it may be more relevant to look at the water absorption during the first hour after the beginning of immersion.

3.1.3 Size distribution

As a result of the crushing and sieving processes, the size distribution of an aggregate fraction is one of the most variable characteristics of this component. Besides manual methods, automated techniques are available, at least for the coarsest range: that is, for sizes higher than 1 mm (Blot and Nissoux, 1994). However, the size distribution obtained by such a method is not directly comparable to that of a classical sieving process.

For sands containing more than some percent passing the 80 μm mesh sieve, it can be of interest to collect an amount of this passing, and to assess its grading as that of a supplementary cementitious material (see section 3.3.2). Hence this fine content plays an important role in the rheological behaviour of mixtures.

Table 3.3 Size distribution: summary of data.

Parameter	Size distribution
Concrete property affected	Rheological properties, air content
Range	1 mm ⩽ maximum size ⩽ 100 mm and more
Variability for a given source	% passing may change by ±5/10% or more
European standard	EN 933-1
US standard	ASTM C 117, 136
French standard	NFP 18-560

3.1.4   Residual packing density

The residual packing density (virtual packing density of a monosize class; see sections 1.1 and 1.2) is governed by the shape and surface texture of aggregate particles. In a given aggregate fraction it is a function of the particle size. It is deduced from the actual packing density, which is measured with a given packing process.

Method of measurement of actual packing density

Following the CPM theory, the actual packing density can be measured using a variety of procedures, all characterized by a compaction index. However, experience has shown that methods producing high packing values (corresponding to high compaction index) are preferable, because their variability is less. We suggest the method explained in Sedran and de Larrard (1994), which corresponds to $K = 9$.

A mass M_D of dry aggregate is put in a cylinder having a diameter $\varnothing$, which is more than five times the maximum size of aggregate. A piston is introduced in the cylinder, applying a pressure of 10 kPa on the surface of the specimen. Then the cylinder is fixed on a vibrating table, and submitted to vibration for 2 min.[1] The actual packing density is calculated from the final height, h, of the sample:

$$\Phi = \frac{4\,M_D}{\pi\varnothing^2\,h\,\rho_D} \tag{3.5}$$

Calculation of the residual packing density

If the aggregate is monosized (which means that it passes entirely through a sieve of size d_i, and is fully retained by a sieve of size d_{i+1}, with $d_i/d_{i+1} \leqslant 1.26$), we have (from equation (1.48)):

$$\bar{\beta} = \left(1 + \frac{1}{K}\right)\Phi \tag{3.6}$$

where $\bar{\beta}$ is the mean residual packing density (affected by the wall effect). The residual packing density in infinite volume β is then (from equation (1.56)):

$$\beta = \frac{\bar{\beta}}{1 - (1 - k_w)[1 - (1 - d/\varnothing)^2(1 - d/h)]} \tag{3.7}$$

where d is the mean size of the particles, and k_w is the wall effect coefficient (see section 1.3.1).

For polydisperse aggregate an hypothesis is necessary for the variation of β with particle size. It is generally assumed that β is uniform among the aggregate grading span. Therefore we have, for each size:

$$\overline{\beta_i} = \{1 - (1 - k_w)\,[1 - (1 - d_i/\varnothing)^2(1 - d_i/h)]\}\beta \tag{3.8}$$

and the actual packing density can be calculated by using equations (1.33) and (1.47). In turn, β is determined so that the theoretical value of ϕ corresponds to the actual experimental packing density given by equation (3.5).

The packing properties of the fine content of sand, if any, should be assessed by the method given in section 3.3.3.

3.1.5   Elastic modulus

The elastic modulus of a crushed aggregate can be measured on rock cores. When no samples of appropriate shape and dimensions are available, the elastic modulus can be determined indirectly from measurement on concrete. The compressive strength and elastic modulus of concrete have to be measured, together with the packing density of aggregate, g^* (unless it is calculated with the CPM). Then the aggregate modulus has to be calibrated in equation (2.97), so that the theoretical value of concrete modulus fits with the experimental one. In this equation the matrix modulus is deduced from the concrete compressive strength by using equation (2.99). For a better assessment of aggregate modulus, and given the error of the model, the process can be repeated with various concretes, and the mean value of aggregate elastic modulus can be calculated.

Table 3.4 Residual packing density (in infinite volume): summary of data (examples are given in Fig. 1.12).

Parameter	Residual packing density (in infinite volume)
Symbol	β
Concrete property affected	All properties, but mostly the rheological ones
Range	From 0.58 (fine rough crushed particles) to 0.74 (smooth spheres)
Variability for a given source	Low (for a constant process: ±0.01 in absolute value)
European standard	–
US standard	ASTM C 29[a]
French standard	–

[a] This method leads to the measurement of dry-rodded packing density, which corresponds, according to Table 1.10, to $K = 4.5$.

Table 3.5 Aggregate elastic modulus: summary of data.

Parameter	Aggregate elastic modulus
Symbol	E_g
Concrete property affected	Elastic modulus, creep, shrinkage, coefficient of thermal expansion etc.
Range	40–100 GPa
Variability for a given source (% relative)	Low (± 5 GPa)
European standard	–
US standard	ASTM D 2845-83
French standard	–

3.1.6 Contribution to compressive strength

As shown in section 2.3.4, and besides the topological effects that are summarized by the maximum paste thickness concept, the aggregate influences the compressive strength of concrete through its bond with cement paste (expressed by p) and its intrinsic strength (expressed by q). The relationship between matrix and concrete strength is given by equation (2.68).

No direct measurement is possible to determine the values of p and q. It is necessary to produce at least two concretes (1 and 2) of very different strength (e.g. a normal-strength concrete and a high-performance concrete), with a cement of known strength. For each concrete the aggregate packing density must be measured (or calculated with the CPM), so that the MPT and the matrix strength can be calculated by equations (2.64) and (2.83). Then a system of two equations with two unknown parameters:

$$fc_1 = \frac{p fc_{m1}}{q fc_{m1} + 1}$$

$$fc_2 = \frac{p fc_{m2}}{q fc_{m2} + 1}$$

(3.9)

leads to the calculation of both coefficients p and q.

When only a single concrete has been previously produced, a relationship between p and q may be assumed if the intrinsic strength of the rock, fc_g, is known. From the data given in Table 2.14, and as a first approximation, the correlation between p/q and fc_g may be expressed by the following equation:

$$\frac{p}{q} \approx 2.14 fc_g$$

(3.10)

from which the expressions for p and q are deduced:

$$p = \cfrac{1}{fc_m \left(\cfrac{1}{fc} - \cfrac{1}{2.14 fc_g} \right)}$$

(3.11)

$$q = \frac{p}{2.14 fc_g}$$

3.1.7 Contribution to tensile strength

This parameter is used in the model that gives the splitting tensile strength as a function of the compressive strength. Its determination, for a given aggregate, requires the knowledge of both strengths, so that

$$k_t = \frac{ft}{fc^{0.57}}$$

(3.12)

(strengths in MPa).

As for the previous parameters, a better precision will be reached if several strength couples are known, dealing with different mixtures produced with the same aggregate.

Table 3.6 Contribution to compressive strength: summary of data (examples are given in Table 2.14).

Parameter	Bond coefficient	Ceiling effect coefficient
Symbol	p	q
Concrete property affected	Compressive strength, tensile strength and deformability properties	Compressive strength, tensile strength and deformability properties
Range	0.5–1.2	0–0.006 MPa^{-1}
Variability for a given source (% relative)	?[a]	?[b]

[a] Large variations of p can be expected, related to the cleanness of aggregates. Compared with washed aggregates, aggregates coated with silt or clay may lose a significant part of their bond with cement paste, especially if no dispersing agent (superplasticizer) is used in the concrete mixture. Intentional reduction of aggregate bond by grease coating has been proven to produce strength reductions of 30% and more (Maso, 1980).
[b] The p/q ratio is probably fairly constant, as it is controlled by the compressive strength of the rock, and provided that no important change in the mineralogical nature of aggregate takes place during production.

Table 3.7 Contribution to tensile strength: summary of data (examples are given in Table 2.18).

Parameter	Tensile strength coefficient
Symbol	k_t
Concrete property affected	Tensile splitting strength
Range	0.32–0.47 MPa$^{0.43}$
Variability for a given source	±0.02/0.03[a]

[a] An excess of silt or clay in aggregates could decrease k_t, as lack of cleanness has more effect on tensile strength than on compressive strength.

Table 3.8 Heat capacity: summary of data.

Parameter	Heat capacity
Symbol	c^{th}
Concrete property affected	Adiabatic temperature rise
Range	0.70–0.90 kJ/K/kg
Variability for a given source (% relative)	Low[a]

[a] As long as the mineralogical nature is constant.

3.1.8 Heat capacity

This parameter is necessary for the evaluation of adiabatic temperature rise. Table 3.8 gives orders of magnitude for siliceous and calcareous aggregates.

3.2 CEMENT

The properties of Portland cement that are used in the models are detailed in this section. Blended cements can be considered as a pure Portland cement as far as the fresh concrete is concerned. However, for the prediction of mechanical properties and heat of hydration they must be split between an amount of pure Portland cement of known composition and another amount of supplementary cementitious materials.

3.2.1 Specific gravity

This is a classical property, generally measured with the pycnometer method. The main cause of variation deals with the secondary components in blended cements.

Table 3.9 Specific gravity: summary of data.

Parameter	Specific gravity
Symbol	ρ_C
Concrete property affected	_a
Range	2.7–3.2 (3.1–3.2 for pure Portland cement)
Variability for a	Very low for Portland cement
given source (% relative)	
European standard	–
US standard	–
French standard	–

a Significant changes in specific gravity of a cement are generally an indication of changes in the proportions of mineral admixtures, which may affect all other properties of the cement, unless the cement manufacturer has intentionally changed this proportion to correct an evolution of the clinker properties.

3.2.2   Grading curve

The grading curve of cement is generally determined with LASER granulometry. Examples of cement grading curves are given in Figure 3.1. The amount of flocculation in the fine part of cement that takes place in fresh concrete is always a matter of uncertainty. As the conventional measurement is performed on an alcoholic suspension of cement, the fineness of agglomerates is certainly overestimated. However, it is only when several binders are mixed in concrete that their

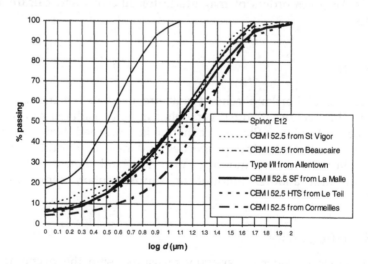

Figure 3.1 Grading curves of some cements, determined by laser granulometry (suspension of cement in ethanol). Tests performed at LCPC Paris. Spinor E12 is a special cement for grouts. La Malle cement is a blended cement containing 8% of silica fume.

grading curve has a significant influence in the mixture optimization process.

The Blaine fineness gives a rough idea of the grading curve. For the purpose of quality control, Blaine fineness monitoring may help in detecting changes in the grading curve of cement.

3.2.3 Residual packing density with and without admixture

We suggest determining the packing properties of cement from the measurement of water demand (Sedran and de Larrard, 1994). If an organic admixture with plasticizing properties is to be used in the final mixture, the water demand measurement should be performed with the same proportion of admixture (as a percentage of cement dosage).

Method of measurement of water demand

This simple test requires only a balance (precision 0.1 g) and a conventional mortar mixer. Taking a mass of 350 g of cement, the aim is to find the minimum water dosage that produces a thick paste. A slightly lower amount should give a humid powder. To blend the two products, first add the water (previously mixed with admixture, if any), and then the cement in the mixer bowl. Mix for 1 min at low speed, scrape the bowl manually, and mix again for 1 min at high speed. The bulk result of the test is the critical water/cement ratio w/c.

Table 3.10 Grading curve: summary of data.

Parameter	Cement grading curve	Blaine fineness
Concrete property affected	Rheological properties, strength	Rheological properties, strength
Range	0.1–100 µm. Mean size around 15 µm except for special cements (see section 5.3.3)	300–500 m²/kg
Variability for a given source (% relative)	Relatively high[a]	±20–30 m²/kg
European standard	–	EN 196-6
US standard	–	ASTM C 204
French standard	–	NFP 15-476

[a] Changes in the grading curve of cement generally reflect changes in the clinker-crushing process. This is one of the main techniques used by the cement manufacturer to maintain a constant strength at 28 days, when changes are detected in the quality of raw materials. But uncontrolled variations in strength at early age and in rheological properties are a possible drawback of this technique.

Calculation of the residual packing density

The actual packing density, Φ, is the solid concentration of the paste, in volume:

$$\Phi = \frac{1}{1 + \rho_C \dfrac{w}{c}} \tag{3.13}$$

where ρ_C is the specific gravity of cement (see section 3.2.1).

To deduce the residual packing density of the cement, the CPM has to be used. If the concrete mix contains only the cement as a binder, a simple hypothesis of uniformity may be taken for the residual packing density. β is then calibrated so that the prediction of the CPM, with $K = 6.7$ (see section 2.1.2), fits with the experimental value coming from equation (3.13). In cases where other binders are to be used, a better precision in the packing calculations will be reached if the distribution of β_i vs. the size of the particles is more thoroughly assessed.[2] As for aggregate, the residual packing density tends to increase with the size of particles, especially below 15 µm (see Table 3.11).

3.2.4 Bogue composition

The Bogue composition of cement is calculated from the chemical analysis. In the models developed in Chapter 2 the Bogue composition is

Table 3.11 Actual packing density (measured in the dry state) of several fractions of a cement (Powers, 1968).

Size group (µm)	2.4	8.3	15.3	30.2	49.2	81.6
Voids content (%)	80	58	46	44	45	45
Actual packing density	0.20	0.42	0.54	0.56	0.55	0.55

Table 3.12 Residual packing density: summary of data.

Parameter	Actual (global) packing density	Residual packing density
Symbol	Φ	β
Concrete property affected	Rheological properties	Rheological properties
Range	Without SP: 0.55–0.58 With SP:[a] 0.60–0.68	Without SP: 0.41–0.44 With SP: 0.45–0.50
Variability for a given source	Without SP: low With SP: may be high	Without SP: low With SP: may be high

[a] At saturation dosage (see section 3.4.2).

Table 3.13 Bogue composition: summary of data.

Parameter	Tricalcium silicate	Dicalcium silicate	Tricalcium aluminate	Tetracalcium aluminoferrite
Symbol	C_3S	C_2S	C_3A	C_4AF
Concrete property affected	Rheological properties and hydration rate			
Range	28–56	19–49	2–14	2–12
European standard	EN 196-2 and 196-4			
US standard	ASTM C 150			
French standard	NFP 15-472			

used to calculate the cement unit heat of hydration, and the contribution of limestone filler to compressive strength.

3.2.5 Strength vs. time

Following the ISO standard, cement strengths are measured with a mortar having relative dosages of $3:1:0.5$ in sand (continuously graded)–cement–water. Prisms measuring $40 \times 40 \times 160$ mm are broken in a flexural test, and then the remaining halves are crushed in compression. The compressive strength is the median of six individual values. Nowadays compressive strength at 28 days may range between 50 and 70 MPa for pure Portland cement produced in developed countries. Lower values are to be expected for blended cements (except when the mineral addition is silica fume). For cements covered by the European standard EN 197, the minimal strength of cements can take the following values: 32.5, 42.5 or 52.5 MPa. In production, cements generally display a mean strength about 10 MPa higher than the minimum value.

The relationship between strengths obtained with the ISO test and the ASTM test has been investigated by Foster and Blaine (1968). From their data, the following correlation is derived:

$$fc_{ISO, MPa} = 1.80 fc_{ASTM, MPa} - 4.4 \tag{3.14}$$

To be accounted for in the compressive strength model developed in section 2.3, the compressive strength at ages different from 28 days must be transformed in kinetics terms, $d(t)$. From the general equations of the model (equation (2.73)), we have

$$d(t) = 0.0522 \left(\frac{Rc_t}{Rc_{28}} - 1 \right) \tag{3.15}$$

where Rc_t is the mortar compressive strength at age t. Typical values of $d(t)$ are displayed in Fig. 2.64. According to equation (3.15), $d(t)$ cannot be lower than -0.0522. However, this is the case for some values in Fig. 2.64. The reason lies in the fact that the parameters given in this figure have been calibrated from concrete data sets in which a number of mixtures contained superplasticizer. But this admixture is known to have a significant retarding effect at early age. Therefore the values of $d(t)$ determined from cement strength tests will be higher, but will lead to overestimations of early age strengths for superplasticized concretes (see remarks at the end of section 2.3).

3.2.6   Contribution to autogenous shrinkage

It has been seen in section 2.5.5 that cements may display significantly different contributions to autogenous shrinkage. These differences are controlled by the porous structure of hardened cement paste (Baroghel–Bouny, 1994), and also by the amount of water consumption during hydration: the finer the structure, or the higher the water consumption, the higher autogenous shrinkage will be. Then, very reactive cements (such as ASTM type III cements) are expected to produce high autogenous shrinkage. Let us recall that this deformation, together with thermal phenomena, is responsible for early age cracking in restrained concrete elements.

Determination of K_c for a cement

When K_c is not available, it can be determined by using the following procedure.[3] A standard ISO mortar (with $w/c = 0.5$) is prepared, and $40 \times 40 \times 160$ mm prisms are cast for shrinkage measurement purposes. The setting time of the mortar is measured with the Vicat needle method (EN 196-3 or ASTM C 191). Just after the end of setting, the prisms are demoulded and the initial length is measured. Then the prisms are carefully wrapped with a double layer of adhesive aluminium (to avoid subsequent water losses), and are kept at 20 °C in a humid environment.

Table 3.14 Strength versus time: summary of data.

Parameter	Compressive strength at i days
Symbol	Rc_i
Concrete property affected	Compressive strength and elastic modulus at i days
Variability for a given source	High at early age, lower at 28 days (some MPa)
European standard	EN 196-1
US standard	ASTM C-109
French standard	NFP 15-471

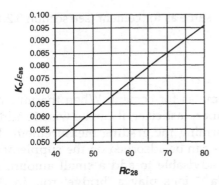

Figure 3.2 Curve for the calculation of K_c (in MPa) from the mortar autogenous shrinkage ($\varepsilon_{as\infty} \times 10^{-6}$) and the compressive strength at 28 days of the mortar cured in water (Rc_{28}, in MPa). This curve has been determined from simulations performed with equations (2.100) and (2.110)–(2.112).

Table 3.15 Autogenous shrinkage coefficient: summary of data.

Parameter	Autogenous shrinkage coefficient
Symbol	K_c
Concrete property affected	Autogenous shrinkage
Range	13–25 MPa
Variability for a given source (% relative)	Presumably high[a]

[a] From several autogenous shrinkage measurements made on the same cement, with a 2-year interval.

The length of prisms is monitored for at least 2 months. Then the final autogenous shrinkage of mortar, $\varepsilon_{as\infty}$, is determined, preferably by numerical extrapolation of the deformations (a simple hyperbolic model can be used). This deformation may range from 100 to 400×10^{-6}. Finally K_c is calculated from the knowledge of $\varepsilon_{as\infty}$ and Rc_{28} (strength of cement; see section 3.2.5), by using Fig. 3.2.

Typical values of K_c are given in Table 3.15.

3.3 MINERAL ADMIXTURES (SUPPLEMENTARY CEMENTITIOUS MATERIALS)

Here we address both pozzolanic materials such as silico-aluminous fly ash (ASTM class F), natural pozzolans, metakaolins or silica fume, and limestone fillers.

3.3.1   Specific gravity

The method of measurement is the same as for cement (see section 3.2.1). Data are given in Table 3.16.

3.3.2   Grading curve

Here again the method is the same as for cement. When the mineral admixture is to be used with a superplasticizer, it can be wise to add a dose of this product when performing the grading measurement. To ensure that the organic molecules can be adsorbed on the supplementary cementitious material, it is advisable to add a small amount of lime ($CaOH_2$) in water, since Ca^{++} ions play a 'bridge' role in the adsorption phenomenon (Buil *et al.*, 1986). Finally, submitting the suspension to ultrasound prior to testing may produce a higher degree of deflocculation.

Silica fume

For silica fume, which is approximately 100 times finer than cement, many laser granulometers are not able to analyse the size distribution. In this case, a sedigraph can be used. In any case, the real state of deflocculation in fresh concrete is largely unknown, as all methods at least require a strong dilution of the suspension, as compared with the product in fresh concrete. For simulation of special mixtures, in which a high dosage of silica fume is combined with a high amount of superplasticizer (see section 5.3.4), it can be important to pay attention to the whole grading of silica fume. Since the finest part is not accessible to most laboratory facilities, a convenient method is to

Table 3.16 Specific gravity: summary of data

	Fly ash	Silica fume	Limestone filler
Parameter	Specific gravity	Specific gravity	Specific gravity
Concrete property affected	Rheological properties[a]	Rheological properties	Rheological properties
Range	2.1–2.4	2.1–2.3[b]	2.6–2.8
Variability for a given source (% relative)	Low	Very low	Very low

[a] Since a change of specific gravity creates changes in the product volume, at constant weight per unit volume of concrete.

[b] The *bulk* density of natural silica fume is very low (0.1–0.2). By densification, this value can be raised up to 0.5–0.6. Higher values can be even attained with *pelletized* silica fume. However, this product is no more appropriate for concrete. Even, its use as a sand replacement may create a strongly disruptive alkali-silica reaction (Marusin and Bradfor Shotwell, 1995).

Table 3.17 Grading curves: summary of data.

Parameter	Fly ash		Silica fume		Limestone filler	
	Grading curve	Specific surface[a]	Grading curve	Specific surface[b]	Grading curve	Specific surface[a]
Concrete property affected	Rheological properties	Rheological properties	Rheological properties	Rheological properties	Rheological properties	Rheological properties
Range	0.1–100 μm	300–500 m²/kg (more if selected)	0.01–10 μm (more if pelletized)	15 000–25 000 m²/kg	0.1–100 μm	200–1000 m²/kg
Variability for a given source	High		Moderate		Very low	Very low

[a] Blaine method.
[b] BET method.

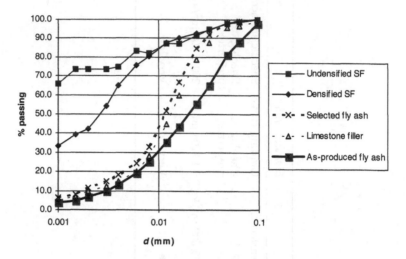

Figure 3.3 Grading curve of some supplementary cementitious materials. Data from Waller (1998).

assume that the 'tail' of the curve is a straight line in a semi-logarithmic diagram (see Fig. 3.4). The slope of the straight line is adjusted so that the specific surface equals the BET value (de Larrard, 1988). It is believed that the error made in the grading curve assessment by using this process has little consequence in the optimization calculations. In fact, cement and other binders are generally much coarser than silica fume. Then, the interaction terms in the CPM that relate the fine part of silica fume to the rest of the granular fractions deal with small size ratios.

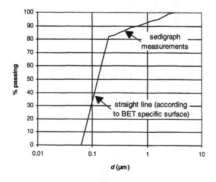

Figure 3.4 Grading curve of a silica fume, with a linear extrapolation in the finest range.

3.3.3   Residual packing density with and without admixture

As for cement (see section 3.2.3), the water demand is first determined. There seem to be some complex interactions between Portland cement and SCM in the presence of superplasticizer. As a result, the *equivalent* virtual packing density of an SCM in the presence of cement may be less than the value measured on pastes containing SCM only. Therefore it is advisable to measure the water demand of mixtures containing 90% SCM–10% Portland cement, then 80% SCM–20% cement, and to extrapolate these results up to 100% SCM–0% cement (Sedran, 1999). With this process the packing calculations carried out with the CPM for superplasticizer mixtures of powders are reasonably accurate (Fig. 2.10).

Then the residual packing density is calibrated by using the CPM. Typical values are given in Table 3.18. Note that, because of the shape of particles, fly ash generally exhibits a higher value than crushed materials such as limestone fillers. In spite of the spherical shape of the silica fume grains, the packing ability of this product is not as high as that of fly ash. The reason probably lies in the difference in size of the products. For diameters lower than some microns, surface forces become important, and overcome the natural tendency of spherical particles to pack closely. Also, silica fume grains could partially agglomerate in the cooling process, forming some grape-like clusters of particles.

3.3.4   Activity coefficients vs. time

This subsection applies only to pozzolanic admixtures, as we have shown that the activity of limestone fillers is essentially controlled by their specific surface (and by the C_3A content of the cement). Then, tests on mortar containing Portland cement and limestone filler reflect as much the compatibility of the cement with filler (hardening rate, C_3A content) as the general capability of the filler to contribute to compressive strength.

The principle of determination of the activity coefficient is to compare the compressive strength of an ISO mortar (see section 3.2.5) with the compressive strength of another mortar where 25% in mass of the cement is replaced by an equal amount of the pozzolanic admixture. The ratio between the two strengths is called the activity index, $i(t)$, where t is the age of mortar when the test takes place. If $i(t)$ is less than 1, the pozzolan is less active than cement. Conversely, if $i(t)$ is greater than 1, the pozzolan has a superior activity (this is generally the case for silica fume).

The result of this test may be partially impaired by the fact that the mixtures do not have the same consistency, so that their air contents can be substantially different. Therefore we recommend using a small amount of superplasticizer to ensure that the two mortars have an

Table 3.18 Residual packing density: summary of data.

Parameter	Fly ash		Silica fume		Limestone filler	
	Actual packing density,[a] Φ	Residual packing density, β	Φ	β	Φ	β
Concrete property affected	Rheological properties	Rheological properties	Rheological properties	Rheological properties	Rheological properties	Rheological properties
Range						
without SP	0.58–0.63	0.47–0.53		0.49–0.63	0.57–0.62	0.47–0.51
with SP	0.62–0.66	0.52–0.60		0.55–0.64	0.61–0.65	0.52–0.58
Variability for a given source	High		Moderate		Low	

[a] Solid concentration in the water demand test.

identically fluid consistency. This process is particularly critical with silica fume, which has the property of decreasing dramatically the workability of a mix when used without superplasticizer.

Another problem in dealing with the conventional mortar test lies in the nature of the specimens. As already mentioned, the prism halves are directly crushed in the ISO test, without capping. Then the test may become less reliable when the compressive strength grows up to 80–100 MPa and more, because the failure is more brittle and more scattered (de Larrard *et al.*, 1994c). Therefore it is advisable, when determining the activity index of silica fume or other very reactive pozzolans, to grind the test faces of the prisms with a lapidary mill; otherwise there is a risk of underestimation of $i(t)$ and K_p.

From the model developed in section 2.3, the activity coefficient K_p is:

$$Kp(t) = -3\ln\left\{1.91 - \frac{2.2}{(i(t)\,[0.0522 + d(t)] - d(t))^{-0.351} - 1}\right\} \quad (3.16)$$

Typical values of K_p versus time are given in Fig. 2.62. A summary is presented in Table 3.19.

For silica fume it may be preferable to limit in the test the amount of cement replaced by the pozzolan to 10–15% of the total weight of binder, a dose that is closer to that used in most high-performance concretes.

Estimation of silica fume activity coefficient at 28 days from SF chemical composition

The activity index (or activity coefficient) of silica fume does not seem to depend very much on the nature of the Portland cement used in the test,[4] but more on the alkali content of the pozzolan (de Larrard *et al.*, 1992).

Table 3.19 Activity coefficient: summary of data.

Parameter	Fly ash		Silica fume	
	Activity index at 28 days	Activity coefficient at 28 days	Activity index at 28 days	Activity coefficient at 28 days
Symbol	$i(28)$	$K_p(28)$	$i(28)$	$K_p(28)$
Concrete property affected	Compressive strength and deformability properties	Compressive strength and deformability properties	Compressive strength and deformability properties	Compressive strength and deformability properties
Range	0.6–1.1	0–1.4	1.4–1.9	2.5–5
Variability for a given source	High	High	Moderate	Moderate

Table 3.20 Compressive strength of mortars of constant mixture proportions, with various types of silica fume.

Silica fume	$Na_2O_{eq.}$ (%)	w/c	fc (MPa)
1	1	0.32	110
2	0.91	0.33	108
3	0.39	0.34	105
4	2.49	0.31	104
5	1.3	0.34	101
6	1.79	0.34	98
7	2.58	0.34	94
8	0.64	0.39	93

Source: Cariou (1988).

We shall use some experimental data (Cariou, 1988) for deriving an empirical formula permitting the evaluation of activity coefficient from a knowledge of equivalent alkali content. In these experiments mortars were produced with 10% silica fume (of the cement weight), and the water content was adjusted to keep the workability constant (see Table 3.20). Taking the activity coefficient as a linear function of the alkali content, and by using the strength model developed in section 2.3 (Féret's type), the following equation for the activity coefficient is found:

$$K_p(28) = 2.8 - 0.7\,Na_2O_{equ} \tag{3.17}$$

where Na_2O_{eq} is the equivalent sodium oxide content. The agreement of this model with regard to experimental data is shown in Fig. 3.5.

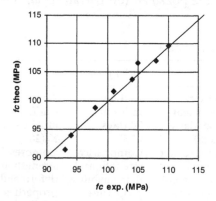

Figure 3.5 Comparison between experimental data and predictions of the strength model, with the activity coefficient of silica fume deduced from equation (3.17).

However, by comparison with Fig. 2.62, it is found that the K_p values in Cariou's data are in the low range of those found in the literature. An explanation could deal with the fact that many silica fumes used in the referred study were densified.

3.4 PLASTICIZERS/SUPERPLASTICIZERS

The products addressed in this section are organic admixtures having the property of either increasing the fluidity of a fresh mixture at constant water dosage, or allowing a reduction of the water dosage at constant fluidity (measured with any usual device). The typical water reduction ranges from 10 to 30% and more, if the Abrams slump is referred to. These admixtures are generally delivered as suspensions in water. The active molecule most often pertains to one of the three following categories:

- naphthalene sulphonate formaldehyde;
- melamine sulphonate formaldehyde;
- lignosulphonate.

The last type of admixture is made from by-products, while the two others are specially produced for use in cementitious materials. In addition to these 'classical' admixtures, new molecules are appearing on the market, such as the polyacrylates. Finally, many commercial plasticizers/superplasticizers are mixtures of several basic molecules. Some of them contain retarding agents. *All* plasticizers/superplasticizers retard the setting to a certain extent.

From a concrete technology viewpoint such an admixture is characterized by its unit weight (or density), its solid concentration, its plasticizing capability (vs. time), and its influence on early strength development. The last point is not detailed in this book, as the compressive strength models do not cover very early ages, nor account for the accelerating/retarding effects of admixtures. Practically speaking, and for comparative purposes, one may perform setting time tests on pastes or mortars in the presence of admixtures, in order to select a product giving an acceptable retardation.

3.4.1 Specific gravity and dry extract

The specific gravity of the liquid is generally given by the producer. Clearly, it depends on the concentration, since the dry extract density is higher than 1. The dry extract content is sometimes more difficult to obtain, but can be easily measured by heating a dose of product at 105 °C and comparing the masses before and after the heating process.

Table 3.21 Specific gravity: summary of data.

Parameter	Density	% of dry extract (in mass)
Range	1.1–1.4	20–40%[a]
Variability for a given source	Very low	±2%
European standard		EN 480-8

[a] Low concentrations are to be avoided because they lead to the handling of a large quantity of water for little active product. Also, in HPC technology, it is generally better to start the mixing process without superplasticizer. But if a large quantity of water is added with the SP, the initial mix becomes very dry, giving inefficient mixing and considerable wear of the mixer blades. Conversely, admixtures that are too concentrated are difficult to disperse in fresh concrete, especially in winter (Le Bris *et al.*, 1993).

3.4.2   Saturation curves of binder–admixture couples

This section deals mainly with high-performance concrete technology. The success of an HPC mix-design, as for any concrete, is the result of a suitable selection of materials, together with a proper proportioning, which accounts for all important mix properties. However, a special feature is the importance of the cement–superplasticizer compatibility (Aïtcin *et al.*, 1994). In the best case, the combination of cement and admixture leads to a mix having at the same time a low water/binder ratio (less than 0.40) and a suitable consistency during the time necessary to cast and finish the concrete structure. In the worse case, either the initial consistency is too dry, or it becomes so very soon (sometimes only 10–20 min after the beginning of mixing). Therefore it is critical when designing an HPC to compare the behaviour of different admixtures with a given binder (a Portland cement with or without supplementary cementitious materials). This is the basis of the AFREM[5] method for mix design of high-performance concrete (de Larrard *et al.*, 1997b). The three criteria will be:

- the fluidifying effect obtained initially;
- the amount of admixture necessary to obtain the fluidification;
- the stability of rheological behaviour within the practical duration of concrete use.

For limiting both labour and material costs, comparative tests are performed on the fine part of concrete (that is, all materials having a size smaller than 2 mm, including free water and organic admixtures). The rheological test chosen in the AFREM method is the Marsh cone, which has the advantage of being simple and widely used worldwide for injection grouts. Other researchers have suggested using the mini-slump test, described in Helmuth *et al.*, (1995), which could be more directly related to a basic rheological property of the grout (namely the yield stress).

Conditions

The usual conditions of use meet the following requirements:

- relative humidity > 65%;
- temperature of production and measurement room 20 °C ± 2 °C;
- temperature of water and materials 20 °C ± 1 °C (unless it is desired to simulate concreting in hot or cold weather).

Apparatus

The following apparatus is required:

- a graduated beaker having a volume greater than or equal to 500 ml to measure the volume of grout that has flowed;
- a Marsh cone (American Petroleum Institute standard API RP 18B);
- a chronometer reading in tenths of a second;
- a thermometer to check the temperature of the mixing water and of the grout at the end of the test;
- a 250 or 500 ml test tube to measure the water;
- a rubber spatula to scrape the sides of the mixer bowl;
- a balance reading to 0.1 g;
- a mixer to CEN standards (EN 196-1), with shaft and pallet in good condition.

Composition of the grout

The initial concrete is formulated with quantities and qualities of binder(s) likely to provide the desired strength, an optimized granular skeleton, an ample proportion of superplasticizer (for example 1.5% dry extract with respect to the weight of cement), and a proportion of water adjusted to give the desired consistency (Abrams cone slump generally greater than 20 cm). The formulation of the grout is deduced from that of the concrete by removing all elements larger than 2 mm, together with the water they absorb. 1.5 litres of grout is prepared.

Production of grout

Dry sand is used, and the steps are as follows:

1 Put the silica fume and then an equal weight of water into the bowl of the mixer.
2 Mix at low speed for 30 s.
3 Put the rest of the water and one-third of the proportion of admixture into the bowl.

4   Mix at low speed for 15 s.
5   Pour in the cement (and any mineral additions other than silica fume) and start the chronometer (t_0).
6   Mix at low speed for 30 s.
7   Pour in the sand during the next 30 s, with the mixer still running at low speed.
8   Mix at high speed for 30 s.
9   Stop mixing for 30 s and scrape the bowl with the rubber spatula for the first 15 s of the stoppage.
10 Add the remaining two-thirds of the admixture.
11 Resume mixing at low speed for 15 s, then at high speed for 1 min 45 s.

Measurement of flow time in Marsh cone

- Choose the nozzle 12.5 mm in diameter.
- At the very end of the mixing cycle, pour 1 litre of grout into the cone (the 1 l level will be marked in the cone after calibration with water).
- Make sure before each measurement that the inside walls of the cone are wet (after rinsing with water, the cone is placed upside down on the lab bench for approximately 20 s.
- The first flow measurement (removal of plug from nozzle) is made at $t_0 + 5$ min.

The flow time of the first 500 ml is measured, and then the temperature of the grout.

To monitor fluidity over time, measurements are made at $t_0 + 5$ min, $t_0 + 15$ min, $t_0 + 30$ min, $t_0 + 45$ min, and $t_0 + 60$ min. This period can be extended as the specification requires. Between measurements the grout is left at rest, and the bowl is covered with a wet cloth or a plastic film. The grout is remixed for 15 s at low speed 1 min before each new measurement.

Determination of the saturation proportion

See Fig. 3.6.

1   Plot the experimental points of the curve of $\log(T)$ versus the percentage of dry extract of superplasticizer with respect to the mass of cement and join these points by a dashed line.
2   Plot a straight line D having slope $\frac{2}{5}$, or for example the hypotenuse of the triangle having base 1% SP and height 0.4 log unit.
3   Plot the straight lines D_i parallel to the foregoing and tangent to the dashed line of the experimental points. The abscissas of the points of tangency of the straight lines D_i are noted (x_i).

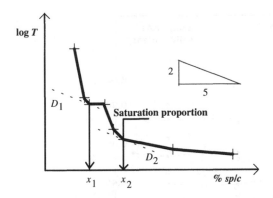

Figure 3.6 Method of determination of saturation proportion. Logarithm of
flow time (in seconds) versus percentage of superplasticizer with
respect to cement (in dry extract).

4   The saturation proportion x_s is then

$$x_s = \sup(x_i)$$

provided that x_s is not equal to one of the limits of the measurement
interval in x.

The value $\frac{2}{5}$ is arbitrary; it was chosen because it leads to reasonable
proportions of admixture.

The saturation proportion is not exactly the proportion of admixture
leading to the maximum fluidity, since this quantity sometimes does not
exist (in other words, the flowing time decreases constantly when the
admixture dose increases). It is more a 'useful' dosage, above which
the increase of efficiency is not worthwhile, given the excessive cost and
the side-effects produced by the product. It is a characteristic of the
binder–admixture couple, with a possible (minor) influence of the sand. It
does not depend significantly on the water/binder ratio (Fig. 3.7). Examples
of saturation curves are given in Fig. 3.8. After round-robin tests between
six laboratories, the determination of saturation dose and the comparison
between admixtures, as performed following the AFREM method, have
been found to be quite reproducible (de Larrard *et al.*, 1997b).

The saturation curves are generally complemented with evolution curves,
in which the flowing time is recorded against elapsed time (Fig. 3.9). These
curves can also be obtained at temperatures other than ambient (either
hotter or colder, to simulate summer or winter concreting respectively).

Finally, let us note that the saturation proportion is affected by the
supplementary cementitious materials contained in the binder. With a

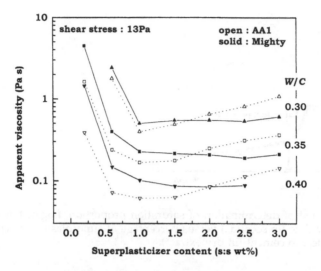

Figure 3.7 Viscosity vs. superplasticizer dosage obtained at different water/
binder ratios (Matsuyama and Young, 1997).

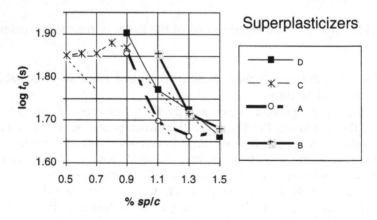

Figure 3.8 Saturation curves obtained with a cement mixed with 10% by
mass of silica fume, and different commercial superplasticizers
(de Larrard *et al.* 1996a). In this case admixture A has been
selected as the one giving the maximum water reduction effect,
with a reasonable amount and a rather stable behaviour (see
Fig. 3.9).

Portland cement/naphthalene type SP combination having a saturation
dose of 1%, the addition of silica fume increased the saturation dose
by 4% of the silica fume mass, resulting in an overall saturation dose of
1.3% for an SF–cement mixture of 10%–90% (de Larrard, 1988).

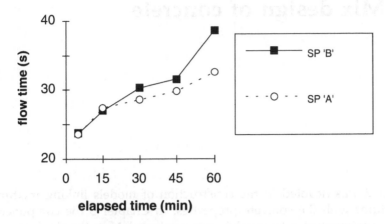

Figure 3.9 Example of evolution curve (at 20 °C). Portland cement plus 10% silica fume. Water contents have been adjusted to give the same flow time at the first measurement (de Larrard *et al*. 1996b).

Table 3.22 Saturation dosage: summary of data.

Parameter	Saturation dosage of an admixture with a given binder (SP dry extract/cement in %)
Range	0.4–1.5%
Variability for a given source	Low

NOTES

1 The result of this test is also influenced by the characteristics of vibration. An acceleration of 4 g is suggested, when the vibrating table allows the user to adjust the frequency or the amplitude.
2 However, the size selection of a powder requires special equipment, not available in most laboratories.
3 Inspired by a private communication from A. Kheirbek, LCPC, 1997.
4 To the author's knowledge, the question is more open for fly ash.
5 Association Française de Recherche et d'Etudes sur les Matériaux et les Structures; French branch of RILEM.

4 Mix design of concrete

Chapter 2 was devoted to the construction of models linking mixture proportions with the concrete properties. In Chapter 3 the component parameters involved in the models were reviewed. Now the aim is to put all the models together, in order to optimize a concrete mixture.

The first step in any method dealing with the formulation of materials is to define the problem: that is, to establish a list of specifications. This may seem simple, but the interactions between the material and the structure specifications mean that an operational list of material requirements can be established only with a full knowledge of all the construction stages.

Knowing the concrete specifications and the constituent characteristics, mix design becomes feasible. Based upon some simplified models developed in Chapter 2, a theoretical solution of the problem is proposed. It permits the rigorous demonstration of a number of rules; most of them are well known by experienced practitioners, without having been already demonstrated (to the author's knowledge). In a second, deeper step of investigations, some of these rules are re-examined in the light of numerical simulations performed with the full set of models. The results sometimes contradict some commonly accepted principles. For example, it appears that the effect of the maximum size of aggregate on the cement demand is very minor, if a concrete of given placeability and strength is sought. Having corrected the set of general rules, a practical mixture-proportioning process is proposed, trying to make the best use of numerical simulations and laboratory experiments. The aim of this part is to show that the approach proposed allows the formulator to investigate a wide range of possible constituents, to limit both labour expenses and delay, and to capitalize on past experience through the determination of material parameters. A real, 'as lived' example of practical mixture proportioning is presented to illustrate this statement.

Following this, classical questions dealing with mixture proportioning are discussed, divided into those that deal with the aggregate skeleton on the one hand, and those that deal with the binding phase on the other. Finally, before reviewing some existing methods in the light of the developed theories, the problem of how to account for production variability at the mix-design stage is investigated.

4.1 SPECIFYING A CONCRETE FOR A GIVEN APPLICATION

Before designing a material it is important to clarify the nature of the problem with regard to the specific application that is aimed for. Most often the goal is to find a combination of constituents that will give a concrete whose properties comply with certain specifications. This is the basis of what is called sometimes a 'performance-based' approach. In addition to the material properties, there may be also conditions emanating from regulations, or from contracts between the partners. Finally, the goal is to determine the mix that meets all the specifications and has the minimum cost (including not only the materials, but also the production cost).

In this section we shall first review the different types of performance that can be measured in a laboratory, and can be included in a set of specifications. We shall try to give orders of magnitude estimates of desirable (or expectable) properties depending on the applications. Finally, we shall make some recommendations for setting up a list of specifications that is both comprehensive and realistic.

4.1.1 Fresh concrete properties

Slump

Except for some special mixes that are not cast by gravity (such as roller-compacted concrete; see section 5.4.1), a range of slump is generally specified for any concrete.[1] This crude test has the advantage of being used more or less worldwide. Moreover, it is related to the yield stress (cf. section 2.1.1), which is a more fundamental property. However, it reflects only indirectly a most critical aspect of fresh concrete: its placeability (see below). In Table 4.1 the range of specified slumps is given for various applications. Generally speaking, a higher slump means an easier placement but a more expensive concrete for a given strength (see section 4.2.1).

The repeatability of the slump, defined as the standard deviation of a population of slumps performed by the same operator on the same concrete, is about 10–20 mm. This is why the slump value of a laboratory study is generally specified with a margin of ± 20 mm, or ± 30 mm in the fluid concrete range. In the production of high-performance concrete such limits are difficult to comply with, as the slump value is very sensitive to the water content. It may also change drastically with a new delivery of

Table 4.1 Range of slump to be specified for various applications.

Application	Desirable slump (mm)
Precast industry: dry concretes for immediate demoulding	0
Slip-formed concrete for pavements	20–50
Normal-strength concrete for reinforced/prestressed structures	80–150
High-performance concrete cast on site	180–250
Self-compacting concrete	> 250 (sump flow[a] > 600 mm)

[a] Mean diameter of the spread of the sample after the slump test.

constituents (e.g. cement). Moreover, for a fixed composition the slump depends on the type of mixing and on the size of the batch: the bigger the batch, the higher the slump. There is often an increase in slump when a mixture designed in a laboratory is produced in a concrete plant.

For most applications there is a delay between the moment when water and cement are in first contact and the moment when the concrete is placed. During this time the slump may decrease because of water absorption by the aggregate, or because of chemical phenomena related to cement–admixture incompatibility (see sections 2.2.1 and 4.2.3). In such a case a margin has to be accounted for at the mix-design stage. It is even common to specify higher slumps in summer than in winter, as a higher temperature promotes faster slump losses.

Finally, let us stress the fact that the slump is not directly related to the mixture placeability (see section 4.2.2). Therefore in most cases it is the placeability that should be specified and not the slump, which should be used only as an easy-to-use quality control tool.

Yield stress

While rheology tools are still seldom used in industrial applications, it is the author's opinion that specifications will refer increasingly to basic rheological parameters. Hence these parameters may be related quantitatively to various concreting operations, while a technological test such as the slump test gives only a qualitative index, which has to be interpreted in terms of the particular materials used for producing the concrete.

The Bingham parameters can be measured for soft-to-fluid concretes only (Hu *et al.*, 1995). Based on the experience gained in the French network of Ponts-et-Chaussées laboratories, target values of yield stress are given in Table 4.2. As for the slump, the likelihood of workability loss has to be estimated prior to fixing a specified initial value.

Table 4.2 Range of yield stress to be specified for various applications.

Application	Desirable yield stress (Pa)
Normal-strength concrete for reinforced/prestressed structures	800–1500
High-performance concrete cast on site	300–1200
Self-compacting concrete	200–500

Too high a value of yield stress means a lack of workability, but too low a yield stress may create some stability problems. For example, the top surface of bridge decks must generally exhibit a given slope, which can be up to 5% and more if required by the longitudinal profile. The critical yield stress $\tau_0^{crit.}$ below which the concrete will flow can be easily calculated with the basic concepts of soil mechanics:

$$\tau_0^{crit.} = \rho g h \sin \theta \tag{4.1}$$

where ρ is the mass of concrete per unit volume, g is the acceleration due to gravity, h is the height between the concrete surface and the bottom form, and θ is the slope of the surface (Fig. 4.1).

If we refer to the case of Pont de Normandie C60 HPC (Monachon and Gaumy, 1996), where the maximum slope was 6%, assuming $h = 200$ mm and $\rho = 2400$ kg/m^3, $\tau_0^{crit.}$ would have yielded a value of 500 Pa. Given the unavoidable variability of the yield stress during production, and to ensure the stability in the case of limited shocks and vibration, a minimum value of 1000 Pa seems advisable in such a case. Then the formulator would have to control the plastic viscosity, so that the shear stress could remain within an acceptable range during concreting operations.

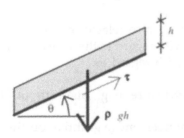

Figure 4.1 Equilibrium of a sloped fresh concrete deck, submitted to gravity. When the shear stress τ equals the yield stress, the concrete flows down.

Plastic viscosity

As already seen in section 2.1.2, this second Bingham parameter may be high as soon as high quantities of superplasticizer are used (as in HPC for example). The plastic viscosity governs the workability of fresh concrete when it is sheared at a high rate, as in a pump (see section below dealing with pumpability).

The plastic viscosity also seems to be related to other aspects of concrete workability as stickiness. Concretes with viscosity[2] higher than 300 Pa.s are difficult to handle, to vibrate, to finish, and to trowel. They often exhibit coarse bubbling at demoulding. We then consider as a good rule of thumb the specification of plastic viscosity in the range 100–200 Pa.s for cast-in-place HPC. However, it should be kept in mind that for a given strength a low plastic viscosity promotes a high paste content, which increases the amount of heat of hydration, shrinkage, and cracking.

Placeability

The concept of placeability, defined in section 2.1.5, is related to the effort needed to compact the mix in a given volume. While the rheological parameters address the flowing properties (intermediate stage between the mixer and the mould), placeability describes the last stage in the casting process (compaction). It is especially useful in the dry concrete range, where the fresh material can no longer be characterized by rheological tools (see section 2.2.1). Recall that placeability is not a directly measurable property[3]. It is a numerical index calculated from the mixture proportions, which can help in designing a mixture that will be compactable in a given container, with a given process.

There is at present little experience in the use of the placeability concept in mixture proportioning. It is therefore difficult to give absolute values that will ensure an easy placement of fresh concrete. However, in the light of calculations performed *a posteriori* about past projects, some orders of magnitude are given in Table 4.3.

The maximum size of aggregate (MSA) is also related to placeability. Most standards give limitations of MSA with regard to the structure to be cast. The MSA should be small enough with regard to:

- the minimum dimension of the structure (e.g. one-fifth of the thickness of a wall);
- the minimum spacing between reinforcement (e.g. three-fourths of this distance);
- the concrete cover (same constraint).

Table 4.3 Orders of magnitude of placeability for several casting processes. The K' value, calculated on de-aired concrete with account taken of the wall effects exerted by the form and the reinforcement, should be less than K^*.

Casting process and type of concrete	K^*
Vibrated, without superplasticizer	6
Vibrated, with superplasticizer	8
Roller compacted	9
Shotcrete (wet process), without superplasticizer	5.5
Shotcrete (wet process), with superplasticizer	7.5
Self-compacting superplasticized concrete	7

Stability

This range of properties cannot be specified in a quantitative manner, as there is no generally accepted method for measuring the stability of fresh concrete. However, the formulator must keep in mind the need for a stable concrete, exhibiting good cohesion, little bleeding, and a limited tendency to segregate. This concern should be especially emphasized in the case of self-compacting concrete (see section 5.4.3). Generally speaking, an even filling diagram must be sought for controlling the stability of fresh concrete (see section 2.1.7). Trying to minimize the segregation potential is a way of ensuring a maximum evenness of the filling diagram. It is also expected that this strategy would lead to better-looking concrete facings.

Air content

Entrapped air is the result of incomplete consolidation, and may range from 0.6 to 2.5% in a concrete without an air-entraining agent. It is undesirable (because it diminishes the mechanical properties of hardened concrete and creates poor concrete facings), but also unavoidable.

However, when the structure is likely to be submitted to freeze–thaw cycles, a certain volume of entrained air is specified. The aim is to incorporate a network of small air bubbles in the concrete so that free water can migrate in these bubbles prior to freezing. The critical physical parameter controlling the system is the mean distance between bubbles, often designated by the symbol $\bar{L}$. According to various researchers (Pigeon and Pleau, 1995), $\bar{L}$ should be below a critical value of 0.20–0.25 mm to ensure good protection against freeze–thaw deterioration. The satisfaction of this criterion leads to the need for a certain volume of total air in concrete. As both *entrapped* air[4] and paste volume increase when the maximum size of aggregate (MSA) decreases, most standards require a minimum air volume related with the MSA (Table 4.4).

The air volume criterion is not fully reliable, because it is the very fine part of the bubbles that provides the most significant protection to concrete (Pigeon and Pleau, 1995), although accounting for only a small part of the total air volume. So for projects where the resistance to freeze–thaw is critical, it is advisable to proceed either with the measurement of $\bar{L}$ on hardened concrete, or with freeze–thaw accelerated tests.

Another remark deals with the specification of air for high-performance concrete. By itself, incorporation of air in concrete is detrimental to concrete mechanical properties, and it can be difficult, if not impossible, to produce an 80 MPa concrete with entrained air. Fortunately, in a number of cases the great watertightness of these materials and the spontaneous self-desiccation provide a natural protection to freeze–thaw, and air-entrainment is no longer necessary (an example is reported in de Larrard *et al.*, 1996a). However, for any actual application this statement has to be checked by relevant accelerated tests.

Finally, at the opposite end of the concrete range, it can sometimes be beneficial to add entrained air in lean concretes, even if the structure environment does not require such a precaution. In this case air bubbles stabilize the fresh concrete, decreasing the amount of bleeding and segregation. This stabilization effect may even counteract the direct effect of entrained air on the voids content, which tends to diminish the mechanical properties. Table 4.5 gives an example of two concretes having the same cement content and the same workability, one without entrained air and one with. In spite of a greater voids/cement ratio, the air-entrained concrete exhibits superior mechanical properties. This tendency is not predicted by the model developed in section 2.3. It probably comes from the fact that, because of excessive bleeding, the aggregate–cement bond is deeply impaired in the first concrete, and the microstructure is flawed.

Specific gravity

When the concrete is produced from normal-weight aggregates, its specific gravity generally varies within a narrow range (from 2.2 to 2.4–2.5). As water is the lightest component of concrete, the specific

Table 4.4 Air content to be specified in a concrete submitted to freeze–thaw, according to European standard ENV 206

Maximum size of aggregate (mm)	8	16	32
Minimum air content (%)	6	5	4

Table 4.5 Effect of incorporating entrained air in a lean concrete.

Mix	Water (l/m³)	AEA (kg/m³)	Cement (kg/m³)	River sand (kg/m³)	CL 0/5 (kg/m³)	CL 5/12.5 (kg/m³)	CL 12.5/20 (kg/m³)	Air (%)	$w+a/c$	Slump (mm)	fc_1 (MPa)	fc_{28} (MPa)	ft_{28} (MPa)	E_{28} (MPa)
M25	193	0	230	446	453	388	619	0.8	0.87	105	5.2	23.6	2.65	32.0
M25EA	160	0.31	230	427	433	454	574	7.2	1.01	110	7.0	28.8	2.84	32.6

AEA, air-entraining agent; CL, crushed limestone; fc, ft and E, compressive strength, tensile splitting strength and elastic modulus respectively.
Source: de Larrard *et al.* (1996b).

gravity for a given source of aggregate is controlled mainly by the water content. Therefore the use of plasticizers or superplasticizers tends to increase the specific gravity of fresh concrete. For a given concrete the specific gravity may evolve slightly from the fresh to the hardened state. When cured in water, concrete generally gains some water, which is absorbed by capillarity as hydration creates internal voids. Conversely, a concrete exposed to an external humidity lower than its internal humidity tends to lose some water. These phenomena may create a variation of specific gravity of less than ±0.1.

The main cases in which the specific gravity of a structural concrete may be specified are concerned with either heavyweight or lightweight concrete. For the latter, the specific gravity may range between 1.5 and 1.9, when a lightweight coarse aggregate (most often expanded clay or expanded shale) is used in combination with normal-weight sand (see section 5.5.1). Intermediate values, between those of lightweight and normal-weight concretes, are simply obtained by mixing the two types of coarse aggregate (Weber and Reinhardt, 1996).

Pumpability

Two phenomena may impair the concrete pumpability in a given installation:

- a separation of the various phases, which creates a blockage;
- a lack of fluidity, which leads to very low pumping rates.

An even filling diagram is a good way to prevent the risk of blockage. The diagram should be calculated in confined conditions (that is, in the pumping pipe), and attention must be paid to the peak that deals with the coarsest aggregate fraction, which should not be higher than the other aggregate peaks. Also, there should be a sufficient amount of fine particles for avoiding a separation between paste and aggregate.

At high strain rate the concrete fluidity is governed mainly by plastic viscosity: the higher the plastic viscosity, the higher will be the pumping resistance (Fig. 4.2), defined as the concrete upstream pressure divided by the flow rate. Here the flow rate is estimated from the number of strokes per minute, for a piston pump.

Therefore the lower the plastic viscosity, the higher the flow rate for a given pressure, or the lower the pressure for a given flow rate. A plastic viscosity lower than 200 Pa.s seems to guarantee easy pumping for moderate pumping distances. In addition to the plastic viscosity requirement, the current state of the art indicates that a slump higher than 50 mm is generally advisable for pumpable concrete (ACI 304). More research is needed to establish more accurate rheological specifications with regard to pumping in a given installation.

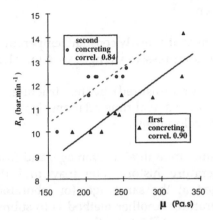

Figure 4.2 Relationship between pumping resistance and plastic viscosity, for two different series of successive batches (de Larrard *et al.*, 1995). R_p (pumping resistance) is defined by the ratio between the maximum pressure in the fresh concrete (in bar) and the number of pump strokes per minute. 1 bar = 0.01 MPa.

4.1.2 Hardening concrete properties

Compressive strength at early age

In most practical cases of mix design, compressive strengths at early age are specified. Table 4.6 gives some figures for common values adopted in various sectors. Here an important distinction must be made between the strength obtained on specimens cured in water at constant temperature and the actual strength attained in the structure, given the temperature history. For estimating the in-situ strength, maturity methods are

Table 4.6 Orders of magnitude for specified compressive strength at early age.

Type of utilization	Purpose	Concrete age (h)	Specified compressive strength (MPa)
Structural concrete for building	Form stripping	14–16	5
	Withdrawal of scaffolding	20–36	10
Prestressed concrete in a bridge built by the cantilever method	Form stripping	14–16	10
	Tension of the first prestressing cables	10–36	15–20

generally used (Carino *et al.*, 1992), the best of which refer to Arrhenius' law. The normal methodology is then

- to measure the strength development at a nearly constant temperature;
- to plot a master curve giving the compressive strength vs. equivalent time (e.g. at 20 °C);
- from the temperature expected (or measured) in the structure, to calculate the in-situ equivalent time vs. real time relationship;
- to deduce the in-situ strength development.

Conversely, from the desired in-situ strength at a given age, and from an estimate of the typical temperature history, one may find the compressive strength to be attained at the same age, for laboratory specimens cured at constant temperature. Another method is to submit laboratory specimens to the temperature history that concrete will experience in the structure, and to measure the obtained strength. This last method is often used in the precast industry.

A rapid strength development after an extended dormant period is one of the most attractive features of HPC. Such concretes may be specified for the strength they exhibit at early age, even if their 28-day strength is in excess of the structural requirements. As an example, it is possible nowadays to design fluid mixtures having an appropriate and stable rheological behaviour 1 hour after mixing, and giving compressive strength in excess of 30 MPa at 16 hours (de Larrard *et al.*, 1996a).

Finally, an early *tensile* strength can be specified in some structures, such as concrete pavements. A typical value for allowing circulation on a young concrete slab is 1.8–2 MPa[5] (splitting strength).

Adiabatic temperature rise

When departing from a mild initial temperature (say around 20 °C), the adiabatic temperature rise may range from 20 to 70 °C, giving final temperatures in the 40–90 °C domain. Temperature rises are of concern in massive structures, which are partially or totally restrained (by their foundations, or by already built parts). Also, the maximum temperature reached during hardening has a detrimental effect on the final in-situ strength of concrete. Furthermore, if the critical temperature level of 60–70 °C is exceeded, some delayed ettringite formation (DEF) may occur later. Then, in a number of cases, it can be worthwhile limiting the adiabatic temperature rise in a structure. While there are processes dealing with construction techniques, such as decreasing the initial temperature of concrete, or cooling the structure during hardening with a network of embedded pipes, the first precaution is to look at the mix design.

When running simulations with the models developed in Chapter 2, it is realized that, *for a given strength*, there is relatively little freedom in changing the adiabatic temperature rise. The use of pozzolans results in minor changes in this respect (Waller *et al.*, 1996), except if high-volume fly ash concrete is being dealt with (see section 5.5.2). The most efficient strategy is to minimize the water content by using superplasticizer and by optimizing the aggregate grading, and/or specifying concrete that is stiffer in the fresh state (such as roller-compacted concrete; see section 5.4.1).

On the other hand, as soon as the concrete is not in pure adiabatic conditions (say, when the thickness of the concrete piece is less than 1 m), a drop is found in the maximum temperature attained. Because of differences in activation energy, this drop is higher when concrete contains pozzolans (see section 2.3). Therefore the *semi-adiabatic* temperature rise can be mastered to a certain extent, essentially by changing the hardening rate of the cement, and by using pozzolans (Waller, 1998).

Autogenous shrinkage

Autogenous shrinkage is another feature of importance in the cracking behaviour of concrete structures at early age. This spontaneous deformation takes place with the progress of cement hydration, and is therefore accelerated by any temperature rise. In massive structures the major part of autogenous shrinkage appears during the first week, and is superimposed on thermal shrinkage when the structure cools down. Compared with drying shrinkage, the amplitude of autogenous shrinkage is generally small (less than 50% of the total shrinkage; see section 2.5.6). However, drying shrinkage does not create restraining effects between several parts of a structure cast at different stages,[6] while autogenous shrinkage does.

Autogenous shrinkage ranges from 0 or so (for high water/cement ratio concretes) up to more than 300×10^{-6} for certain silica-fume HPCs. For the latter it is strong enough to create cracking in restrained pieces, even at constant temperature (see section 5.1). So, when the aim is to limit cracking in a restrained structure made of HPC, it is wise to limit the amount of autogenous shrinkage. This may be done by optimizing the aggregate grading, by a relevant choice of cement (see section 2.5.3), or by using ternary blends of binders (see section 5.3.2).

4.1.3 Hardened concrete properties

Compressive strength

There is no inferior limit on specified compressive strength as, in some applications, concrete may only have a role of filling material. In everyday building construction it is common to ask for a compressive strength of 20–30 MPa. Higher strengths are often required in the prefabrication industry, especially for prestressed structural elements.

For bridges, where the mechanical function is predominant (while acoustic insulation may prevail in some housing constructions), the specified compressive strength at 28 days ranges from 30 to 80 MPa, according to present regulations of various developed countries. Higher values are still exceptional, and can be reached in production only if superior materials, facilities and workmanship are involved.

Since the compressive strength of mature concrete is the main requirement for structural stability, this property must be carefully specified and assessed. A safety margin has to be kept between the design strength and the mean value obtained in the preliminary laboratory study. A reasonable order of magnitude is 10–20%. However, some regulations give precise requirements in this respect, accounting for the maximum drop of cement strength that is likely to occur during production.[7] For a rigorous approach of *target* strength as a function of *design* strength, an assessment of the expectable strength drops in production is desirable (see section 4.5).

Tensile strength

A tensile strength may be specified for some applications where concrete is not reinforced (such as pavement slabs or precast tetrapods), or when a superior resistance to cracking is aimed for. Because it is easy to perform and fairly repeatable, the splitting test is generally preferred to the direct tension test, although it may slightly overestimate the real strength by 5–12% (Neville, 1995). At 28 days the splitting cylinder strength of structural concrete may range from 2 to 6–7 MPa. Values in the range 2.7–3.3 MPa are generally required for road or airport concrete pavements[8]. The splitting strength exhibits little improvement after 28 days. Even some retrogression may happen in the case of silica-fume HPC (de Larrard *et al.*, 1996b). During the mix-design study, the control of concrete tensile strength is essentially performed in the same way as for compressive strength, since these two properties are closely related with each other. However, following the nature of the aggregate, the correlation may change slightly (see section 2.4). Therefore it can be worth selecting an aggregate for the purpose of having a high tensile/compressive strength ratio.

Flexural strength is higher than splitting strength. There is no universal coefficient for deducing the former from the latter. However, the order of magnitude of flexural strength is about 4/3 of the splitting strength, for compressive strengths lower than 50 MPa (Raphaël, 1984; cited by Neville, 1995).

Elastic modulus

The elastic modulus of hardened normal-weight concrete may range from 25 to 55 GPa. A high elastic modulus may be aimed for if structural deflections are to be limited, as for example in some high-rise buildings submitted to wind effects. To increase this parameter, the strategy is to increase the specified strength, to optimize the aggregate grading and, if the previous attempts fail in reaching the desired level, to replace the original aggregate by a stiffer one.

It can sometimes be tempting to limit the elastic modulus to reducing stresses due to restrained deformations. However, when the restrained deformations come from concrete shrinkage (either autogenous or total), it should be kept in mind that every action aiming to decrease the elastic modulus tends to increase the concrete shrinkage (see section 2.5.7), and sometimes to diminish the tensile strength of concrete.

4.1.4 Long-term concrete properties

Shrinkage

The shrinkage addressed in this section is the total free deformation that takes place from the setting to the long term, at constant temperature. There is a single mechanism causing shrinkage,[9] which is the lowering of internal humidity. However, this drop may come either from exchanges with the structure environment, or from self-desiccation (autogenous shrinkage).

Ultimate shrinkage (drying + autogenous shrinkage) of structural concrete at 50% R.H. may range from 300–400 to 800–1000 × 10^{-6} and more. The lowest values are obtained with carefully optimized HPCs (see section 5.3), while the highest amplitudes can be found in concretes in which an excessive water dosage, which would normally bleed out of the material, remains by the action of viscosity agents. At equal strength, the nature of the binder has little influence on concrete shrinkage. Optimization of aggregate grading and limitation of the paste content remain the most efficient ways of reducing total shrinkage; some admixtures that have recently appeared on the market could change this situation in the future.

For humidity higher than 50%, the total shrinkage likely to appear is roughly proportional to the difference between the initial and the final internal humidity (Granger *et al.*, 1997a,b). Therefore, the prediction of the model developed in section 2.5.6 must be multiplied by the factor

$$\frac{100 - RH}{100 - 50} = 2 - \frac{RH}{50}$$

(where *RH* is the external relative humidity in per cent).

Creep

The creep deformation of a given concrete depends on various factors, such as the age of concrete at loading, and the environmental conditions (temperature and humidity). Its development is very slow, and some authors even state that it does not tend towards a finite limit (Bazant *et al.*, 1991–92). As we did in section 2.5, we shall conventionally deal with the creep deformation that develops between loading (at 28 days) and 1000 days. In Table 4.7 the orders of magnitude of creep are given, calculated from the French code BPEL 97 (Le Roy *et al.*, 1996).

The creep deformations can be controlled in the same manner as for the elastic modulus. However, unlike elastic modulus, creep is sensitive to the type of binder used for reaching the desired strength. Silica fume reduces the amount of creep sharply, whatever the humidity conditions.

Durability specified through performance criteria

Apart from freeze–thaw-related properties, concrete durability may be specified through a variety of performance tests. Here is a non-restrictive list of the properties measured:

- permeability to water or gases;
- water absorption;
- diffusion coefficients for various ions, including chlorides;
- electrical resistivity of hardened concrete;
- depth of carbonation after exposure to carbon dioxide;
- account of internal microcracking after various curing regimes;
- alkali–silica reactivity tests;
- abrasion tests, etc.

Ideally, the results of the tests, together with knowledge of the environmental conditions, would allow one to evaluate the length of

Table 4.7 Range of typical creep deformations after 1000 days, for a 150–160 mm diameter cylinder.

Strength grade (MPa)	Basic creep strain (10^{-6}/MPa)	Basic creep coeff.	Total creep strain (at 50% R.H., 10^{-6}/MPa)	Total creep coeff. (at 50% R.H.)
C 20–C40	20–50	0.8–1.3	50–170	2.3–4.5
HPC without SF	15–40	0.8–1.3	40–80	2–3
HPC with SF	8–20	0.4–0.8	13–35	0.7–1.4

the structure service life (Schiessl *et al.*, 1997). In turn, the specification of a given service life would then correspond to target values for the durability-related performances. Today, owing to a lack of sufficient feedback from the field, and given the incomplete knowledge of the degradation mechanisms of concrete, most specifications of this type present the following characteristics:

• They often appear arbitrary, and depend rather significantly on country and regulations.
• The specified property, which is seldom a material property, cannot be predicted from the mix design. Therefore the classical trial-and-error approach prevails.

Durability specified through the prescription of mix-design ratios

Some standards and regulations, such as the European standard ENV 206 or the French standard NFP 18 305, propose a simplified approach to specifying durability, through some mix-design ratios. An equivalent cement dosage is first defined, accounting for the Portland cement content plus the presence of supplementary cementitious materials, weighted by 'k-coefficients'. In this calculation a maximum cementitious addition/Portland cement ratio is allowed. Then the specifications generally include a minimum binder (equivalent cement) content requirement, and a maximum water/binder ratio requirement. The thresholds of these ratios are fixed with regard to the environment to which the structure will be submitted.

A minimum content of fine materials ($d < 75 - 80$ µm) is physically necessary to prevent excessive bleeding and segregation (see section 2.1.5). But the requirement that a majority of this fine content be cement is more, in the author's opinion, a result of the power of the cement lobby than a will for better concrete. Day (1995) demonstrated how the constraint of minimum *cement* dosage tends to overcome the natural incentive for a concrete producer to make a more optimized and economical concrete, with a reduced standard deviation.

The water/(Portland) *cement* ratio could be a good indicator for the depth of carbonation in a given test (Rollet *et al.*, 1992), especially for mixtures containing pozzolans. But the water/binder ratio or water/ equivalent cement ratio are more usual in the code prescriptions. Prescribing a maximum water/binder ratio is more or less equivalent to specifying a minimum compressive strength. However, compressive strength is sensitive to changes in such secondary parameters as cement strength, reactivity of pozzolans, or paste–aggregate bond (which is affected by a lack of cleanness of aggregates). These parameters also influence durability, but are not reflected in the water/binder ratio.

Moreover, the assessment of water/binder ratio requires the knowledge of fresh concrete water content, for which no fully reliable method exists nowadays.

The maximum addition/cement ratio accounted for can be scientifically justified, since too high a proportion of pozzolans leads to a low fraction of remaining portlandite in hardened concrete, which could be detrimental to the prevention of reinforcement corrosion. However, this concern could be more a chimera than a real hazard, except for normal-strength mixtures containing more than 20–25% of silica fume as a proportion of the cement weight (Wiens *et al.*, 1995).

Finally, as suggested by Day (1995), specification of a minimum compressive strength that would depend on the environmental conditions could be a better way to ensure an optimal level of quality and durability. This approach would account for the link between strength and durability (see section 2.6). It could be complemented by a minimum strength requirement at the end of the curing period (most often at form stripping) for avoiding too rapid a carbonation. This performance-based approach would tend eventually to promote responsible behaviour of all partners, without impairing the rational use of available resources.

Durability specified through the selection of constituents

When concrete is to be submitted to aggressive environments, it is usual to define categories of acceptable constituents. For cold climates, aggregates resistant to freezing and thawing have to be selected. In marine environments, only some Portland cements can be used, complying with specifications dealing with their chemical composition (C_3A content, alkalis etc.). This type of specification appears in the national standards; thus they will be detailed no further in this book.

4.1.5 Some rules for setting up a list of specifications

Before proceeding more deeply with the mix-design process, it will be useful to set down a series of seven rules, which should always be considered in practical mix design.

Rule 1: Refer to all relevant standards and regulations

Regulations that apply in a mix-design study can deal either with the concrete produced or with the materials used. For the latter, standards are sometimes implicitly referred to, but they can be mandatory in some countries or specific contexts.

Rule 2: Consult every partner involved in the construction

They can be:

- the material suppliers (Are the chosen constituents the most appropriate for this application? Will they be available for the whole project duration?)
- the concrete producer (Is he able to produce the desired mixture, with a relevant quality-control organization?)
- the transporter (Is the time after which the slump is specified long enough? Does the concrete need to be pumpable? If yes, in which installation?)
- the contractor (Do the requirements dealing with fresh concrete and early age properties fit with its own needs and organization?)
- the site engineer (Is the quality-insurance plan appropriate with regard to the mixture to be used?)
- the design team (Are the mechanical properties well specified with regard to the structural calculations, at all stages of the site schedule?)
- the owner (Are the durability requirements strict enough to ensure a sufficient service-life duration? Is the environment of the structure accounted for in the durability provisions?)
- the architect (Is the concrete likely to produce satisfactory facings, with the required colour and homogeneity?) etc.

Rule 3: Carefully check each step of concrete/structure utilization, in the light of past experience

This item corresponds to the review of the whole list given in sections 4.1.1–4.1.4.

Rule 4: Do not set up unrealistic specifications

If a long time (six months or more) is available for performing the study, a challenging set of specifications can be proposed, and amended in case of failure. But in the opposite case it can be wise to review previous cases, taking advantage of local experience, for specifying reasonable characteristics likely to be attained. On the other hand, it is quite common to see cases where people do not dare to specify high-strength concrete, presumably because of poor local aggregates. Hence, with modern admixtures and supplementary cementitious material, it is now exceptional to fail in producing an HPC formula with strength in excess of 60 MPa at 28 days[10], at least in developed countries.

Rule 5: Do not set up conflicting specifications

Contractors generally ask for early strength, for early form removal, and for acceleration of the site schedule. But this requirement is contradictory with the need for an extended dormant period, which allows more freedom in the site organization, and lightens the pressure on work teams in the case of hot weather concreting. Also, a high early strength often means higher cement dosages, or higher hardening rate of cement, which is a factor for more cracking.

Rule 6: Do not change the specifications during the mix-design study

Such a problem may happen when the structural design and the mix-design study are performed in parallel (while it is of course more appropriate to perform the latter after the former). Then, tests have to be restarted, which results in a more expensive and less focused study. However, let us point out that, with the approach proposed in this book, a wide variety of concretes can be simulated and quickly designed, as soon as the material characteristics have been duly determined (which is the first step of the study).

Rule 7: Start the mix-design study soon enough with respect to the site schedule

In the absence of an accelerated acceptance test for compressive strength, any mix-design study has an 'uncompressible' duration of 5–6 weeks. However, as soon as durability tests are involved, this minimum increases to several months. Creep tests, if scheduled, should last at least six months if a reasonable extrapolation is aimed for.

4.2   SOLUTION OF THE MIX-DESIGN PROBLEM

In this section we shall show how the models developed in Chapter 2 can be used to formulate a concrete composition. Rather than state some arbitrary principles that allow us to find a theoretical mixture composition, we shall demonstrate that a strictly mathematical optimization process leads to the optimum mixture; the only error comes from the limited precision of the models used for predicting the concrete properties.

We shall first use some simplified models to solve the basic problem by analytical means. This will allow us to derive a series of mix-design rules, most of which are well known by competent practitioners, but seldom rationally demonstrated. Some of these rules will be discussed in more detail by using more precise models, with the help of spreadsheet software. Finally, we shall show how it is possible to take maximum advantage of the knowledge contained in the models, together with the experimental data generated by the user from trial batches.

4.2.1   Analytical solution and general relationships

The mixtures considered in this section will be made up with a maximum of six materials:

- a coarse aggregate (volume per unit volume of concrete, CA);
- a fine aggregate (FA);
- a cement (C);
- an inert[11] filler having the same size as the cement (F);
- water (W);
- a plasticizer (volume per unit volume of *cement*, P).

No special hypothesis is made about the nature of the plasticizer, which may have the properties of a *super*plasticizer as well. The volumes of air and plasticizer will be neglected at the batch level (Fig. 4.3).

The following developments are assumed to be valid in the range of all cementitious materials made up with this set of components, having a common consistency (excluding mixtures that are too dry or too fluid).

Simplified models for concrete properties

Any strength model presented in section 2.3.2 means that the first parameter controlling concrete compressive strength is the cement concentration in the matrix, which is most commonly expressed by the water/cement ratio (in mass). As this ratio is proportional to the water/cement ratio in volume, we may assume the following model for

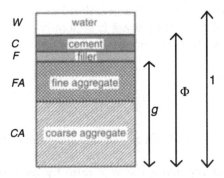

Figure 4.3 Symbols for the composition of a unit volume of concrete.

compressive strength:

$$f_c = \varphi(W/C) \tag{4.2}$$

where φ is a decreasing function of the water/cement ratio. f_c is the compressive strength at a given age.

As for the workability, we shall use one of the two following models. In the paste/aggregate model, the apparent viscosity is given by

$$\eta_a = \Psi_1(P, W/C) \cdot \Psi_2(g/g^*) \tag{4.3}$$

which is a generalized form of equation (2.29). η_a is the apparent viscosity, Ψ_1 is a decreasing function of both variables, and Ψ_2 is an increasing function of g/g^*, with $\Psi_2(0) = 1$. g is the volume of aggregate per unit volume of concrete, and g^* is the packing density of aggregate.

The solid concentration model is of the following form (based on equations (2.27) and (2.28)):

$$\eta_a = \Psi_3(P) \cdot \Psi_4(\Phi/\Phi^*) \tag{4.4}$$

Ψ_3 is a decreasing function of the plasticizer content; Ψ_4 is an increasing function of Φ/Φ^*, where Φ is the total solid volume in a unit volume of concrete; and Φ^* is the packing density of the dry materials (including filler and cement).

Solution of the basic mix-design problem

In this paragraph, we shall assume that no filler or plasticizer is used. The specifications are reduced to a required apparent viscosity η_a and a required strength f_c. We shall, for the general case, use the strength model (equation (4.2)) together with the paste/aggregate model for the viscosity (equation (4.3)). As for the optimization criterion, we shall assume that the most economical mixture is the one that has the least cement content. As the unit cost of cement by mass is generally *about* 10 times that of aggregate fractions, this statement appears reasonably acceptable.

Inverting equation (4.2), we have

$$W/C = \varphi^{-1}(f_c) \tag{4.5}$$

where φ^{-1} is the inverse function of φ, and is a decreasing function of the strength.

By inverting equation (4.3), and replacing the w/c ratio by its value from equation (4.5), we have

$$g = g^* \cdot \Psi_2^{-1} \left\{ \frac{\eta_a}{\Psi_1[0, \varphi^{-1}(f_c)]} \right\} \tag{4.6}$$

The volume balance equation means that the total volume of the batch is equal to unity, so that

$$W + C + FA + CA = W + C + g = 1 \tag{4.7}$$

From this equation, the cement content is calculated as follows:

$$C \cdot (1 + W/C) + g = 1$$

$$\Rightarrow C = \frac{1 - g}{1 + W/C} \tag{4.8}$$

Replacing the aggregate volume and the water/cement ratio by expressions coming from equations (4.5) and (4.6) respectively, we obtain

$$C = \frac{1 - g^* \cdot \Psi_2^{-1} \left\{ \dfrac{\eta_a}{\Psi_1[0, \varphi^{-1}(f_c)]} \right\}}{1 + \varphi^{-1}(f_c)} \tag{4.9}$$

The aggregate packing density, g^*, is controlled by the mutual proportions of coarse and fine aggregates (see section 1.4.2). The general shape of the curve is shown in Fig. 4.4. g^*_{max} is the optimum value of g^*. x is the optimum ratio between coarse and fine aggregate. These two parameters are characteristics of the aggregate, and can be determined either by testing or by using models such as the CPM (described in Chapter 1).

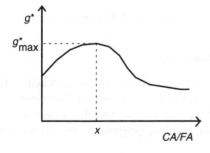

Figure 4.4 Relationship between the aggregate packing density and the coarse/fine aggregate ratio.

Let us now complete the solution of the problem. Since we want to minimize the cement content, it is clear from equation (4.9) that we have to maximize the aggregate packing density, and then to choose x for the coarse-to-fine aggregate ratio. The optimum mix design is therefore given by the following equations. For the cement content, we just replace g^* by its maximum value in equation (4.9), which gives:

$$C = \frac{1 - g^*_{max} \cdot \Psi_2^{-1}\left[\dfrac{\eta_a}{\Psi_1(0, \varphi^{-1}[f_c])}\right]}{1 + \varphi^{-1}(f_c)} \tag{4.10}$$

To deduce the water content, the cement dosage is multiplied by the w/c ratio, which comes from equation (4.5):

$$W = \frac{\varphi^{-1}(f_c)}{[1 + \varphi^{-1}(f_c)]} \cdot \left(1 - g^*_{max} \cdot \Psi_2^{-1}\left[\frac{\eta_a}{\Psi_1(0, \varphi^{-1}[f_c])}\right]\right) \tag{4.11}$$

The aggregate volume, g, is obtained by replacing g^* by g^*_{max} in equation (4.6). Then, as the sum of fine and coarse aggregate equals g, we have for the fine aggregate volume

$$FA = \frac{g^*_{max}}{(1 + x)} \cdot \Psi_2^{-1}\left[\frac{\eta_a}{\Psi_1(0, \varphi^{-1}[f_c])}\right] \tag{4.12}$$

from which we deduce the coarse aggregate volume by a simple multiplication by x (the optimum coarse/fine aggregate ratio):

$$CA = \frac{x \cdot g^*_{max}}{(1 + x)} \cdot \Psi_2^{-1}\left[\frac{\eta_a}{\Psi_1(0, \varphi^{-1}[f_c])}\right] \tag{4.13}$$

Finally, all the specified properties are matched at the minimum cost.

Some rules for concretes without admixture

Let us now enumerate a number of rules that help in understanding the behaviour of the 'concrete system'. The first 10 rules are demonstrated within the frame of the theory developed in the previous section, for concretes made up with a single source of aggregate (in two fractions), cement and water.

Rule 1: The mutual aggregate proportions in an optimum concrete maximize the packing density of the aggregate skeleton. They do not depend on the specified workability and strength

This is a direct consequence of the previous calculations. In particular, this rule is still valid whatever the number of aggregate fractions (which can yield four or five in some applications).

Rule 2: The cost of an optimized concrete is an increasing function of both workability and strength

For demonstrating this rule, let us refer to equation (4.10), which is

$$C = \frac{1 - g^*_{max} \cdot \Psi_2^{-1} \left[\dfrac{\eta_a}{\Psi_1(0, \varphi^{-1}[f_c])} \right]}{1 + \varphi^{-1}(f_c)}$$

When the specified workability increases, the apparent viscosity decreases, and so the term $\Psi_2^{-1}[\eta_a/\Psi_1(0, \varphi^{-1}[f_c])]$ decreases (since Ψ_2^{-1} is a decreasing function, as the inverse of a decreasing function), and then the cement volume increases.

Similarly, when the strength increases, $\varphi^{-1}[f_c]$ decreases (since φ^{-1} is a decreasing function, as the inverse of a decreasing function). Then the denominator in the cement equation decreases. At the same time the term $\Psi_2^{-1}[\eta_a/\Psi_1(0, \varphi^{-1}[f_c])]$ decreases, so the numerator in the cement equation increases. Finally, the cement volume C can only increase. As we assumed that the concrete unit cost was mainly a function of the cement content, the rule is demonstrated.

Rule 3: The optimum mutual aggregate proportions can be found by searching experimentally those that lead to the maximum workability, at constant cement and water contents

Let us recall equation (4.3), which expresses the apparent viscosity of the mix:

$$\eta_a = \Psi_1(P, W/C) \cdot \Psi_2(g/g^*)$$

From any initial composition, if the water and cement contents are kept constant, and if the mutual aggregate proportions are changed, only the term g^* will vary (since $g = 1 - W - C$). Then the workability will be optimum for a minimum value of the apparent viscosity, which corresponds to a maximum value of g^*. This rule is the basis of a popular French method called the Baron–Lesage method (Baron and Lesage, 1976); see section 4.6.4.

Rule 4: For a specified workability and a specified compressive strength, the cheapest concrete will have the lowest possible water content at the fresh state

The cheapest concrete is assumed to have the minimum cement content. As the w/c ratio is determined by the specified strength, the water content will be also minimum.

Rule 5: For a specified workability and a specified compressive strength, the cheapest concrete will have the lowest proneness to segregation (in an infinite volume)

If we assume that a separation between aggregate fractions is the most common type of segregation, we have seen in section 1.5.1 (fact 2) that the aggregate proportions leading to the maximum packing density also minimize the risk of segregation of either the fine or the coarse aggregate.

Rule 6: For a specified workability and a specified compressive strength, the cheapest concrete will have the lowest deformability and heat of hydration

By lowest deformability, we mean that the concrete will have the highest elastic modulus, the lowest creep and the lowest shrinkage.

Again, referring to the paste–aggregate model for workability, as given by equation (4.3):

$$\eta_a = \Psi_1(P, W/C) \cdot \Psi_2(g/g^*)$$

let us point out that the specified strength fixes the w/c ratio. Then, within the range of concretes having the same strength and the same workability, the only free parameters are g and g^*, their ratio being fixed by the previous equation. If we now recall the triple-sphere model, thanks to which we evaluated the deformability properties, it turns out that the *total* volume of the two inner phases is determined, since it is equal to g/g^* (Fig. 4.5).

Moreover, the ratio of the inner matrix volume to the aggregate volume is equal to $(1 - g^*)/g^*$, which is a decreasing function of the aggregate packing density, g^*. Finally, when we want to minimize the cost of the concrete, we are led to maximize g^*, then minimizing the volume of the inner matrix sphere, the rest of the geometry being constant. Thus it is clear that this process tends to minimize the deformability of the triple sphere,[12] which demonstrates the first part of the rule.

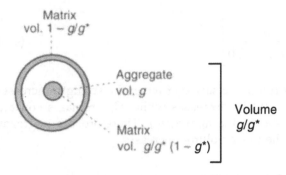

Matrix
vol. $1 - g/g^*$

Aggregate
vol. g

Matrix
vol. $g/g^* (1 - g^*)$

Volume
g/g^*

Figure 4.5 The triple-sphere model for the calculation of deformability properties.

As far as the heat of hydration is concerned, we have seen that at equal w/c ratio the total heat produced during hydration is proportional to the amount of cement. So minimizing the cement content minimizes the heat of hydration.

Rule 7: For a given source of aggregate, the higher the maximum size of aggregate (MSA), the more economical and undeformable the optimum mixes will be at equal specifications

From equation (4.10), the cement volume in the optimum mix appears to be a decreasing function of the optimum aggregate packing density, g^*_{max}. We have seen in Chapter 1 (section 1.4.4) that increasing the grading span of an optimized granular mix tends to increase its optimum packing density. As increasing the MSA produces an augmented grading span, we see that the cement dosage and hence the cost will tend to decrease when the MSA becomes higher. For the last part of the rule, we have seen in the demonstration of the previous rule that increasing g^*_{max} at equal strength and workability tends to decrease the volume of the central phase in the triple-sphere model, thus diminishing the overall deformability of the composite. Therefore, again the rule is demonstrated.

A consequence of this rule is that a mortar will always be more expensive and deformable than a concrete having the same workability and the same strength.

Rule 8: At constant specified workability, an increase of the specified strength leads to an increase of the paste content

Hence from equation (4.6) the aggregate volume in an optimized concrete is

$$g = g^*_{max} \cdot \Psi_2^{-1}\left[\frac{\eta_a}{\Psi_1(0, \varphi^{-1}[f_c])}\right] \tag{4.14}$$

In the demonstration of rule 2 we saw that when the strength increases, the term $\Psi_2^{-1}[\eta_a/\Psi_1(0, \varphi^{-1}[f_c])]$ decreases while g^*_{max} remains constant (since it is a characteristic of the aggregate). Therefore the aggregate volume decreases, and the paste volume increases.

Rule 9: At constant specified workability, the maximum strength is attained with a pure cement paste

Let us first notice that all basic equations apply for pure cement paste, for which $g = 0$. If any concrete with $g > 0$ had a strength higher than that of the cement paste, this would violate the previous rule. So the maximum strength can be attained only with the pure paste.

While this statement is true (see section 2.3.4), remember that the thorough determination of cement paste compressive strength involves a number of experimental precautions (listed in section 2.3.1).

Rule 10: At constant specified workability, an increase of the specified strength leads to a non-monotonic variation of the water demand: it first decreases, reaches a minimum, and then increases

To demonstrate this rule we have to use the second workability model (which we called the solid concentration model). Inverting equation (4.4), we obtain

$$\Phi = \Phi^* \cdot \Psi_4^{-1}\left[\frac{\eta_a}{\Psi_3(P)}\right] \tag{4.15}$$

However, the volume balance gives the following equation (Fig. 4.3):

$$W + \Phi = 1 \tag{4.16}$$

From the two previous equations, we can then calculate the water content:

$$W = 1 - \Phi^* \cdot \Psi_4^{-1}\left[\frac{\eta_a}{\Psi_3(P)}\right] \tag{4.17}$$

From Chapter 1 we know that, if the aggregate has been optimized, the solid material packing density, Φ^*, is controlled by the aggregate/cement ratio. The relationship between these two parameters is analogous to the one that exists between the *aggregate* packing density and the coarse/fine aggregate ratio (Fig. 4.6).

Now, from rules 1 and 8 we know that when the required strength increases, the cement dosage, C, increases, and the aggregate volume, g, decreases. Therefore the aggregate/cement ratio decreases, and the curve in Fig. 4.6 is covered from the right to the left side. So the solid material packing density increases up to a maximum value, g^*_{max}, and then decreases towards the value of the cement. The pattern for water is just opposite to the Φ^* pattern, because of the minus sign in equation (4.17).

Rule 11: At constant specified workability, the water demand is minimum for a given cement dosage, corresponding to a critical aggregate/cement ratio. This ratio does not depend on the workability value

The validity of this statement is visible if one looks at Fig. 4.6.

Rule 12: At constant specified workability, if the specified strength is lower than a given threshold, it is beneficial to add an inert filler to the mix

Introducing a new component in the system, let us rewrite the basic equations. We have two volume balance equations, plus two dealing with the specified properties. The strength equation remains unchanged, since the filler is assumed to be inert. As for the previous rule, the solid concentration model is chosen for describing workability.

$$1 = W + \Phi \tag{4.18}$$

$$\Phi = C + F + g \tag{4.19}$$

$$f_c = \varphi(W/C) \tag{4.20}$$

$$\eta_a = \Psi_3(P) \cdot \Psi_4(\Phi/\Phi^*) \tag{4.21}$$

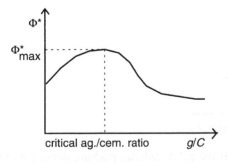

Figure 4.6 Relationship between the solid material packing density and the aggregate/cement ratio.

Combining these equations, the cement dosage becomes

$$C = \frac{1 - \Phi^* \Psi_4^{-1} \left[\dfrac{\eta_a}{\Psi_3(P)} \right]}{\varphi^{-1}(f_c)} \tag{4.22}$$

If the same specifications are maintained, the cement dosage evolves only through the changes in the solid material packing density, Φ^*. This parameter is now controlled by the aggregate/fine ratio, where the fine volume includes both cement and filler (Fig. 4.7).

So, if the initial concrete (without filler) has an aggregate/cement ratio higher than the optimal ratio, the fact of adding a filler into the components allows the solid material packing density, Φ^*, to increase up to its maximum possible value, Φ^*_{max} (Fig. 4.7). From equations (4.22) and (4.18) it is clear that this process tends to diminish both the cement and water contents. The final concrete is therefore more economical than the initial concrete (if the unit cost of filler is neglected). Moreover, since the solid material has the highest possible packing density, the likelihood of segregation is minimized.

Additional rules applying when a plasticizer is added

Rule 13: All rules demonstrated for non-plasticized mixtures also apply when a plasticizer is added at a fixed percentage of the cement content. In particular, the optimum mutual aggregate proportions remain the same

Hence it suffices to remark that all demonstrations of rules 1–12 remain valid with a non-nil value for P. In particular, since the plasticizer/cement ratio is constant, it is still acceptable to minimize the cement content to obtain the most economical mix.

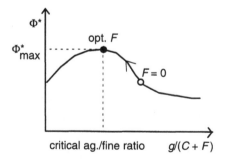

Figure 4.7 Relationship between the solid material packing density and the aggregate/fine ratio.

Rule 14: The addition of plasticizer reduces both water and cement demands at constant specified workability and strength

From the previous developments (using the paste/aggregate model for workability), we have

$$C = \frac{1 - g^*_{max} \cdot \Psi_2^{-1} \left[\dfrac{\eta_a}{\Psi_1(P, \varphi^{-1}[f_c])} \right]}{1 + \varphi^{-1}(f_c)} \tag{4.23}$$

$$W = \frac{\varphi^{-1}(f_c)}{[1 + \varphi^{-1}(f_c)]} \cdot \left(1 - g^*_{max} \cdot \Psi_2^{-1} \left[\frac{\eta_a}{\Psi_1(P, \varphi^{-1}[f_c])} \right] \right) \tag{4.24}$$

Ψ_1 is a decreasing function of P. However, let us remember that Ψ_2 and Ψ_2^{-1} are increasing functions. Finally, when a plasticizer is added into the system, the effect on the various terms is as follows:

$P \uparrow$

$$\Rightarrow \Psi_1(P, \varphi^{-1}[f_c]) \downarrow$$

$$\Rightarrow \frac{\eta_a}{\Psi_1(P, \varphi^{-1}[f_c])} \uparrow$$

$$\Rightarrow \Psi_2^{-1} \left[\frac{\eta_a}{\Psi_1(P, \varphi^{-1}[f_c])} \right] \uparrow$$

$$\Rightarrow - g^*_{max} \cdot \Psi_2^{-1} \left[\frac{\eta_a}{\Psi_1(P, \varphi^{-1}[f_c])} \right] \downarrow$$

$$\Rightarrow C \downarrow, W \downarrow$$

Note that the cost of concrete decreases only if the cement reduction is not overcome by the additional cost of the plasticizer.

Rule 15: The addition of plasticizer reduces the heat of hydration and the deformability at constant specified workability and strength

The effect on heat of hydration is fairly obvious, as the cement content is reduced at constant w/c ratio. With regard to the concrete deformability, let us refer to Fig. 4.5. The fact of adding a plasticizer tends to increase the aggregate volume, g, at constant g^*. So the external layer decreases, the aggregate phase increases, and the inner matrix phase increases in proportion to the aggregate phase. This means that the deformability of the two inner-sphere combinations remains constant. Therefore, if the system is analysed as the combination of a composite phase (the two inner layers) and a matrix phase (the external layer), the proportion of the matrix phase is decreasing while the specific deformability of the two phases remains constant. It is then clear that the deformability of the whole system decreases.

Rule 16: The addition of plasticizer leads to a lower critical aggregate/cement factor

The critical aggregate/cement factor corresponds to the amount of cement that best fills the aggregate voids. As seen in section 2.1, the plasticizer increases the packing density of the cement, thus making it possible to insert more cement between aggregate grains. Therefore the aggregate/cement ratio is lower.

Rule 17: The strength below which it is beneficial to add an inert filler is higher with plasticizer than without

This threshold is attained when the solid material packing density is equal to its maximum value, Φ^*_{max}, for an aggregate/cement ratio equal to its critical value. It is then easy to calculate the w/c ratio:

$$W = 1 - \Phi^*_{max}$$

$$C + g = \Phi^*_{max}$$

$$\Rightarrow C = \frac{\Phi^*_{max}}{1 + g/C}$$

Then

$$W/C = \frac{(1 - \Phi^*_{max})}{\Phi^*_{max}} (1 + g/C) \qquad (4.25)$$

We have seen in the demonstration of the previous rule that an addition of plasticizer tends to increase Φ^*_{max} and to decrease g/C. It is then clear from the previous equation that the water/cement ratio can only diminish, and the strength increases at the critical aggregate/cement ratio.

A practical consequence of this rule is the following. If it is desired to use a plasticizer at a significant dosage in the range of low-strength concretes, it will also be interesting to combine the plasticizer with an addition of a supplementary cementitious material (see section 4.4).

4.2.2 Numerical solution: discussion of the previous relationships

In this section, some of the rules demonstrated in section 4.2.1 are discussed more deeply, in the light of the refined models developed in Chapter 2. The other rules, not referred to in this section, have not been found invalid in the light of the author's practical experience, nor from the results of numerous numerical simulations.

All the simulations presented in this section have been performed with the following reference set of constituents:

- a crushed limestone coarse aggregate (two fractions having extreme sizes of 5/12.5 and 12.5/20 mm);
- a river sand (0/4 mm);
- a correcting sand (0/0.4 mm). A value of 50 FF/t[13] has been taken for the unit cost of all aggregate fractions;
- a CEM I 52.5 Portland cement with a compressive strength at 28 days of 64 MPa and a moderate aluminate content of 5% (unit cost: 500 FF/t);
- water (unit cost: 5 FF/t).

In some mixtures, the following additional constituents were used:

- a limestone filler having a specific surface of 400 m^2/kg (unit cost: 100 FF/t);
- a superplasticizer of the melamine type (unit cost per kg of dry extract: 20 FF);
- a silica fume delivered in the slurry form (unit cost: 1500 FF/t).

To calculate the price of a unit volume of concrete an overhead cost of 200 FF/m^3 has been taken, irrespective of the type of mixture. Otherwise, let us note that in the range of normal-strength concretes the coarse aggregate has better strength characteristics (in terms of the coefficients p and q; see section 2.3.4) than the sand from the Seine. This is not a general feature, but only a peculiarity of the materials chosen in the simulations.

Now, let us recall the first rule.

Rule 1: The mutual aggregate proportions in an optimum concrete maximize the packing density of the aggregate skeleton. They do not depend on the specified workability and strength

The question is to know whether there exists one single optimum grading curve for a set of aggregate fractions, as has been demonstrated in the light of the simplified models for workability and strength. To investigate this problem, a series of simulations has been performed with the full set of refined models. Constant cement and free water contents have been adopted, and the coarse/fine aggregate ratio has been changed from 1.0 to 3.5. Obtained slump varied from 0 to 16 cm. The results of the simulations are given in Fig. 4.8.

These findings conflict with Lyse's rule, as very different slump values are obtained with a constant water content (see section 2.1.8). Clearly, this rule is valid only in a limited range of aggregate proportions.

The *CA/FA* ratio giving the maximum aggregate packing density is about 1.75. It also gives the maximum placeability of the concrete mixture. However, the lowest plastic viscosity is attained for a value of 2, while the lowest yield stress requires an even higher value (2.5). As for the plastic viscosity, the shift of optimum *CA/FA* ratio comes from the fact that it is the 'squeezed concrete' that is considered in the calculation of the concrete packing density (Φ^* in equation (2.7)). Since this system is more compact than the de-aired concrete, the amount of fine material needed to fill the voids of the coarse materials is lower (see section 1.4.2). This is why less sand is required to optimize the plastic viscosity than for the placeability. Concerning yield stress and slump, it has been shown that the contribution of granular fractions increases when the particle size decreases, as described by equation (2.13). So, when the minimum yield stress is sought, the sand content is less than in the case of the other fresh concrete properties.

Air content does not display any minimum in the range of *CA/FA* ratios investigated. Finally, although water/cement ratio is constant, compressive strength increases markedly with the amount of coarse aggregate. In the present case, there are two main causes: first, the decrease of air content; and, second, the better intrinsic quality of the crushed limestone,[14] as compared with the parent rock of the river sand.

From these simulations, one may conclude that there is no optimum size distribution, if the preferred fresh concrete property is not claimed. An optimum placeability will lead to more 'sandy' mixtures, and optimum slump will promote a higher amount of coarse aggregate. If only compressive strength and slump are considered, the result will be a mixture with a very high coarse/fine aggregate ratio. But this mixture will exhibit a poor placeability (Johnston, 1990) and in some cases[15] a high proneness to segregation. Conversely, a mixture having a minimum placeability in addition to a given combination of slump and compressive

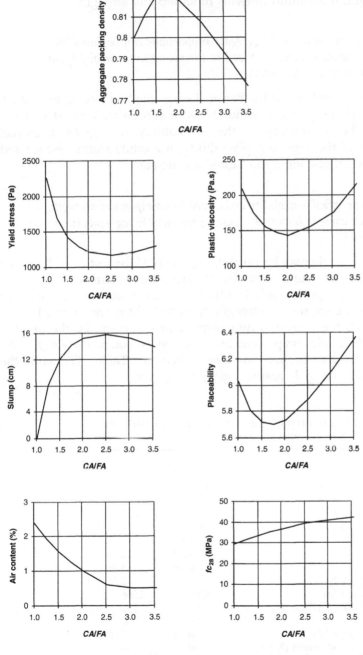

Figure 4.8 Effect of coarse/fine aggregate ratio in a concrete with a cement content of 350 kg/m^3, and a water/cement ratio of 0.56 (simulations).

strength can be more costly than the previous mixture, which gives the strict economical optimum in terms of slump and strength.

Rule 3: The optimum mutual aggregate proportions can be found by seeking experimentally those that lead to the maximum workability, at constant cement and water contents

This strategy is the basis of the Baron–Lesage method (see section 4.6.4). In the light of the previous simulations, it can be anticipated that the result will be satisfactory if the workability test gives a correct combination of rheology and placeability. If a crude slump test is used there is a risk of obtaining too 'stony' a concrete.

Rule 5: For a specified workability and a specified compressive strength, the cheapest concrete will have the lowest proneness to segregate (in an infinite volume)

Here again this rule may be invalid, especially if the test considered for workability is the slump test. To illustrate this statement, additional simulations were performed. The basic specifications were a slump of 100 mm and a compressive strength of 30 MPa. Mixtures A and B were formulated on the basis of a minimum cement content. Furthermore, a constraint of a minimum placeability was imposed on mixture B ($K' \leqslant 6$). Obtained mixture proportions are given in Table 4.8, and the corresponding filling diagrams appear in Fig. 4.9.

Table 4.8 Mixture proportions and properties of mixtures A and B (simulations).

	Mixtures	
	A	B
Cement (kg/m^3)	279	316
Water (l/m^3)	186	189
w/c	0.66	0.60
CA/FA	2.73	1.59
Air (%)	0.50	1.49
Specific gravity	2.40	2.36
Yield stress (Pa)	1569	1566
Plastic viscosity (Pa.s)	302	255
Slump (mm)	100	100
Placeability (= K')	6.75	6.00
Compressive strength (MPa)	30	30
Elastic modulus (GPa)	36	33
Adiabatic temperature rise (°C)	44	49
Total shrinkage (10^{-6})	511	638

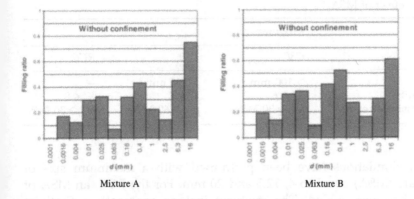

Figure 4.9 Filling diagrams of mixtures A and B. For mixture A, the cement peaks are low (which means a risk of bleeding) and the coarse aggregate peak is very high (which means that there is a risk of segregation and poor concrete facings).

The placeability constraint leads to a significant increase of cement content. Also, the elastic modulus of the concrete is lower, and the dimensional stability (as expressed by adiabatic temperature rise and shrinkage) is less satisfactory. But the fresh concrete properties are better: the placeability is improved, the plastic viscosity is decreased, and the likelihood of bleeding and segregation is minimized (Fig. 4.9). This is an example of the conflict that often appears between fresh concrete and hardened concrete properties at the mix-design stage. However, the formulator has to bear in mind that if he does not pay attention to fresh concrete, the mechanical and durability-related performance of in-situ concrete will be far below that measured in the laboratory.

Finally, it appears that optimization with regard to slump may produce mixtures that are too rich in coarse aggregate. It is therefore essential to look at the filling diagram, and to correct the grading by specifying a minimum placeability.

Rule 7: For a given source of aggregate, the higher the maximum size of aggregate (MSA), the more economical and undeformable the optimum mixes will be at equal specifications

For checking this rule, three types of concrete have been simulated: a normal-strength concrete (C30), a high-performance concrete (C60), and a roller-compacted concrete (RCC). The chosen specifications are given in Table 4.9. In these examples all aggregate fractions are supposed to come from the same quarry (crushed limestone aggregate) in order to avoid a bias due to differences in their mechanical characteristics. For each type

Table 4.9 Specifications applied to the simulated mixtures for investigating the effect of MSA.

	Mixtures		
	C30	C60	RCC
Slump	$\geqslant 100$ mm	$\geqslant 200$ mm	–
Placeability	$K' \leqslant 6$	$K' \leqslant 7$	$K' \leqslant 8$
Compressive strength	$fc_{28} \geqslant 30$ MPa	$fc_{28} \geqslant 60$ MPa	$fc_{90} \geqslant 30$ MPa

of mix, simulations have been performed with a maximum size of aggregate (MSA) equal to 4, 12.5 and 20 mm. For the RCC, an MSA of 40 mm has been added. The mixtures include aggregate, a Portland cement, and water. Superplasticizer is added in the C60 series, while RCC contains a fly ash in addition to the cement. Optimizations have been carried out on the basis of compliance with the specifications at a minimum cost.

In Fig. 4.10, the influence of the MSA on various properties is illustrated. As a general remark, it can be seen that the most important effect occurs when the MSA increases from 4 to 12.5 mm. Subsequent improvements are only marginal. The main reason is the following: in the simplified approach given in section 4.1, the compressive strength is supposed to be controlled only by the water/cement ratio. So any increase of the grading span provokes a decrease of the water demand, which gives a decrease of cement content at constant specified strength. But, in reality, two more factors influence the compressive strength: the air content, which is high for mortars (because of their high sand content), and the maximum paste thickness (MPT; see section 2.3.3), which is partially controlled by the MSA. As a result, the increase of MPT between 4 and 12.5 mm is overcome by the decrease of entrapped air, and there is a clear benefit as far as the cement content is concerned. But, with further increases of MSA, the drop in water demand and the increase of MPT balance each other, and the cement consumption is nearly constant.

Technically speaking, in most practical concretes there is a benefit in increasing the MSA above 10–15 mm only if the coarsest aggregate fraction has better properties[16] than the adjacent one. Hence economic considerations promote the use of all available fractions, which is a reason for having MSA up to 20–25 mm. Also, when crushed aggregate is used, coarser fractions induce a lower energy consumption. Finally, when the active constraint in the mix design is not the compressive strength but the water/cement ratio (see section 4.1), the simplified approach still holds, and a higher MSA leads to a reduction of the binder content. Similarly, if the placeability is not considered and slump is the

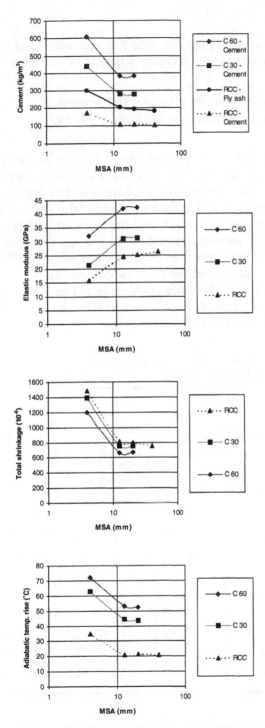

Figure 4.10 Effect of the maximum size of aggregate on the properties of optimized mixtures (simulations).

only fresh concrete criterion accounted for, rule 7 is still valid. But the placeability of optimum mixtures will decrease when MSA increases.

Rule 12: At constant specified workability, if the specified strength is lower than a given threshold, it is beneficial to add an inert filler to the mix

To validate this statement, two series of simulations have been carried out, the specifications for which are given in Table 4.10. Each concrete was optimized to meet the specifications at minimum cost. The materials used were the same as those indicated at the beginning of this section. In the first series no superplasticizer was used, and the price ratio between the filler and the cement was assumed to be 0.30. In the second series a superplasticizer was added to the mixtures (at a constant rate of 0.3% of the cement weight in dry extract), and the price ratio was 0.12.

In Fig. 4.11 it can be seen that the filler content decreases when the required strength increases. However, there is no threshold above which there is no interest in adding a filler. This comes from the fact that the

Table 4.10 Specifications applied to the simulated mixtures for investigating the effect of MSA.

	Mixtures	
	Series 1	Series 2
Slump	$\geqslant 100$ mm	$\geqslant 200$ mm
Placeability	$K' \leqslant 6$	$K' \leqslant 6.5$
Compressive strength	20–50 MPa	20–60 MPa

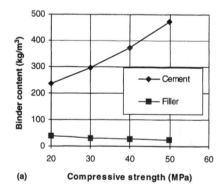

(a)

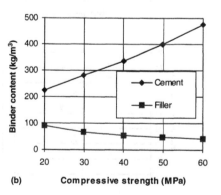

(b)

Figure 4.11 Relationship between specified compressive strength and binder contents, for the two series (simulations): (a) without superplasticizer; (b) with 0.3% of superplasticizer.

marginal effect of the filler is especially significant at low dosage (see section 2.3.7). However, not only is the filler not inert; it is also seldom free of charge. This is why the optimum dosage depends on the unit cost of the product, as compared with that of cement. The sand used in the computations was assumed to have a negligible fines content. Following the simulations, it is clear that the presence of some 10% of fine limestone particles in a fine aggregate can be beneficial for most common concrete applications.

4.2.3 Practical mixture-proportioning process

After some analytical and numerical investigations intended to increase familiarity with the concrete system, it is time to present the general approach that comes logically from the models developed in Chapter 2. The mixture-proportioning process requires the use of a computer, equipped with a spreadsheet package. It is the combined use of simulations and laboratory experiments that allows the best and the quickest optimization.

Choice of components

After having written a series of specifications, the formulator has to select the constituents. This selection is carried out with regard to the following issues:

- *Local availability.* This criterion is critical for aggregates and water, and applies to a lesser extent to the binders, while organic admixtures can normally be shipped from (almost) any part of the world at low additional cost.
- *Technical suitability.* As an example, if a high-performance concrete (HPC) is to be designed, a superplasticizer is mandatory (not an 'old-fashioned' plasticizer). Likewise, if a compressive strength greater than 80 MPa is required, a good mineral admixture (generally a silica fume) will be helpful. As far as durability is concerned, a sulphate-resistant cement should be chosen if the concrete is intended for a marine environment.
- *Technical compatibility.* This criterion is especially relevant for the cement–superplasticizer couple. In HPC mix design, it is best to carry out preliminary tests to select a suitable combination of products (see sections 3.4.2 and 5.5). Another example in which compatibility between constituents is critical is the prevention of alkali–silica reaction. A potentially reactive aggregate can be used, provided that the amount of alkalis brought by the binders[17] is low. Conversely, a cement which a high alkali content should be used only with a non-reactive aggregate.

- *Compliance with standards.* If the standards are satisfactory, the technical suitability and compatibility should normally ensure compliance with standards. However, risks affecting durability may have been forgotten in the technical examination, or some standards may not have been updated to reflect recent advances, so that attention to standards may lead to rejection of some constituents that are technically attractive.
- *Economy.* This aspect should be examined last, but clearly is of great importance, as concrete remains basically a low-cost material that is used in vast quantities.

Let us stress that one of the greatest advantages of the approach presented in this book is the possibility of considering alternative choices of materials. Various types and origins of constituents can be envisaged in parallel, the selection being done on the basis of simulations – if the differences between the results are significant – or by a limited number of laboratory tests.

Supplying the software with component data

The models developed in Chapter 2 and summarized in the Appendix have to be implemented in software. A convenient way is to use a spreadsheet package, having an optimization function.

Chapter 3 gives the details and experimental procedures for obtaining the technical data necessary to run the software. Depending on the properties dealt with in the specifications, only some parameters have to be determined. In any case, the size distribution and packing density of the granular materials must be measured, together with the cement strength.

Running simulations to find the theoretical solution

Once the models have been implemented with relevant values of the various adjustable parameters, a solution should be sought that complies with the specifications, at minimum cost. A first possibility is to perform a series of simulations, by 'manually' changing the various mix-design ratios up to the point where all the specifications are fulfilled. This process takes some time, but gives the user a good understanding of the behaviour of the concrete system. Another solution is to let the machine find a numerical optimum *automatically*. Since such an optimum is sought by an iteration technique, without a special mathematical analysis, there is no assurance that it is *the* real (theoretical) optimum. In principle, it could be only a *local* optimum. However, in the author's experience, when the problem is correctly formulated the solver generally finds a solution that cannot be significantly improved by further 'manual' simulations.

When a theoretical mixture has been obtained, the filling diagram has to be examined. If a risk of excessive bleeding/segregation appears, the mix has to be modified in order to obtain a more even filling diagram (see section 2.1.7). Sometimes a composition with an excess of coarse aggregate is suggested by the computer. This comes from the fact that the air content is an increasing function of the sand content. Therefore, when the curve linking coarse/fine aggregate ratio (CA/FA) and workability is rather flat in the area of optimum value, the computer will tend systematically to suggest a high CA/FA value, reducing the air content and the cement demand, for a given required strength. In such a case, the concrete is too 'stony'. Moreover, in the case of changes of CA/FA during the production process, the effect on workability will be significant.

To avoid this problem, the optimization process can be performed in three steps:

1   a series of 'manual' simulations, for evaluation of the cement demand;
2   the search for the maximum placeability, at constant paste content and composition;
3   final optimization of the paste (cement content, water/binder ratio) to match the specifications.

In this way the mixture will exhibit a minimum segregation risk (see section 2.1.7) and a maximum stability in the industrial process (section 4.5.1).

Making trial batches and re-optimizing the mixture

Let us stress a very important reservation dealing with the approach proposed in this book: **any specified property of the mixture must be measured on a trial batch, to make sure that the specifications are fulfilled.** While the models have been checked as much as possible, the validation is most often of a statistical nature. This means that a small proportion of mixtures may deviate from the models, as a result of their (at least partially) empirical nature. It is therefore essential to check that the mixture that is going to be produced at a large scale for a real application does not pertain to the 'outliers' category. The only specified property that cannot be easily verified experimentally in the laboratory is the placeability. The K^* value used in the optimization process has to be modified if the placeability *in the structure* is not satisfactory (for example if the concrete facings display coarse bubbling or honeycombing when the form is removed).

Once a trial batch has been produced in the laboratory, the tests dealing with the various specifications are performed. Therefore, two possibilities may occur:

- The concrete matches the specifications, but then a preliminary field test[18] has to follow, before production can be started.
- At least one property is not satisfactory, so the mixture has to be adjusted. Software has to be recalibrated in order to predict the property as measured, and the mixture proportions are changed so that the new predictions fit with the specifications. In the recalibration process a change has to be performed in the material parameters that are the most likely cause of the error in the system.

Final adjustment of the formula

When only one property is not satisfactory, the formulator may use a direct technique to adjust the formula without changing the other properties. Let us assume that a mixture has to meet specifications dealing with yield stress (or slump), plastic viscosity and compressive strength:

- If the yield stress has to be changed, the amount of superplasticizer can be modified at constant water content. There will be very little effect on the other two properties.
- If the plastic viscosity is not correct, the paste volume has to be adjusted (at constant water/cement ratio), and the superplasticizer content re-dosed to maintain a constant slump (or yield stress).
- Finally, if the rheological properties are correct but the strength is inadequate, the binder amount must be changed at constant water ratio. Following Lyse's rule, the effect on fresh concrete properties will be minor.

Checking the 'yield'

In any case, the specific gravity of the mixture has to be measured, for the sake of the real composition of one unit volume of concrete. If the theoretical value γ is not matched by the measurement γ', while the specific gravity of all constituents and the water absorption of aggregate have been duly determined, then the air content is responsible for this deviation. Here the correct composition will be obtained by multiplying all the individual masses by the ratio γ'/γ. At this last stage the mutual proportions of constituents are unchanged, but the correct contents per unit volume are calculated, which will form a sound basis for economical evaluation.

4.2.4   Example

In this section a real example of the use of the method is given in detail.[19] After the presentation of specifications, measurements dealing with the constituents are given. Then the numerical optimization leads to two formulae that are adjusted on the basis of tests at the fresh state. At 28 days, compressive strength measurements allow the formulator to determine the formula that complies exactly with the specifications. Emphasis will be put on the efficiency of the overall approach in terms of the number of tests and the time necessary to arrive at an optimum composition.

Specifications

A structural concrete is to be designed, with the following specifications: slump between 100 and 150 mm ('very plastic' consistency), good placeability ($K' \leqslant 7$), and a design compressive strength at 28 days of 45 MPa. The requirements dealing with durability, as given by the French standard NFP 18 305, appear in Table 4.11. These requirements are very close to those given by European standard ENV 206.

We shall show that because of the uncertainty in the material parameters and the precision of dosages in the industrial process, the target strength in the mix-design study is 54 MPa. Given the delay between trial batches and compressive strength measurements, it is decided to design two concretes with target compressive strength of ±5% compared with the goal. This strategy allows the formulator to derive the optimal mixture by interpolation from the results of the two trial batches. Therefore the aim is to design two concretes, B1 and B2, having target compressive strengths at 28 days of 51.3 and 56.7 MPa respectively.

Note that in this example the goal is to design a formula that is assumed to remain constant during the whole construction period. This is why the strength margin is so large. One could imagine using the system of equations in order to adjust the recipe continuously in production, on the basis of tests at early age. This would lead to a more economical and consistent concrete (in terms of properties) (Day, 1995). However, many specifiers, such as the French Ministry of Transportation, still do not allow this approach.

Table 4.11 Durability requirements for a prestressed concrete placed in a humid environment, with a moderate probability of freeze–thaw, according to NFP 18 305.

Class of environment	2b$_2$
Minimum cement content	300 kg/m^3
Maximum water/cement ratio	0.55
Minimum compressive strength	30 MPa

Material parameters

The cement is a Portland cement CEM I 52.5, the characteristics of which are given in Table 4.12 and Fig. 4.12. It is proposed to use a superplasticizer of the melamine type, which has in the past shown a good compatibility with this cement.

The aggregate is a rounded gravel available in three fractions, complemented with a correcting fine sand. The characteristics of the various fractions appear in Table 4.13 and Fig. 4.13. The compressive and tensile strength parameters (p, q, and k_t) have been taken equal to those of another aggregate of similar type (a river gravel). In a past concrete study

Table 4.12 Characteristics of the Portland cement/superplasticizer couple.

Specific gravity of cement (section 3.2.1)	3.168
Compressive strength at 28 days[a] (section 3.2.5)	66 MPa
Minimum strength in production (claimed by the manufacturer)	60 MPa
Specific gravity of superplasticizer (section 3.4.1)	1.2
Solid content in mass (section 3.4.1)	30%
Saturation dose in combination with the cement (section 3.4.2)	0.8% (dry extract)
Water demand of the cement without SP (section 3.2.3)	$w/c = 0.228$
Residual packing density of the cement without SP (section 3.2.3)	0.4869
Water demand of the cement with SP[b] (section 3.2.3)	$w/c = 0.199$
Residual packing density of the cement with SP[b] (section 3.2.3)	0.5188

[a] For a previous delivery of the same cement
[b] At the saturation point.

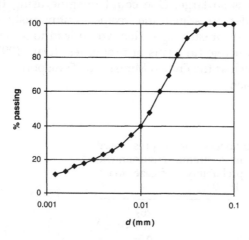

Figure 4.12 Grading curve of the cement, determined with a LASER granulometer.

Table 4.13 Characteristics of aggregate fractions.

	Aggregate fractions			
	8/16	3/8	0/3	0/1
Specific gravity in dry conditions (section 3.1.1)	2.540	2.546	2.561	2.536
Water absorption (section 3.1.2)	1.1%	1.1%	0.8%	0.1%
Packing density ($K = 9$) (section 3.1.4)	0.6304	0.6547	0.08/0.315: 0.6700 0.315/1.25: 0.6582 1.25/4: 0.6490	0.08/0.315: 0.6216 0.315/1.25: 0.6685
Residual packing density[a] (section 3.1.4)	0.6737	0.6696	0.08/0.315: 0.6668 0.315/1.25: 0.6622 1.25/4: 0.6728	0.08/0.315: 0.6108 0.315/1.25: 0.6803
Bond coefficient p (section 2.3.4)			1.08	
Ceiling effect coefficient q (section 2.3.4)			0.0058 MPa^{-1}	
Tensile strength coefficient k_t (section 2.4)			0.44	

[a] Calculated with account of the wall effect given by the mould.

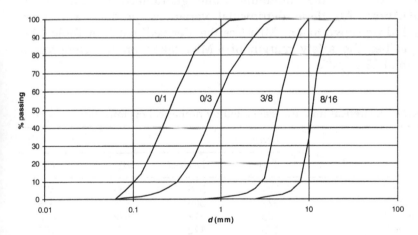

Figure 4.13 Grading curves of aggregate fractions.

involving the same materials (Sedran, 1997) these values were found to give fair predictions of the experimental compressive strengths.

Numerical optimization

After having fed the computer with the material data, a first optimization is performed by looking for the formula that gives the minimum cost, while matching the specifications (compressive strength equal to 54 MPa). The minimum slump with regard to the plastic concrete range is sought, since large concrete plant batches generally exhibit a softer consistency than small laboratory batches. The 'target' composition found by the computer appears in Table 4.14. Note that the computer selected a mixture with a small amount of superplasticizer. Hence the marginal rheological effect of such a dose is so large that it is most often interesting to use it, except when the constraint of minimum cement content is already active in the absence of superplasticizer. On the other hand, the saturation dosage is seldom the optimum dosage as far as economy is concerned, because the marginal effect of superplasticizer decreases with its dose. However, the prevention of rapid slump loss is often a more important issue than the strict cost aspect. At this stage the dose of superplasticizer is maintained, but the need to check the rheological stability is kept in mind.

In order to check the robustness of this theoretical composition a simulation is performed based on the possible variations during production, as given in Table 4.21. This corresponds to the worst case, in which the effects of all possible errors in weighting are cumulated. Moreover, it is assumed that a strength drop is experienced with the cement, down to the minimum value guaranteed by the cement manufacturer. All these detrimental variations produce a concrete strength drop down to 45.1 MPa (see the 'derived' batch in Table 4.14). This justifies the safety margin of 9 MPa taken on the compressive strength at 28 days[20].

Table 4.14 Theoretical mixtures generated by numerical optimization.

	Batches			
	Target	Derived	B1	B2
8/16 (kg/m^3)	584	578	589	577
3/8 (kg/m^3)	372	369	375	368
0/3 (kg/m^3)	623	617	628	616
0/1 (kg/m^3)	173	172	175	172
Cement (kg/m^3)	403	391	384	424
Superplasticizer (kg/m^3)	1.90	1.76	1.84	1.97
Water (l/m^3)	179.2	189.5	178.9	180.0
w/c	0.409	0.448	0.427	0.391
CA/FA	1.20	1.20	1.20	1.20
Specific gravity	2.335	2.317	2.331	2.337
Slump (cm)	10.0	16.5	10.0	10.0
fc_{28} (MPa)	54.0	45.1	51.3	56.7
ft_{28} (MPa)	4.3	3.9	4.2	4.4
Cost (FF/m^3)	417.42	410.38	407.63	428.44
K'	7.00	6.46	7.00	7.00

Then, as explained in the introduction, two concretes are designed, corresponding to a compressive strength below and above the target strength (B1 and B2). Note that both mixtures comply with all specifications, including minimum cement content and maximum water/cement ratio.

Trial batches

Ten litres of B1 concrete are produced in a small mixer, giving a slump of 150 mm (instead of 100 mm). The difference between the theoretical and experimental slump is of the same order of magnitude as the model precision (see section 2.1.4). Therefore a direct experimental adjustment is performed. As the aim is to reduce the slump while keeping the compressive strength unchanged, it is decided to reduce the super-plasticizer dosage. Halving the initial dosage leads to a slump of 95 mm, which is considered satisfactory. Based on the experience of the B1 mixture, B2 is then produced in the same quantity, with a reduced dose of superplasticizer. The measured slump is 90 mm. At this stage, three small batches have been sufficient to adjust two mixtures matching the required fresh concrete properties.

Larger batches are then prepared (70 l), in order to perform the whole series of tests (given in Table 4.15). Because of a scale effect, the slumps

Table 4.15 Results of tests performed on 70 l batches of concretes. The mixture compositions have been corrected by accounting for the experimental specific gravity. The numbers in italics are predictions from computer simulations.

Batches	B1	B2	Final mixture
8/16 (kg/m^3)	610	596	595
3/8 (kg/m^3)	389	380	379
0/3 (kg/m^3)	651	636	635
0/1 (kg/m^3)	181	177	177
Cement (kg/m^3)	398	438	443
Superplasticizer (kg/m^3)	0.95	1.01	1.02
Water (l/m^3)	185.3	186.0	186
w/c	0.426	0.390	0.386
CA/FA	1.20	1.20	1.20
Air (%)	1.5	1.6	*1.6*
Specific gravity	2.415	2.414	*2.414*
Slump (mm)	140	100	*100–150*
Expected compressive strength (MPa)	*51.3*	*56.7*	
Measured compressive strength (MPa)	48.9	52.4	
Re-calibrated theoretical strength (MPa)	48	53.4	*54*
Expected tensile strength (MPa)	4.2	4.4	
Measured tensile strength (MPa)	4.0	4.2	
Re-calibrated tensile strength (MPa)	4.0	4.2	*4.1*

are higher than in the previous batches, but fit with the range of specifications.

The rheological stability of the mixtures is checked with the BTRHEOM rheometer (see section 2.1.1), with the same sample kept in the rheometer for 40 min (Fig. 4.14). In the absence of a rheometer, this stability would have been checked by slump tests performed on successive samples of the same batch. It is verified that no cement/superplasticizer incompatibility takes place. Finally, the specific gravity is determined for each concrete. The measured values γ' being different from the theoretical ones γ, the 'yield' of the mixture is assured by multiplying all quantities by the ratio γ'/γ. The correct compositions are finally indicated in Table 4.15.

The aspect of the fresh concrete is satisfactory, without any visible segregation or excessive bleeding. For checking the placeability, the CES test[21] is carried out for the two mixtures. This test consists in filling with fresh concrete a cubic mould, one vertical face of which is transparent and equipped with a steel mesh standing for reinforcement (Fig. 4.15). Here, fresh concrete is placed in two layers, crush rodded 25 times, and a 10 s vibration is applied to the box. Then the aspect of concrete facing is observed through the transparent face. As can be seen in Figs. 4.16 and 4.17, the obtained placeability is satisfactory: after vibration, the reinforcement is fairly coated with fresh concrete, and no bubbles nor honeycombing appear[22].

Tests at 28 days: final formula

The compressive strengths obtained at 28 days are given in Table 4.15. The results are somewhat lower than expected, probably because of a change in the cement strength since the previous delivery. For the calculation of the final formula, the software is re-calibrated (by changing

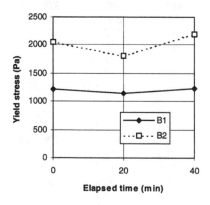

Figure 4.14 Evolution of yield stress, for concretes B1 and B2. No significant evolution takes place during the test period.

Figure 4.15 CES test device (empty).

the cement strength) in order to predict compressive strengths close to those measured. Then the definitive cement dosage is obtained by searching for a predicted strength equal to the 54 MPa target strength. The final formula is very close to that of the B2 concrete.

Finally, a suitable mixture composition has been developed with minimal laboratory work, within a short period. Depending on the local regulations, either the final mixture can be batched in the laboratory, in order to check that the compressive strength at 28 days fits with the specification, or the formulator can move to the next step, which deals with adjustment of the mixture proportions in the real conditions of production.

4.3 QUESTIONS RELATING TO THE AGGREGATE SKELETON

In this section questions relating to the aggregate phase are emphasized. We have seen that the aggregate influences the concrete properties through a number of parameters. Characteristics of the parent rock, such as water absorption, compressive strength, bond with cement paste and elastic modulus, are important. Another series of parameters is controlled by the production process: the packing density of monosize fractions (which incorporates shape effects), and size distribution. All

Figure 4.16 CES test device (after filling with mixture B1, before vibration).

Figure 4.17 CES test device (after vibration).

these parameters are included and quantitatively accounted for in the models developed in Chapter 2.

The question now is to investigate the influence of aggregate proportioning on the final result of concrete optimization. The complexity of optimizing the grading curve has been studied in sections 1.4 and 1.5.3 (for dry mixtures) and 4.2.2 (for concrete). Let us review some classical questions in the light of the previous theoretical developments.

4.3.1 Choice of the maximum size of aggregate (MSA)

There is first a limitation of MSA dealing with the structure to be cast (see 'Placeability' in section 4.1.1). Also, if the concrete is to be pumped, it is usual to keep the MSA under one-fifth to one-third of the pipe diameter. Taking these 'external' constraints into account, the classical theory states that the highest MSA is desirable (rule 7 in section 4.2.1). The author himself has long believed in the validity of this rule, and has tried to gain experimental data supporting it in the range of high-performance concrete (de Larrard and Belloc, 1992).

However, this principle was questioned as long ago as the early 1960s (Walker and Bloem, 1960). We have shown in this work that the negative effect of the maximum paste thickness (MPT) counteracts the lowering of water demand, when the MSA increases. It turns out that there is little gain above 10–15 mm, at least if concrete is specified by a minimum placeability and a minimum strength criterion, and provided that these constraints are active (see section 4.2.2). Therefore aggregate fractions with a size higher than 10 mm should be technically compared in terms of their intrinsic characteristics, other than the size. On the other hand, there is no reason to exclude *a priori* a coarse aggregate having an MSA of 20–25 mm if the aim is to design a high-performance concrete. A good HPC with a compressive strength up to nearly 130 MPa at 28 days may be produced with such a coarse aggregate (de Larrard *et al.*, 1996b).

4.3.2 Rounded vs. crushed aggregate

In classical concrete technology rounded aggregates are often preferred, because they are assumed to produce concrete with a better workability than when crushed aggregates are used. On the other hand, crushed aggregates are known to develop a better bond with cement paste, thanks to their roughness. When the two types of aggregate are compared on the basis of identical concrete specifications, which is the best? Of course there is no general answer to this question because, as already stated, the aggregate intervenes in the concrete system through several independent parameters.

We shall now perform a comparison between two coarse aggregate sources as an illustration of the power of simulations. But the conclusions will be valid only for these materials. The crushed aggregate is a semi-hard limestone, used by the author in a project dealing with HPC for a nuclear power plant (de Larrard *et al.*, 1990; see section 5.3.2). The rounded aggregate is a rounded flint from the Channel. Aggregate characteristics are summarized in Table 4.16. Basically, the rounded aggregate has a good shape and good intrinsic strength, but develops a poor bond with the cement paste. The crushed aggregate has characteristics in terms of shape and strength that are less satisfactory, but has an excellent adhesion with the cementitious matrix, due to its roughness, porosity and chemical nature.

For the two series of mixtures, concretes with three strength levels will be formulated: 30, 60 and 90 MPa. The specifications forming the basis of the simulations are given in Table 4.17. Apart from the coarse aggregate, the mixtures will be made up with a well-graded river sand and a Portland cement. In addition, the C60 and C90 mixtures will contain a superplasticizer. Finally, a silica fume will be added in the C90 concretes.

The results of simulations are displayed in Fig. 4.18. As expected, the water demand is higher with the crushed aggregate, because of differences in water absorption and packing density. However, as soon as the mixture is rich enough, there is no more effect of the aggregate

Table 4.16 Characteristics of the two coarse aggregate sources used in the simulations.

	Aggregate	
	Rounded (Crotoy)	*Crushed (Arlaut)*
Minimum and maximum dimensions	5/25 mm	5/25 mm
Water absorption	0%	1%
Virtual packing density, β_i, (see section 1.2)	0.702–0.714	0.644–0.656
Bond coefficient, p (see section 2.3.4)	0.583	0.96
Strength coefficient, q (see section 2.3.4)	0	$0.004\ MPa^{-1}$

Table 4.17 Specifications of the three types of mixture (SP = superplasticizer; SF = silica fume).

	Strength grades		
	C30	*C60*	*C90*
Optimization criteria	Minimum cement	Minimum cement SP/C = 0.5%	Minimum cement SP/C = 1.3% −SF/C = 10%
Placeability	$K' \leqslant 6$	$K' \leqslant 6.5$	$K' \leqslant 7$
Slump	$\geqslant 100$ mm	$\geqslant 200$ mm	$\geqslant 200$ mm
Compressive strength at 28 days	30 MPa	60 MPa	90 MPa

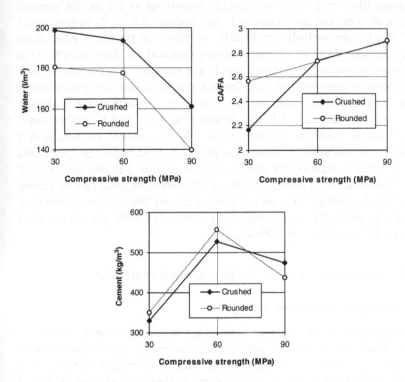

Figure 4.18 Effect of specified strength on various parameters (simulations).

shape on the optimal coarse/fine aggregate ratio. Finally, the cement consumption is higher with the rounded aggregate, because of its low bond with cement paste, except in the HPC range, where the aggregate high strength becomes beneficial. As already found by Day (1995), it appears that the ranking between two types of aggregate may change depending on the type of concrete to be proportioned (see section 2.3.4).

Finally, if a concrete is specified by placeability and strength criteria, there is no general ranking between rounded and crushed coarse aggregate. However, if a water/cement ratio constraint is active, a mixture produced with a rounded aggregate will require less cement.

As far as the sand fractions are concerned, let us point out that the two types of particle (rounded or crushed) can be equally suitable for the formulation of any type of concrete. All concretes produced in some French *départements* (in the south of the Massif Central) are made up entirely with crushed aggregate, as here are no more available sources of alluvial sands.[23] Crushed sands have generally good size distributions in the 0.1–4 mm range, while river sands often lack particles in the fine fractions (see next section). The bad side of crushed sands generally lies in their high fines content. We have shown that an amount of about 10%

of limestone filler in a sand is generally beneficial as far as the general properties of concrete as concerned (see section 4.2.2). A non-washed crushed sand may include up to 20% or more of particles finer than 80 μm, which provokes an excessive water demand for the concrete. This is why many commercial crushed sands are washed. Unfortunately, the washing process removes not only the fine particles, but also a number of particles in the 0.1–0.5 mm range, which are essential for concrete stability (see next section). A better solution can be to use a crushed sand with its 'natural' size distribution, with a lower cement content and a water reduction obtained through the use of a superplasticizer, but this practice is often restrained by standards that impose a minimum cement dosage. Alternatively, a sieved fraction of the washed sand can be added to the original product. This technique permitted the author to produce an excellent 100 MPa self-compacting concrete with an all-crushed aggregate (de Larrard *et al.*, 1996a).

4.3.3   Continuously graded vs. gap-graded concretes

Gap between fine and coarse aggregates

Since the early times of concrete technology there have been numerous disputes about the merits of gap-graded concretes as compared with continuously graded mixtures. In France, Valette advocated a method in which a compact mortar was first designed, then mixed with a gravel much coarser than the sand. The opposite school, led by Caquot, supported the use of continuously graded aggregates. Intuitively, gap-graded materials fit better with the natural idea of forming a packing of particles by successive introductions of aggregate fractions, the i fraction being small with regard to the $i - 1$ fraction.

In section 1.4.4 we investigated the problem of optimizing the size distribution for the sake of maximum packing density. While the maximum *virtual* packing density is found with a discontinuous distribution, such a feature has not been found in the numerical optimizations dealing with actual packing density. However, when the number of monosize fractions permitted for designing a mix is gradually increased, the overall granulometry seems to tend towards a continuous distribution, except for two peaks corresponding to the minimum and maximum sizes of the grading span (Fig. 1.43). Differences between good gap-graded and continuous distributions are minor, and often smaller than the precision of the packing model (CPM).

Concrete simulations have been performed with the following constituents: two coarse fractions of crushed limestone (5/12.5 mm and 12.5/20 mm), a Portland cement, and water. As far as the fine aggregate was concerned, the continuously graded concrete (A) incorporated the two fractions of rounded sand given in Fig. 4.19. The gap-graded

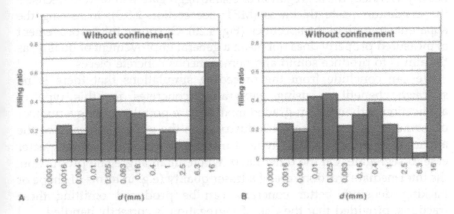

Figure 4.19 Filling diagrams of mixtures A (continuously graded) and B (gap-graded).

concrete (B) included the same fine sand, but a coarser sand having a lower MSA, so that there were no particles between 2.5 and 12.5 mm. Concretes were optimized in order to fit with specifications typical of a dry concrete for precast products: a placeability criterion ($K' \leqslant 6.5$) and a compressive strength at 28 days equal to 50 MPa. Obtained mixture proportions are given in Table 4.18. In this example it can be seen that concrete A needs a little less cement than concrete B, while the latter has a much higher tendency to segregate (Fig. 4.19).

Actually, the difference in cement requirement comes only from the difference in sand content at the optimum. As there is more sand in concrete B, the air content predicted by the model is slightly higher, and the water/cement ratio slightly lower for the same theoretical strength. However, if a concrete sample is produced with the B composition, and

Table 4.18 Mixture proportions of concretes (simulations).

	A (continuously graded)	B (gap-graded)
Crushed limestone 12.5/20 mm (kg/m³)	688	1230
Crushed limestone 5/12.5 mm (kg/m³)	637	0
River sand (kg/m³)	328	507
Correcting sand (kg/m³)	197	98
Portland cement (kg/m³)	379	389
Water (kg/m³)	182	181
w/c	0.48	0.47
Slump (mm)	0	0
Air (%)	0.6	1.0
Segregation potential	0.89	0.97

heavily vibrated, the segregation of coarse aggregate will tend to decrease the maximum paste thickness (MPT; see section 2.3.3), increasing the apparent strength of the specimen (Fig. 4.20). Of course this improvement of measured property does not mean a general improvement of quality, as the aim is to produce sound and homogeneous concrete pieces.

We can conclude from the preceding simulations that there is no *systematic* benefit in using gap-graded concretes for this type of application, although gap-graded mixtures are popular among a number of formulators in the precast industry. It is the author's belief that the apparent superiority of this type of mixture may come from its greater ease to segregate. On the other hand, when, in an aggregate production, the intermediate fractions are of a lesser quality (e.g. in terms of shape or packing density), better concretes can be produced omitting these fractions, provided that the risk of segregation is correctly handled.

Gap between binders and fine aggregate

Utilization of river sands in which fine particles are lacking is a common situation (see section 4.3.2). This is due to the process of either formation or extraction, where the fine particles are entrained by water. In such a case the fine aggregate is improved by inclusion of a correcting sand, with a grading span that is intermediate between those of fine aggregate and cement.

To illustrate the benefit of this practice, let us perform some simulations, dealing with a low-quality concrete for building. The minimum cost is sought with a placeability of 6 and a compressive strength of 25 MPa at 28 days. The constituents are the same as those used in the previous simulations. The grading curves of fine materials are displayed in Fig. 4.21.

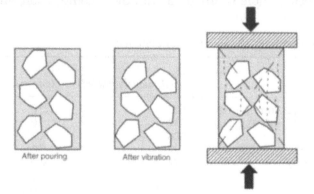

After pouring   After vibration

Figure 4.20 Segregation-induced increase of apparent strength. In the central zone of the specimen, where failure occurs, the maximum paste thickness tends to decrease when the coarse aggregate sinks.

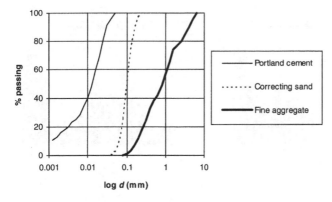

Figure 4.21 Size distribution of cement and fine aggregate fractions used in the simulations.

The result of the simulations is given in Table 4.19, and the filling diagrams of the concretes appear in Fig. 4.22. Interestingly, the lack of particles around 0.1 mm leads to a high fine aggregate content, which in turn diminishes the amount of intermediate aggregate (5/12.5 mm) in the optimal grading (mixture A). The use of a correcting sand leads to a slight reduction of the cement demand (about 12 kg/m³), and brings the 5/12.5 mm fraction to a higher level, giving a lower segregation index (mixture B). A third mixture has been designed with the addition of a limestone filler. A further reduction of cement is gained, with an even lower segregation index. Finally, it appears that the use of a variety of aggregate fractions, in order to obtain a continuous and optimized

Table 4.19 Results of simulations.

	Mixture		
	A	*B*	*C*
Crushed limestone 12.5/20 (kg/m³)	774	697	690
Crushed limestone 5/12.5 (kg/m³)	436	585	591
Fine aggregate (kg/m³)	670	390	387
Correcting sand (kg/m³)	0	246	230
Portland cement (kg/m³)	276	264	245
Limestone filler (kg/m³)	0	0	36
Water (l/m³)	196	188	189
Segregation potential	0.93	0.88	0.87
Slump (mm)	180	140	130
Compressive strength at 28 d. (MPa)	25	25	25
Placeability, K'	6	6	6
Unit cost per m³ (FF)	433	429	422

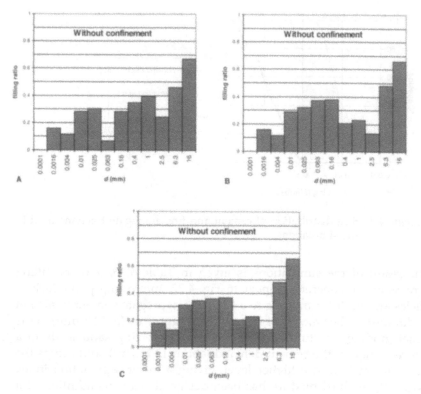

Figure 4.22 Filling ratios of the mixtures.

grading curve, is a factor of both quality *and* economy. In practice, the availability of aggregate fractions or silos can limit this optimization process.

4.4   QUESTIONS RELATING TO THE BINDERS

Having reviewed the various questions dealing with the aggregate phase, let us now investigate the features of cement and cementitious materials.

As far as Portland cement is concerned, its significant properties have been reviewed in section 3.2, and their influence appears quite obviously in the models developed in Chapter 2. The question of binder influence is less trivial when OPC (ordinary Portland cement) is combined with a supplementary cementitious material (SCM). The technical literature is full of examples of pure Portland cement concretes, compared with other mixtures where the cement has been partially replaced by an SCM, but this type of comparison is of little use for the engineer, because none of the properties are equal.

One interest of a computer system, fed with suitable models, is to allow relevant comparisons of mixtures fitting the same basic specifications. The cost and the secondary properties can then be compared. Finally, it becomes possible to highlight the types of mixture where the benefit of using a SCM is the highest.

Let us point out that in many cases SCMs are already mixed with clinker in the form of blended cements. The following developments also apply to this type of product. The interest in blended cements is to allow the use of two binders with only one silo. On the other hand, the user cannot modify the mutual proportions of cement and SCM, which is a limitation for reaching a mixture that is optimum both technically and economically.

4.4.1 Use of limestone filler

We have already seen that an inert filler is theoretically of interest in lean concretes, because it allows us to complement the cement action in filling the aggregate voids (rule 12 in section 4.2.1). However, as a limestone filler is neither inert nor free of charge, a small amount of filler is always beneficial as soon as the cement contains some aluminates. The optimum amount of limestone filler increases when the amount of superplasticizer increases, and when the target strength decreases (see sections 2.3.7 and 4.2.2). Also, incorporation of limestone filler is more attractive when the C_3A content of the cement is higher.

As far as the fineness of filler is concerned, we have seen that the acceleration effect is controlled by the specific surface of the product (section 2.3.7). But without superplasticizer a finer filler will require more water in terms of yield stress and slump, because the contribution of granular fractions to the yield stress increases when the grain size decreases (see section 2.1.3). Thus a very fine limestone filler (with a mean diameter as small as 1 μm or less) is only of interest if a large amount of superplasticizer is used. But in the range of mineral admixtures for high-performance concrete (HPC), limestone fillers have to compete with pozzolanic products such as silica fume, which are much more active (de Larrard, 1989).

In order to realize the various effects of incorporating a limestone filler in a concrete mixture, we shall examine some simulation results. The specifications dealt with are the following: a minimum placeability ($K' \leqslant 6$), a minimum slump of 100 mm, and a compressive strength at 28 days ranging from 20 to 50 MPa. The mixtures will be formulated with the reference set of constituents. The filler cost will be assumed to be 16% of the cement cost.

In the first series the mixtures contain only cement as a binder. In the second series a limestone filler is added, while in the third one the mixtures may also contain superplasticizer. The minimum cost is sought for all mixtures. The results of the simulations are displayed in Fig. 4.23.

Following these simulations, it appears that incorporation of limestone filler allows the user to save a moderate amount of cement[24], giving a slight economy on the overall composition. It is only in the highest strength range that the interest of filler is significantly enhanced by the superplasticizer. As far as total shrinkage is concerned (see section 2.5.6), the use of filler leads to a slight increase except in the right-hand part of the diagram. Finally, the strength development is hardly modified by the limestone filler (Fig. 4.24). The accelerating effect given by the additive equilibrates the increase of water/cement ratio, which is *per se* a factor for a slower strength development (see section 2.3.5).

Apart from the already mentioned effect on shrinkage, and when it is not overdosed, limestone filler hardly influences the overall material durability (at equal workability and strength) (Detwiler and Tennis, 1996) except for its scaling resistance (Caré *et al.*, 1998).

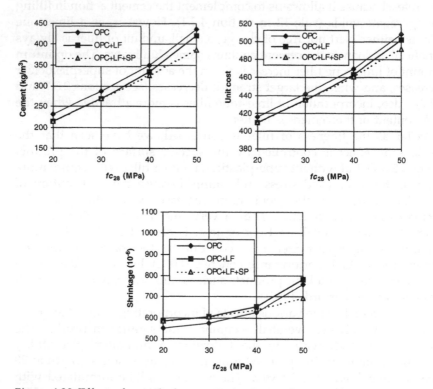

Figure 4.23 Effect of specified compressive strength on various concrete ratios (simulations). LF = limestone filler; SP = superplasticizer.

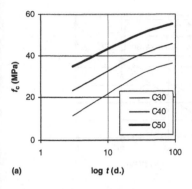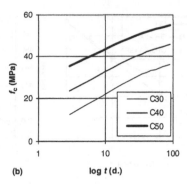

Figure 4.24 Strength development of mixtures (a) without and (b) with limestone filler.

4.4.2 Use of fly ash

Class F fly ash is the main pozzolan used in concrete technology. The concrete technologist is duty bound to use it as much as possible, because its disposal is increasingly being forbidden, or at least discouraged for environmental reasons.

This product can be a beneficial admixture for concrete in several respects. But as a by-product, fly ash has sometimes varying properties for a given source, especially when the type of coal changes. Keeping in mind the question of variability, which must be assessed for any actual application, simulations have been performed to highlight the effect of incorporating fly ash in concrete. The same approach as that used for limestone filler has been applied to fly ash. A 'standard' fly ash has been used in the simulations, with a unit cost of 80 FF/t. The residual packing density was 0.50 without superplasticizer, and 0.54 with superplasticizer. The values of K_p (activity coefficient) at $1, 2, 3, 7, 28$ and 90 days were $0, 0, 0.1, 0.2, 0.5$ and 0.7 respectively. The results of the simulations appear in Fig. 4.25.

Without superplasticizer, the benefit of using fly ash is higher than in the case of limestone filler, in the low strength range. However, in the high-strength range this benefit vanishes, and the pure Portland cement is more competitive. A slight amount of superplasticizer restores the economical interest of fly ash in the whole range.

Since the activity of fly ash does not display so rapid a saturation as limestone filler,[25] the search for economy may lead to a very high fly ash dosage, which gives the concrete a high paste content. This phenomenon is detrimental to all deformability properties, including shrinkage. Another adverse effect is the lowering of hardening rate,

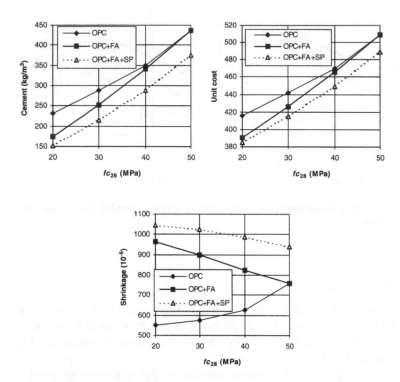

Figure 4.25 Effect of specified compressive strength on various concrete ratios (simulations).

visible in Fig. 4.26. In a competitive economical context the use of fly ash has to be limited by a global management of the mix design, including delayed deformations and strength at early age. Or, if a maximum amount is to be used, a 'high-volume fly ash' concrete, incorporating a large dose of superplasticizer, has to be developed (see section 5.5.2).

As far as durability of fly ash concrete is concerned, if such a concrete is compared with a control mixture incorporating Portland cement only, and having the same workability and the same strength:

• the need for careful curing is more stringent;
• the depth of carbonation may be higher, as there is less portlandite to form calcium carbonate, which tightens the porous network and slows a further penetration of carbon dioxide;
• provided that the curing is satisfactory, the permeability of hardened concrete is less;
• the resistivity of hardened concrete is better;
• the likelihood of alkali-silica reaction development is lower;

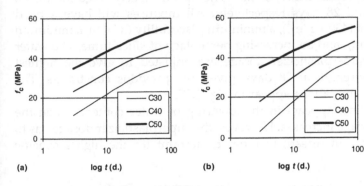

Figure 4.26 Strength development of mixtures (a) without and (b) with (second series) class F fly ash.

- the resistance to scaling under the application of deicing salts tends to increase (Neville, 1995).

Finally, as most fly ash contains some percentage of free carbon, the colour of the concrete will be darker.

4.4.3 Use of silica fume

Silica fume presents numerous similarities with class F fly ash: the grains are spherical, and the product displays pozzolanic properties. However, silica fume is much more reactive than fly ash, and also much finer. Fine, selected fly ash with a maximum size of 10–15 μm tends to be an intermediate product (Naproux, 1994).

Filler effect of silica fume

Because of its size distribution, silica fume exerts a filler effect in concrete, even when the cement dosage is high (Bache, 1981). However, this effect is significant only when a large amount of superplasticizer is used together with the mineral admixture. Also, the magnitude of the filler effect depends on the state of deflocculation of silica fume in fresh concrete, which is partially controlled by the form of the product (as-produced, in slurry, densified, crushed with the cement etc.). There is at present no experimental method to assess the real size distribution of silica fume in the presence of cement and admixtures.

In order to realize what is the filler effect of silica fume in high-performance concrete (HPC), some simulations will be performed. The constituents are the same as in the previous sections, except for a silica fume (in slurry) having a mean size of 0.1 μm and a packing density[26] (in presence of superplasticizer) of 0.65. The K_p values accounted for are 1.6 and 4, at 3 and 28 days respectively. All mixtures will have a fixed cement content (430 kg/m³), a minimum placeability ($K' \le 8$), a minimum slump (200 mm), and an increasing percentage of silica fume. The water content and superplasticizer dosage are adjusted to give the highest compressive strength at 28 days, given the previous constraints. The results of the simulations appear in Fig. 4.27.

The filler effect appears in the lowering of water content, when the silica fume dosage increases. However, the superplasticizer dosage has to be continuously increased, in order to account for the higher specific

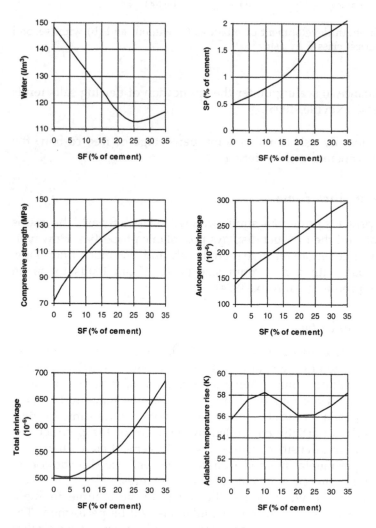

Figure 4.27 Effect of silica fume incorporation on the proportions and properties of an HPC having a fixed cement content.

surface of the mixture[27]. The filler effect continues up to about 25% of silica fume, which also appears to be the optimum dosage for compressive strength. Above this dose any further pozzolanic effect is overcome by the increase of water dosage. Note that, thanks to silica fume, the compressive strength at 28 days is nearly doubled.

As expected, the autogenous shrinkage also increases in the same proportion. This is a factor that favours early cracking in restrained structures. However, the adiabatic temperature rise hardly evolves. Even if the total binder dosage increases, representing a growing *potential* heat development, the decrease of water/binder ratio diminishes the final degree of transformation of cement and silica fume (see section 2.2). Total shrinkage increases only in a moderate way. Therefore the unique property of silica fume is that it allows the formulator to reach the very-high-strength domain without excessively impairing the secondary properties of concrete.

Comparison between SF and non-SF concretes

Another way to investigate the effect of silica fume is to compare SF and non-SF concretes based upon the same specifications. Here, the same placeability and slump requirements will be maintained (with regard to the previous simulations). Instead of looking for the maximum strength, the binder, water and superplasticizer dosages will be optimized in order to minimize the adiabatic temperature rise. Compressive strength at 28 days will range from 60 to 100 MPa, for concretes having no silica fume and 10% SF respectively. The results of the simulations are displayed in Fig. 4.28.

The mixture of Portland cement with 10% SF constitutes a binder with superior properties: that is, a high strength together with a low water demand. This is why the amount necessary to reach a given strength grade is always lower than in the corresponding pure Portland cement concrete. Interestingly, the cost of SF concrete is higher in the moderate strength domain (60–75 MPa) and lower in the very-high-strength domain. This explains why non-SF concrete is often used for C 60 HPC, while concretes at a higher strength grade are always formulated with an SCM, most often a silica fume (see section 5.3.1). In the cost computations, silica fume is assumed to be three times as costly as cement.

However, the use of silica fume has an adverse effect: lower strength at early age. In turn, the cracking tendency is reduced, since both adiabatic temperature rise and total shrinkage are lower. Finally, the creep deformation is greatly diminished by the use of silica fume. This is why, in the appendix of the French code BPEL applying to HPC, the creep coefficients for non-SF and SF HPCs are 1.5 and 0.8 respectively (Le Roy *et al.*, 1996).

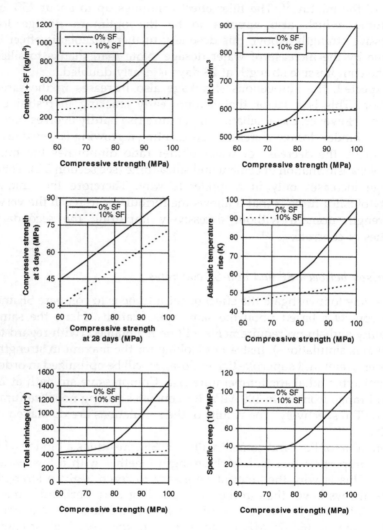

Figure 4.28 Comparison between SF and non-SF mixtures, at equal specifications.

Finally, silica fume appears to be a unique mineral admixture for the mixture proportioning of HPC. Most durability properties of concrete are improved by the incorporation of silica fume: the permeability is reduced (whatever the fluid), the resistivity of hardened concrete is increased, the resistance to sulphates and acids is improved, and the risk of alkali–silica reaction development is greatly reduced (Neville, 1995). However, it must be kept in mind that not all silica fumes have the same qualities (see section 3.3). In particular, the extent of filler effect, related to the amount of deflocculation, is critical.

4.5 STABILITY OF CONCRETE IN AN INDUSTRIAL PROCESS

When a concrete mixture is designed, it is important to anticipate the variability of properties in the context of industrial production. As soon as the material is to be produced within a certain duration, changes in both quality and quantity of the various constituents are to be expected. Facing this reality, two basic approaches are possible: the classical approach and the 'modern' approach.

In the classical approach the target formula is kept constant. Then a sufficient margin should exist between the target and the design strength. In many countries the concrete technologist is forced to adopt this approach, because of stringent regulations that are more prescription-based than performance-based. However, the actual composition will fluctuate, because of the limited accuracy of the weighing process.

In the 'modern' approach an attempt is made to limit the variability of properties by continuously adjusting the target composition. In such a case, a lower difference between the target and the design strength can be adopted. This is the rational approach, because it is clearly preferable to have a mixture of variable composition but constant (main) properties, rather than have a constant composition with variable properties. This type of technique is detailed by Day (1995), who advocates the utilization of CUSUM (cumulative sums) monitoring of the various measurements performed in a concrete plant.

While the model of the concrete system presented in Chapter 2 would be suitable for this purpose, the quality control and production control aspects of concrete variability are not covered in this book. However, when establishing the theoretical composition of a concrete mixture devoted to a given production, it is important to design a mixture that is not too 'sharp'.[28] Moreover, when a composition is to be selected, a quantification of the expected variability can be carried out, either experimentally or by simulation. These issues will be addressed in the following sections.

4.5.1 Strategy for designing a mixture of minimum variability

Let us first review the various causes of variability of a concrete mixture that is to be produced in a concrete plant.

Variability in the constituents

As far as the characteristics of the constituents are concerned, the changes that can be expected are listed in Table 4.20. In the following developments, only workability and strength will be considered as concrete properties.

Table 4.20 Factors of variation of concrete properties concerned with the characteristics of constituents.

Constituent	Parameter	Effect on workability	Effect on strength
Aggregate	Grading curve	**	–
	Shape[a]	**	–
	Cleanness	*	**
	Water content	* **	***
Portland cement	Strength	–	***
	Fineness	**	** (at early age)
	C_3A and sulphates	*** (in presence of SP)	** (at early age)
	C_3S/C_2S	–	* (at early age)
Supplementary cementitious materials	Fineness	*	* (at early age)
	Activity index	–	**

[a] As assessed through packing density measurements; see section 3.1.4. * Low, ** medium, *** high.

Based on this table, the formulator may select the constituents of minimum variability, so that a more consistent concrete will be produced. This concern may be relevant when choosing a quarry: if a strict quality assurance programme is adopted, the likelihood of important variations in the aggregate production is minimized. The major effect of changes in aggregate water content may induce the selection of an aggregate of low water absorption. Likewise, when selecting the source of cement, an examination of statistical data dealing with the production in recent years can be of interest. As far as supplementary cementitious materials are concerned, attention to low variability may lead the formulator to prefer the use of a limestone filler, which is a product, rather than to a fly ash, which is a *by*-product (unless the fly ash production is controlled and a rigorous quality assurance programme is carried out).

Variability in the proportions

In most industrial processes the quantities of constituents are either weighed (aggregate, powders) or measured by volume (admixtures, water). The aggregate fractions are often added in a cumulative way, and errors in a given weight can be corrected by the next weighing, so that there is better control of the *total* aggregate content than there is for each particular fraction. Therefore it can be of interest to minimize the effect of coarse/fine aggregate ratio on the workability: that is, to choose the optimal proportions in this respect (see sections 4.2.1 and 4.2.2). It is common to see mixtures in which there is an excess of sand, either because it makes concretes with better 'trowelability', or because the fine aggregate is cheaper locally than the coarse one. In such a case one should realize that the mixture will be more variable in terms of workability. Furthermore, as the water content of the sand is generally much higher than that of gravel, the likelihood of changes in *concrete* water content will be higher. Finally, the more sandy the mixture, the more sensitive will be the strength to changes in workability. Hence, if we refer to the air content model presented in section 2.1.6, it is clear that the effect of slump on air content greatly increases with the sand content of the mixture, and this air content will in turn affect the compressive strength (see section 2.3).

Another important factor with respect to variability is the admixture dosage. If a high-performance concrete (HPC) of a given workability and strength is to be designed, there is a certain margin of freedom between a mixture with a lower binder content and a saturation dosage of superplasticizer on the one hand (see sections 3.4.2 and 5.5), and a mixture with more binder and less superplasticizer on the other hand. As we shall see below, the latter will have a lower plastic viscosity. However, adopting a superplasticizer amount that is lower than the saturation dosage may introduce two types of problem:

- Changes in the amount of superplasticizer, or incomplete mixing, will have a greater effect on workability (since the curve linking workability with superplasticizer amount is steeper in the region of low dosages; see section 3.4.2).
- Changes in the nature of the cement (C_3A content, form of gypsum etc.) will also produce stronger effects in terms of rheological properties. Hence it is clear that a part of the superplasticizer is quickly consumed by early hydration reactions in fresh concrete. Therefore it is important to keep a significant part of the initial dosage in solution. Otherwise, if nearly all the superplasticizer poured into the mixture is adsorbed on the cement grains, a slight acceleration of the early hydration will have a tremendous effect on workability. Rapid slump losses are to be expected.

In conclusion, when fixing the concrete proportions, a lower variability of the workability will be obtained if the aggregate proportions are near the optimum ones (in terms of the specified fresh concrete property), and if the superplastizer is dosed near the saturation proportion (when such a product is of interest).

4.5.2 Assessment by testing

If the uncertainties on the various characteristics and amounts of constituents are estimated, it is possible to assess the variability of concrete properties in a given production process. Even a refined probabilistic approach could be carried out.

The simplest approach is to measure the effect of each critical parameter by changing the value of this parameter, the other parameters being constant. Under the French regulations such a study is mandatory if a mixture is devoted to a bridge owned by the state (Fascicule 65 A). Then a linear combination of the various effects gives the overall variation of the property, when all parameters are supposed to change simultaneously. We performed such a study to compare the reliability of three mixtures: a normal-strength concrete (NSC), a high-performance concrete (HPC), and a very-high-performance concrete (VHPC) (de Larrard, 1988; de Larrard and Acker, 1990). The investigated properties of the mixtures were the slump and the compressive strength at 28 days.

Given the fact that the aggregate skeletons had been optimized, the critical parameters for slump and strength were the water content, the cement content, the silica fume content, and the superplasticizer dosage. Expected changes of these parameters in industrial production are given in Table 4.21. Trial batches were produced in which one parameter was changed, while the others were kept constant. In these tests the variation of the selected parameter was amplified, in order to decrease the uncertainty in the obtained effect (given the intrinsic error of the test).

The effects of the various changes on slump and strength appear in Figs 4.29 and 4.30 respectively. As far as the slump is concerned, it can be seen that the water content is the main influencing factor, as the saturation dosage of superplasticizer had been adopted in the HPC. For compressive strength, water is still the main factor, followed by the cement content. Compared with NSC, HPC and VHPC have a greater sensitivity to water, but a change in the cement content has only minor effects for these mixtures. Finally, the cumulative effect of weighing errors has been calculated, on the basis of the likely changes in the mixture proportions. In relative terms the strength decrease was 12.2%, 10% and 13.8% for the NSC, HPC and VHPC respectively. Contrary to expectations, HPC and VHPC are not significantly 'sharper' than NSC. This statement is supported by the feedback obtained from certain bridge sites. As an example, the standard deviation[29] obtained in the quality

Table 4.21 Typical variations of mixture proportions in concrete production. The signs of variations are taken in order to give a strength drop.

Parameter	Expectable variation in concrete production	Variation adopted in trial batches
Excess of water	$+3\% + 5 \ l/m^3$	$+15 \ l/m^3$
Lack of cement	-3%	-6%
Lack of silica fume	-5%	-10%
Lack of superplasticizer	-5%	-10%

Source: de Larrard (1988).

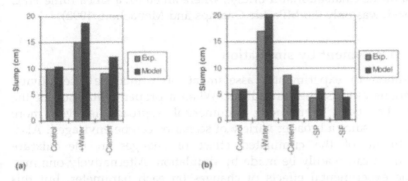

(a)   (b)

Figure 4.29 Effect of changes of the mixture proportions on the slump of
mixtures: (a) normal-strength concrete; (b) high-performance concrete. Experimental data from de Larrard and Acker (1990).
SP = superplasticizer; SF = silica fume.

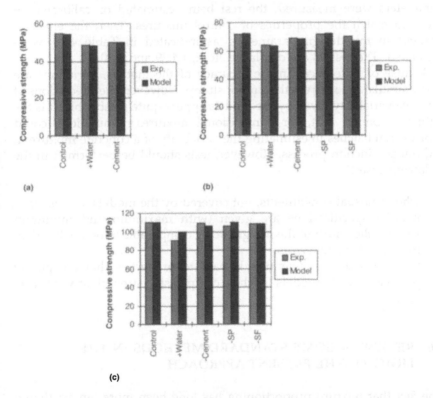

(a)   (b)

(c)

Figure 4.30 Effect of changes of the mixture proportions on the compressive
strength of concretes: (a) normal-strength concrete; (b) high-performance concrete; (c) very-high-performance concrete. Experimental
data from de Larrard (1988) and de Larrard and Acker (1990).
SP = superplasticizer; SF = silica fume.

control of the Chateaubriand bridge, where an 80 MPa silica fume HPC was used, was only 2.6 MPa (de Champs and Monachon, 1992).

4.5.3   Assessment by simulation

In addition to experimental assessment of a mixture's sensitivity, simulations can also be carried out. When a proper calibration of the models has been performed, a numerical approach is even more satisfactory, since a broader variety of scenarios can be envisaged. Also, the estimate of the cumulated effect of changes in the mixture proportions can readily be made by simulation. Alternatively one may add the experimental effects of changes on each parameter, but this addition may lead to wrong results, as the models linking mix proportions and properties are far from linear. To validate such an approach, the experiments described in the previous section have been simulated with the models developed in Chapter 2. Most constituent parameters were measured, the rest being estimated or calibrated in order to match the properties of control mixtures. Then the effect of changes in mix-design parameters, as indicated in Table 4.21, were simulated. The results are displayed in Figs 4.29 and 4.30.

Because of differences in the accuracy of the models, variations are better simulated for strength than for slump. However, the predictions of concrete variability given by the models appear quite reasonable. It is the author's opinion that, for conventional mixtures, simulations could replace trial batches in estimating the variability of a concrete mixture in a given production process. However, tests should be performed in the following cases:

* when unusual constituents, not covered by the models, are used;
* if strict specifications are given (with maximum and minimum values), the width of the margin being comparable to or smaller than the typical model errors;
* more generally, if attention is paid to properties that are poorly predicted by the models (such as compressive strength at very early age).

4.6   REVIEW OF SOME STANDARD METHODS IN THE LIGHT OF THE PRESENT APPROACH

The fact that mixture proportioning has long been more 'an art than a science' (Neville, 1995) is illustrated by the variety of methods encountered worldwide. Not only are there deep differences in the strategy carried out to solve the problem by the various authors, but a quantitative comparison between these methods leads to impressive differences in terms of formula, cement content and cost for the same set of specifications, as shown by the international working group RILEM TC 70-OMD. This is due to the empirical nature of the models used, which are moreover calibrated on the basis of local data.

We shall now review some of these methods, in order to point out their positive and negative aspects. We shall see that the models and theoretical developments of this book allow the reader to understand and to analyse the hypotheses and simplifications that support these methods. Of course, it is the author's hope that the reader will be convinced of the superiority of the scientific approach, made possible by both theoretical advances and computer technology.

4.6.1   US method (ACI 211)

This method (ACI 211) is probably one of the most popular worldwide. It is based mainly on the works of American researchers (Abrams and Powers).

Successive steps of the method

Step 1

Choice of slump.

Step 2

Choice of maximum size of aggregate (MSA).

Step 3

Estimate of water content and air content. The water content is supposed to be controlled by the MSA and the required slump, while the air content would be determined by the MSA.

Step 4

Selection of water/cement ratio. This selection is performed with regard to durability (specified maximum value) and compressive strength. For the latter, a table is given for deducing the water/cement ratio from the required strength.

Step 5

Calculation of the cement content. This calculation is simply done from the water content and the water/cement ratio previously determined.

Step 6

Estimate of coarse aggregate content. The bulk volume of coarse aggregate, placed by a standard method (dry rodded), is first measured. This amount is then multiplied by a tabulated coefficient, which depends on the maximum size of aggregate and on the fineness modulus of the sand.

Step 7

Calculation of the fine aggregate content. This quantity is calculated by difference between the unit volume of the batch and the other volumes previously calculated.

Step 8

Final adjustments.

Critical examination

Like most currently available methods, ACI 211 was developed at a time when concrete was essentially produced with aggregate, cement and water. It is therefore not directly and easily applicable when organic or mineral admixtures are used. However, it does cover air-entrained concretes. In the restricted field of concretes without admixtures, the method is based upon some underlying hypotheses that are worth discussing.

At step 3, water content is determined by MSA and slump. The MSA is related to the grading span of fresh concrete, since the finest particles are controlled by the cement. Thus this simplification refers implicitly to Caquot's law (see section 1.4.4, equation (1.69)), which states that the porosity of a granular mix is mainly controlled by its grading span, and to Lyse's rule, which assumes that the workability of a concrete made up with a given set of constituents is controlled by its water content. We have seen previously that these two 'laws' are valid only in the vicinity of optimum combinations. Therefore deviations are to be expected for lean or rich mixtures. Moreover, the effect of particle shape is not accounted for in the prediction of water demand.

The relationship between water/cement ratio and compressive strength (step 4) is assumed to be unique. Hence if the diversities of aggregate nature[30] and cement strength are cumulated, the compressive strength obtained for a given water/cement ratio may range from 1 to 2, in relative terms. Therefore, the prediction of water/cement ratio appears very crude. This is especially unfortunate, as a large mistake in the determination of water/cement ratio may take 28 days to be detected, while an inaccurate prediction of water demand can be corrected within some minutes.

At step 6 it is assumed that the coarse aggregate content does not depend on the cement and water content. We shall evaluate this hypothesis by examining some simulations. A series of mixtures have been simulated on the basis of the reference set of constituents. In this first series the ratios between 12.5/20 and 5/12.5 fractions on the one hand, and between fine aggregate and correcting sand on the other, have been kept constant. The specifications appear in Table 4.22.

First, it is shown in Fig. 4.31 that Lyse's rule applies quite well in the range of mixtures without superplasticizer. Also, the choice of keeping the coarse aggregate content constant, in lieu of the fine aggregate content, seems justified.

However, the assumption that the coarse aggregate volume does not depend on the workability has the following adverse effect: if the specified workability increases with the same set of materials, the water content will increase, and so will the cement content (because the water/cement ratio is kept constant). As the sand is calculated from the difference between the total volume of the batch and the volume of the rest of the constituents, it can only decrease, and the coarse/fine aggregate ratio will increase. On the other hand, we have seen that the less compacted the mixture, the higher the proportion of fine materials (see section 1.4.4 and Fig. 1.45). This weakness of the ACI 211 method is recognized in a footnote, where it is noted that the proportion of coarse aggregate can (must) be increased for a dry concrete, or decreased for 'more workable concrete'.

In another series of simulations, the effect of the sand fineness modulus on coarse aggregate volume has been investigated. Keeping the same constituents, and a fixed cement content of 350 kg/m^3, the fineness modulus was varied by changing the ratio between the two fine aggregate fractions. As predicted by ACI 211, the coarse aggregate volume decreases when the fineness modulus increases. In terms of packing concepts, this trend is to be expected, because an increase of fineness modulus means a coarser sand, which exerts a higher loosening effect on the coarse aggregate arrangement (see section 1.1.3). However, the values found in the simulations are significantly higher than the ones proposed by ACI 211 (Fig. 4.32).

Table 4.22 Specifications for the mixtures with increasing cement dosage.

Mixtures	Without superplasticizer	With superplasticizer
Cement	From 200 to 550 kg/m^3	From 350 to 550 kg/m^3
Superplasticizer	–	1% of the cement weight
Placeability	$K' \leqslant 6$	$K' \leqslant 7$
Slump	$\geqslant 100$ mm	$\geqslant 200$ mm
w/c	Minimum	Minimum

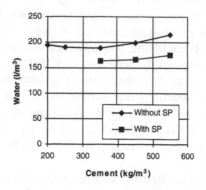

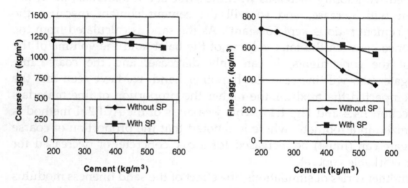

Figure 4.31 Results of simulations dealing with mixtures of increasing cement dosage.

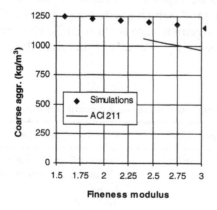

Figure 4.32 Effect of fineness modulus on the coarse aggregate content.

On the whole, the prediction of coarse/fine aggregate ratio provided by the ACI 211 tables seems reasonable, though it could be improved. This method is actually one of the very few that emphasize the packing ability of aggregate fractions. It is only a pity that this approach (the measurement of the coarse aggregate packing density) is not extended to the fine aggregate fraction.

However, there is no way to manage more than two aggregate fractions, nor to detect a lack of particles in some areas of the grading span. Some authors have even stated that this limitation is a cause of durability problems throughout the USA (Shilstone, 1993). Finally, the prediction of water/cement ratio is definitely too inaccurate.

4.6.2   British method (BRE 1988)

Successive steps of the method

The method of the Department of the Environment, revised in 1988 (BRE, 1988) and cited by Neville (1995), comprises the following steps.

Step 1

Determination of water/cement ratio. A first constraint is imposed by durability concerns (maximum value). Then the water/cement ratio is determined by the required compressive strength, for ages ranging from 3 to 91 days. The type of cement (normal hardening rate vs. rapid hardening) and aggregate (crushed vs. rounded) is also listed in the table and the nomographs.

Step 2

Determination of water content. The principle is very similar to that adopted in ACI 211 (see section 4.6.1). A further refinement is brought by paying attention to the shape of particles (crushed vs. uncrushed), which is assumed to govern the water demand, together with the MSA and required slump.

Step 3

Calculation of cement content, by dividing the water content by the water/cement ratio. At this step, a possible inferior limit dealing with durability is accounted for.

Step 4

Determination of total mass of aggregate. The user can read the 'fresh density of concrete mix' as a function of water content and aggregate specific gravity. From this density, the masses of cement and water are subtracted.

Step 5

Determination of the fine aggregate/total aggregate ratio. A series of tables is given, in which this ratio is given by the required workability (expressed in terms of either slump or Vebe time), the MSA, the water/cement ratio, and the amount of particles in the fine aggregate passing through the 0.6 mm sieve. When the coarse aggregate is delivered in several fractions, fixed percentages of the various fractions within the total coarse aggregate mass are proposed in another table.

Critical examination

The management of water/cement ratio with regard to compressive strength is clearly the most advanced within the four standard methods investigated in this section. It has the advantage of considering ages other than 28 days, and it also recognizes the influence of the constituents. However, we have shown in section 2.3 that not all cements have the same strength development, even within the same category (Fig. 2.64). Likewise, not all crushed aggregate gives the same contribution to compressive strength. Nevertheless the British method at least permits a reasonable guess at the necessary water/cement ratio from a visual appraisal of the constituents.

As for the granular skeleton, the level of sophistication is also higher than in ACI 211, as more than two fractions can be managed. The sand proportion in the total aggregate increases:

- when the cement content decreases (as in ACI 211);
- when the slump increases (this is a further advantage over the US method);
- when the MSA decreases:
- when the amount of particles smaller than 0.6 mm decreases. In this case the sand becomes coarser, so that the loosening effect exerted on the gravel increases, which in turn makes the sand content rise. This corresponds to the influence of fineness modulus in ACI 211.

Therefore all trends regarding the evolution of fine aggregate content are consistent with the packing theories developed in Chapter 1. However, the weakness of the British method is the absence of packing measurements on aggregate fractions. In particular, while the difference between crushed and rounded aggregate is accounted for in the prediction of water/cement ratio and water demand, it is not dealt with in aggregate proportioning.

4.6.3 French method (Dreux 1970)

This method is basically of an empirical nature, unlike the previous Faury's method (Faury, 1944), which was based upon Caquot's optimum grading theory (Caquot, 1937). Dreux made an extensive enquiry to collect data about satisfactory concretes. By plotting the grading curves obtained, he could derive an empirical approach for determining a reference grading curve, having the shape of two straight lines in a semi-logarithmic diagram (Fig. 4.33).

Successive steps of the method

Step 1

Examination of the structure (dimensions, spacing between rebars, concrete cover etc.).

Step 2

Determination of the target compressive strength. It is calculated from the design strength, and from the expected coefficient of variation.

Step 3

Selection of fresh concrete consistency. This specification is performed in terms of slump, or with regard to the CES test[31].

Step 4

Selection of MSA. This is done with the help of a table fixing maximum values for the MSA, as functions of the structure dimensions analysed in step 1.

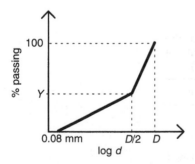

Figure 4.33 Reference grading curve according to Dreux (1970).

Step 5

Calculation of the cement dosage. The cement/water dosage is first calculated from the target strength, by using a Bolomey law (see section 2.3.2). This law incorporates the cement strength, plus an adjustable aggregate factor. A minimum cement dosage is then computed, for durability purposes. This dosage is assumed to be inversely proportional to the fifth root of the MSA. Finally, the cement dosage is determined from a nomograph, as a function of cement/water ratio and slump.

Step 6

Calculation of the (total) water content. It is calculated from the knowledge of cement content and cement/water ratio. At this step, a correction can be made with regard to MSA (the water content increases when MSA decreases). Then the water to be added in the mixer is evaluated from the previous value, by accounting for the water contained in the raw materials (aggregate fractions).

Step 7

Examination of the sand. The fine aggregate is either accepted or rejected on the basis of its cleanness. Then the grading curve is plotted on a standard diagram. If the fineness modulus is too high (that is, the sand is too coarse), a correcting fine sand is added.

Step 8

Plotting of the reference grading curve. The Y parameter (Fig. 4.33) is determined by the following list of factors: MSA, cement dosage, amount of vibration, particle shape, and fineness modulus of the sand. Table 4.23 summarizes the variation of Y as a function of these parameters.

Table 4.23 Parameters accounted for in the determination of Y (ordinate of the break point in Fig. 4.33).

Parameter	Y increases when
MSA	↓
Cement dosage	↓
Vibration	↓
Particle shape	rounded to crushed
Fineness modulus	↑

Step 9

Determination of the solid volume of the fresh mixture. This quantity is tabulated as a function of MSA and consistency.

Step 10

Determination of the amounts of aggregate fractions. The mutual proportions (in volume) are determined graphically, with the help of the reference grading curve plotted at step 9. Then, for each fraction, the volume per unit volume of concrete is calculated by mutiplying the proportion by the solid volume. Finally, multiplying the volume of aggregate fraction by the specific gravity leads to the content (in mass) of the aggregate fraction.

At this step the theoretical composition is known, and must be tested by trial mixtures and then adjusted (if necessary).

Critical examination

Unlike the British method, Dreux's method does not take account of compressive strength requirements at an age other than 28 days. There is only a 'rule of thumb' giving an approximate ratio between the strength at 7 days and the strength at 28 days. However, the prediction of the latter is rather refined, while not accounting for the presence of SCM (like all standard methods).

The determination of cement content on the basis of consistency is rather surprising. And since the cement/water ratio is already known at the end of step 5, the water dosage could have been determined by a nomograph as well.

In the author's opinion, the most interesting part of the method lies in the determination of the masses of aggregate fractions. It has been shown in section 1.4 that there is no optimal grading curve, irrespective of the shape of particles. Also, in section 1.5, even when all aggregate fractions have the same individual packing ability (residual packing density), Dreux's distribution does not appear as the optimum curve, from either the packing density or the segregation index viewpoints. On the other hand, relying upon a reference grading curve has the advantage of facilitating the management of more than two aggregate fractions. Moreover, it allows the detection of segregation hazard related to a missing fraction in the overall grading span.

As far as the coarse/fine aggregate ratio is concerned, it is determined by the value of the Y parameter (Fig. 4.33). Note that all trends given in Table 4.23, which are the result of extensive practical experience, are consistent with the prediction of the packing theory developed in Chapter 1:

- When the MSA decreases, the equivalent number of monosize fractions decreases (see section 1.4.1); then the relative proportion of a fixed number of classes (corresponding to the sand grading span) increases. Finally, the coarse/fine aggregate ratio decreases, and Y increases.
- When the cement dosage decreases, there is less loosening effect exerted on the fine aggregate (see section 1.1.3). Then the proportion of sand rises, and the coarse/fine aggregate ratio decreases.
- When less vibration is applied, the compaction index, corresponding to the placeability of the mixture, decreases (see Table 1.10 in section 1.2.3), which tends to increase the Y value (according to Fig. 1.45).
- When the particle shape shifts from rounded to crushed, this corresponds to a lower residual packing density (β_i parameter). This is again a factor for a rising breaking point in Fig. 4.33, according to Fig. 1.46 in section 1.4.
- Finally, when the fineness modulus increases (that is, the sand becomes coarser), we have seen in Fig. 4.32 that our model predicts a lower coarse/fine aggregate ratio. From the simulations detailed in section 4.6.1, we have extracted the curve given in Fig. 4.34. Here again, sophisticated models based upon packing concepts confirm the empirical experience condensed in Dreux's method.

4.6.4   Baron and Lesage's method (France)

This method is not *per se* a mix-design method, but more a technique for optimizing the granular skeleton of a concrete (Baron and Lesage, 1976).

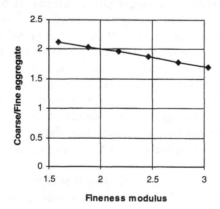

Figure 4.34 Effect of fineness modulus on optimum coarse/fine aggregate ratio (simulations).

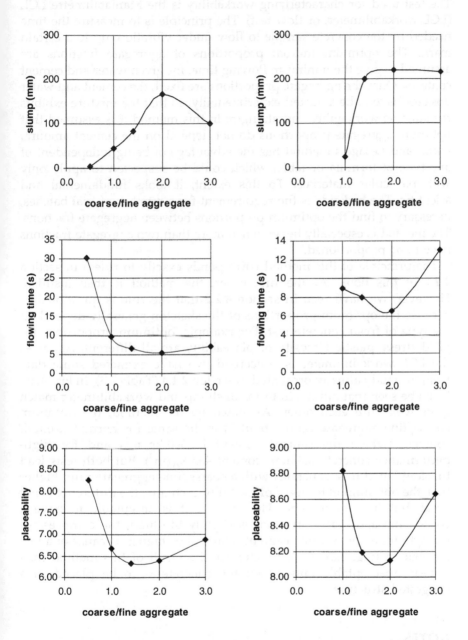

Figure 4.35 Effect of coarse/fine aggregate ratio on fresh concrete properties, at constant cement and water contents: left, normal-strength concretes; right, high-performance concretes. Mixture proportions given in Table. 2.3. Slumps and flow times are experimental values, while the placeability is calculated with the CPM (see section 2.1.5).

The test used for characterizing workability is the Maniabilimètre LCL (LCL workabilimeter, or flow test). The principle is to measure the time needed by the concrete sample to flow under vibration, up to a certain mark. The optimum mutual proportions of aggregate fractions are assumed to give the minimum flowing time, for given water and cement contents. Once the aggregate proportions are fixed, the cement and water dosages have to be adjusted experimentally, so that the mixture exhibits the suitable workability and strength. In this method it is assumed that optimum aggregate proportions do not depend on the cement amount. Baron and Lesage's method has the advantage of being independent of any type of formula or table, which could be suspected to apply only with particular materials. To this extent, it looks fundamental and scientific. The bad side is the requirement for a number of trial batches, necessary to find the optimum proportions between aggregate fractions. The method is especially heavy when more than two aggregate fractions have to be proportioned.

The principle of this method corresponds exactly to rule 3 in section 4.2.1 of this book. At the first order, the method is then justified. However, we have seen in section 4.2.2 that the rule is questionable, because the optimum proportions of the skeleton generally depend on the type of fresh concrete test. For example, optimum proportions for yield stress, plastic viscosity or placeability are all different. As far as the LCL workabilimeter is concerned, we have extracted some data from the study already presented in section 2.1.4 (Table 2.3). In Fig. 4.35, it can be seen that optima in terms of slump and workabilimeter match quite well with each other. As stated by the method, the optimum coarse/fine aggregate ratio (about 2) is the same for normal-strength concrete (where the cement content is 340 kg/m^3) and for high-performance concrete (cement content 435 kg/m^3). But both tests lead to 'stony' mixtures: concretes with a coarse/fine aggregate ratio higher than the optimum in terms of placeability. Actually, the difference is not great, but it is systematic. However, if it is acceptable to have a coarse/fine aggregate ratio that is slightly too high, the compressive strength for a given water/cement ratio will be higher, because of a low air content (Fig. 4.8). As a conclusion, the result of this method may appear unacceptable only when the concrete is to be placed in a congested structure.

NOTES

1 Except in countries where this test is not used, as in Germany. Here the DIN flow table replaces the slump test.
2 Here the plastic viscosity of fresh concrete after vibration is addressed (as predicted by equation 2.1.8, in section 2.1.2).

3 The VB test is close to a placeability measurement, as it induces compaction more than flowing, but it is sensitive only in the range of dry-to-plastic consistencies.
4 Which does not by itself give any protection to the material.
5 Charonnat, Y., private communication, LCPC, 1997.
6 Provided that the delay between successive lifts is not in excess of some weeks.
7 Fascicule 65A from the French Ministry of Transportation requires that the following inequalities be satisfied:

$$\bar{f}_c \geqslant 1.1 f_{c28}$$
$$\bar{f}_c \geqslant f_{c28} + \lambda(c_E - c_{min})$$

where $\bar{f}_c$ is the mean strength (obtained in the tests), f_{c28} is the specified compressive strength, c_E is the strength of the cement used in the tests, and c_{min} is the lower bound of c_E, according to the cement standard.
8 Charonnat, Y., private communication, LCPC, 1997.
9 With the possible exception of *carbonation* shrinkage (Neville, 1995).
10 If a relevant approach is used.
11 By inert it is meant that the filler has no effect on the compressive strength of concrete.
12 If we assumed that the matrix is more deformable than the aggregate.
13 French francs per tonne.
14 Through the bond with cement paste.
15 For example if a high slump is specified, or a discontinuous grading is used.
16 In terms of packing density, bond with cement paste and intrinsic strength.
17 And by the rest of the constituents, including the aggregate itself.
18 For in-situ cast concrete.
19 This mix-design study took place at LCPC in December 1997/January 1998, as related.
20 Hence this approach is rather conservative, since the probability that all negative variations will occur at the same time is very small. One could compute a safety margin in compressive strength following a probabilistic approach. The choice of the method depends on the regulatory context.
21 Centre d'Essai des Structures. This test was designed and advocated by Dreux (1970).
22 Actually, this test has been performed in order to prove the validity of the placeability concept. The author does not recommend a systematic use of the CES test in everyday mixture-proportioning studies.
23 J.M. Geoffray, Private communication, LCPC, 1995.
24 As already said, the cement saving depends also on the C_3A content of the cement.
25 As can be seen by comparison of Figs 2.63 and 2.67.
26 Overall packing density in the water demand test (see section 3.3.3).
27 The superplasticizer demand of the silica fume is assumed to be four times that of cement (de Larrard, 1988).
28 That is, which is not too sensitive to marginal variations of the proportions of constituents.
29 As calculated on the population of *results*, a result being the mean strength of three crushed cylinders.
30 See section 2.3.4.
31 See section 4.2.4.

5 Applications: various concrete families

The previous chapter was dedicated to concrete mixture proportioning, a process based upon the use of the models developed in Chapter 2, fed with material data detailed in Chapter 3. In this last chapter the aim is to apply the overall approach in a number of typical cases. Hence one ambition of this book is to propose a strategy covering a broad range of cementitious materials, since a scientific approach is always governed by a maximal economy of the basic concepts.

In the first section, an attempt will be made to embrace the 'concrete landscape', from low-strength, everyday concrete to very-high-strength mixtures. Then special cases dealing with various typical industrial contexts will be developed. For each concrete family, a typical set of specifications will be given. Then the result of numerical optimizations based upon these specifications will be presented and discussed. For most simulated mixtures, the reference set of materials will be used. Finally, the theoretical mix design will be compared with an actual concrete, taken from the literature or from the author's experience. Note that no efforts were made to use the real material data in the simulations, since all these 'real' concretes were designed and used before, or independently from, the development of the present method. As a consequence, in most cases some basic parameters[1] were lacking to permit relevant simulations.

5.1 PRELIMINARY SIMULATIONS: FROM NORMAL-STRENGTH TO VERY-HIGH-STRENGTH CONCRETES

A broad series of concrete mixtures will be optimized, made up with the same aggregate fractions, but with different binder/admixture combinations. The goal of these simulations is to realize the comparative interest of various technological solutions, depending on the strength range. Secondary properties such as deformability and cracking tendency will be stressed, together with economical aspects.

Constituents and specifications

In a first series of concretes, only Portland cement as a binder is used, and the target strength ranges from 20 to 50 MPa. Then a second series covering the same strength range is formulated with a combination of the same Portland cement and 20% of class F fly ash. This corresponds to what could be obtained with a blended cement. In a third series, the proportion of fly ash, together with a superplasticizer amount, is optimized for compressive strength ranging from 30 to 70 MPa. Finally, a series of silica fume high-performance concretes (HPC) completes the simulations, having a fixed percentage of 10% silica fume (with regard to the Portland cement dosage), and a target strength ranging from 60 to 120 MPa. For each particular mixture, the cheapest formula is sought, matching with the rheological specifications given in Table 5.1.

Results

Let us first have a look at some mix-design ratios (Fig. 5.1). The water content increases with the compressive strength, except in the HPC series, where a minimum appears around a compressive strength of 100 MPa. Therefore it turns out that in the range of practical mixtures, the cement dosage is above the critical cement content (see rule 10 in section 4.2.1). Note that the use of fly ash without superplasticizer is not a factor of reduction of the water demand. Although fly ash particles are spherical, the total volume of binder required to reach a given strength is higher, so that the water demand increases when fly ash is introduced into the system. Superplasticizer is an efficient tool for reducing the water demand, especially in combination with silica fume.

The amount of binder always increases with the target compressive strength. In the range of normal-strength concrete, pure Portland cement remains the most efficient product (in terms of MPa/kg). However, because of its filler and pozzolanic effects, silica fume in combination with superplasticizer substantially reduces the binder demand for a given strength level. The paste volume follows the same trend as that of binder content. But the effect of incorporating silica fume and

Table 5.1 Rheological specifications for the concrete series.

	Series			
	1 (OPC)	2 (OPC + FA)	3 (OPC + FA + SP)	4 (HPCs)
Placeability	$K' \leqslant 6$	$K' \leqslant 6$	$K' \leqslant 7$	$K' \leqslant 8$
Slump	$\geqslant 100$ mm	$\geqslant 100$ mm	$\geqslant 150$ mm	$\geqslant 200$ mm

OPC, ordinary Portland cement; FA, fly ash; SP, superplasticizer; HPC, high-performance concrete.

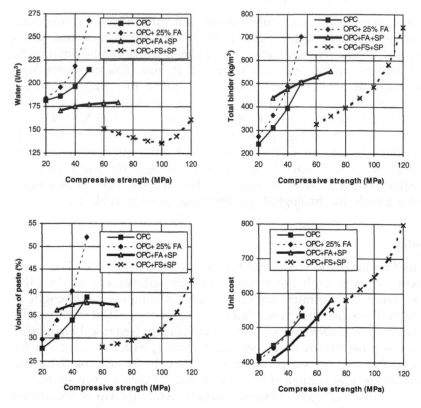

Figure 5.1 Effect of target strength on mix-design ratios.

superplasticizer is even more dramatic: with these products, one can prepare concretes having a compressive strength of 60–80 MPa, with the same paste volume as that of low-strength ordinary mixtures. This fact, among others, explains the success enjoyed by high-performance concrete technology in most sectors of concrete applications.

Finally, as far as cost is concerned, with the particular hypotheses of these simulations it is realized that the 10–120 MPa is covered only by doubling the unit cost of concrete. The four concrete categories tend to coincide with a single (cost, strength) curve, with a slight advantage, at equal strength for mixtures containing pozzolan and superplasticizer.

Unlike most code-type models, which assume that elastic modulus is a monotonic function of compressive strength, the simulations suggest the existence of local maxima in elastic modulus. For higher strength, the elastic modulus would drop according to Fig. 5.2. This is explained for a given type of binder by the 'exponential' evolution of the paste volume (Fig. 5.1). This prediction is consistent with the fact that the elastic

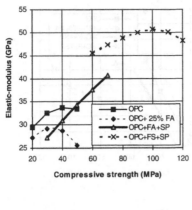

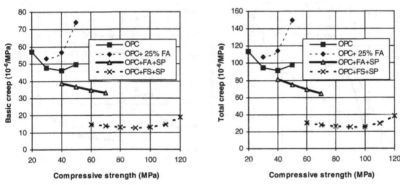

Figure 5.2 Effect of target strength on elastic modulus and creep.

modulus of 200 MPa ultra-high-performance materials hardly reaches 50–60 GPa (see section 5.3.4), a range already attained for 100 MPa mixtures. Creep, expressed in terms of deformation per unit stress, exhibits a trend opposite to that of elastic modulus. As already stated in section 2.5.3, it is drastically reduced by incorporation of silica fume combined with a superplasticizer. This is an advantage when a maximum rigidity of structural elements is aimed for, but this reduction can be detrimental as far as tensile stresses due to differential shrinkage are concerned.

Let us now examine the concrete properties concerned with free deformations. The adiabatic temperature rise increases rapidly with the target strength. This is mainly an effect of the increase in binder content (Fig. 5.3). Surprisingly, in the range of mixtures without superplasticizer, the use of fly-ash-blended cement does not promote a drop in adiabatic temperature rise at equal strength. However, recall that in semi-adiabatic conditions the activation energy also plays a role in controlling the

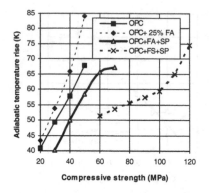

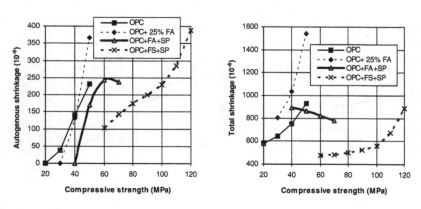

Figure 5.3 Effect of target strength on adiabatic temperature rise and shrinkage.

maximum temperature rise. As the activation energy of the pozzolanic reaction is greater than that of cement hydration, the drop in maximum temperature rise due to heat diffusion increases with the proportion of pozzolans in the binder. Again, silica fume combined with super-plasticizer shows a unique capability in decreasing the maximum temperature rise. A well-proportioned silica fume HPC produces a heat quantity comparable to that of a conventional Portland cement structural concrete.

Autogenous shrinkage roughly follows the same trend as that of adiabatic temperature rise. The main difference lies in the fact that autogenous shrinkage is non-existent or very minor for low-strength concrete, while any concrete exhibits a heat of hydration. On the other hand, total shrinkage (appearing in a dry environment) remains of the same order of magnitude over almost the whole range of compressive strength, if the least shrinking concrete is chosen at each strength level. This is because there is a balance between autogenous and drying shrinkage: the former increases with the amount of self-desiccation, while the latter decreases. Total shrinkage increases only when the paste volume becomes large.

Based upon simulations, it can be of interest to compare the various mixtures in terms of cracking tendency. For this purpose, we shall assume that a concrete piece is totally restrained, and we shall try to compare the tensile stress created by this restraint with the tensile strength of the material. We shall then define a *cracking index*, equal to the ratio between stress and strength. Theoretically, a crack is likely to occur when the cracking index is greater than unity.

In sealed and isothermal conditions, autogenous shrinkage creates only tensile stress (case 'without drying' in Fig. 5.4). This corresponds to the case of a thin, well-cured slab-on-grade of large dimensions. We have

$$Ic_{sealed} = \frac{\varepsilon_{as}\,E_d}{ft} \tag{5.1}$$

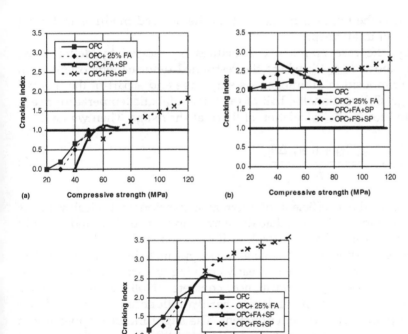

Figure 5.4 Effect of target strength on cracking tendency: (a) without drying; (b) with drying; (c) massive structure.

where Ic_{sealed} is the cracking index in sealed conditions, ε_{as} is the autogenous shrinkage, E_d is the delayed modulus (including the effect of basic creep), and ft is the tensile strength of concrete. Of course, this definition is a rough approximation of the reality, since creep at early age is much larger than in the case of loading at 28 days. A more realistic evaluation of the stress would require an incremental calculation, which would necessitate knowledge of creep and shrinkage evolutions from setting. Equation (5.1) is therefore used mainly for comparative purposes, not for a sound evaluation of the cracking hazard. The same reservation applies to the definition of the two next cracking indices.

In drying conditions the total shrinkage should be accounted for (case of the same slab-on-grade without curing, and submitted to a dry environment). Then the cracking index in dry conditions, Ic_{dry}, is

$$Ic_{dry} = \frac{\varepsilon_{ts}\, E_d'}{ft} \tag{5.2}$$

where ε_{ts} is the total shrinkage, and E_d' the delayed modulus including the effect of total creep.

Finally, the case of massive structures is assessed by defining the cracking index $Ic_{massive}$, where the restrained deformations originate in the superimposition of autogenous shrinkage and a part of the thermal shrinkage. Conventionally, half of the total adiabatic temperature rise is accounted for in the calculation of thermal shrinkage. Then we have

$$Ic_{massive} = \frac{\varepsilon_{as}E_d + 0.5c_{te}\Delta\theta}{ft} \tag{5.3}$$

where c_{te} is the coefficient of thermal expansion (a typical value is $10^{-5}\,K^{-1}$), and $\Delta\theta$ is the adiabatic temperature rise. This index could stand for the case of a long, thick wall cast on a stiff foundation.

In sealed conditions the cracking index appears to be mainly controlled by the target strength. Actually, early cracking of cured, restrained slabs is a feature of high-performance concrete (Holland *et al.*, 1986). Conversely, the cracking index is hardly influenced by strength for concrete placed in a dry environment. The slight increase appearing in Fig. 5.4 is hardly significant, given the simplifications underlying the calculation of Ic_{dry}. Finally, in massive conditions the trend observed for the sealed conditions still applies: the higher the strength, the higher the likelihood of cracking. To some extent, there is a risk, when using high-performance concrete, of having an excellent and durable material between the cracks, as claimed by Mehta (1997). Fortunately, we shall show in section 5.3.2 that it is possible to design 60 MPa HPC with a reduced tendency to crack, even in massive conditions. The only thing is that economical issues must be put in a second position in the optimization process, unlike the present simulations.

5.2 NORMAL-STRENGTH STRUCTURAL CONCRETE

This section deals with normal-strength structural concrete: that is, the most common type of concrete. The cases of C40 concrete for bridges and C25 concrete for building will be covered.

5.2.1 C40 for bridges

Concrete is present in most bridges built nowadays. Either the structure is made of reinforced/prestressed concrete, or the main bearing elements are made of steel. But even in this last case the deck is often made up with reinforced concrete. We shall investigate in this paragraph the case of a concrete having a design compressive strength at 28 days equal to 40 MPa, intended for the structural parts of a bridge (that is, not for the foundations, where a leaner mix should be preferred).

Specifications

Concrete is often heavily reinforced in some parts of bridges, especially in the anchorage zones of prestressed structures. Therefore it should exhibit a good placeability in the fresh state. Moreover, the mixture should be pumpable, if this technique is to be used. In construction, it is common to specify high early compressive strength when the bridge is to be built by the cantilever method. Such a property allows the contractor to build a segment per day, which requires a sufficient strength to put the first prestressing cables in tension before 24 hours.

In service, bridges have to support their dead load plus important loads due to traffic and/or wind. Thus the mechanical function is predominant, while in housing acoustic insulation often governs the wall/slab thickness. This is why compressive strengths in the upper range of normal-strength concretes are generally specified for bridge construction. As far as durability is concerned, it is common to expect a 100-year lifespan for bridges, which are exposed to all weather perturbations, including freeze-and-thaw and deicing salt attacks. Therefore most regulations include stringent limitations in terms of minimum binder, or maximum water/binder ratio. Also, all precautions dealing with prevention of alkali–silica reaction must be taken. As an example, we shall look for a mixture complying with the specifications given in Table 5.2.

344   *Applications: various concrete families*

Table 5.2 Specifications for a C40 concrete for bridges.

Criterion	Value
Placeability	$\leqslant 6.5^{a}$
Slump	$\geqslant 150$ mm
Plastic viscosity (for pumpability)	$\leqslant 200$ Pa.s
Compressive strength at 24 hours	$\geqslant 20$ MPa
Compressive strength at 28 days	$\geqslant 40 \times 1.2 = 48$ MPa
Cement content[b]	$\geqslant 384$ kg/m^3
Cost	Minimum

[a] According to section 4.1.1 a value of 6 is recommendable for mixtures without superplasticizer. In the present case it is common to use a limited amount of superplasticizer, so that the maximum K' value can be increased somewhat.
[b] According to Fascicule 65 A, the French code for state-owned prestressed bridges, the cement content in kg/m^3 should be higher than $700/\sqrt[3]{D}$, where D is the maximum size of aggregate in mm.

Simulations and comparison with a real case

Only Portland cement is used as a binder, according to the present state of the art in France (for this type of concrete). A superplasticizer is added. After some preliminary simulations, it turns out that a cement content close to 400 kg/m^3 is necessary. So this value is taken tentatively, with a water/cement ratio of 0.5, and the proportions of aggregate fractions are optimized in order to maximize the placeability, according to the procedure described in section 4.2.3. The last step consists in maintaining these aggregate proportions and letting the computer solve the problem by changing the cement, water and superplasticizer contents. This process leads to the mixture presented in Table 5.3. Note that a theoretical compressive strength of 30 MPa is sought at 1 day, in order to account for the fact that the compressive strength model overestimates the early strength development of superplasticized concretes (see section 2.3.8).

In the optimized mixture, both constraints dealing with fresh concrete are active, while the early strength requirement determines the water/binder ratio and cement content. This is a general trend for this type of concrete, where the two critical issues are the fresh concrete properties (workability, pumpability, and quality of concrete facings) and the compressive strength development. For the latter, it is increasingly common to use maturity techniques, in order to monitor the real in-place compressive strength development (Carino *et al.*, 1992; Chanvillard and Laplante, 1996). The filling diagram of the simulated mix appears in Fig. 5.5. In spite of a certain gap between coarse and fine aggregate, the mixture should be reasonably pumpable, as the fine (cement) peaks are high enough, and the coarse aggregate is not overdosed.

Table 5.3 Theoretical mixture matching the specifications and real project concrete.

Constituents	C40 (simulations)	C40 from the Viaduc de Carrières
12.5/20 aggregate (kg/m^3)	708	1041
5/12.5 aggregate (kg/m^3)	342	(4/20 mm)
0/4 River sand (kg/m^3)	640	743
0/1 Correcting sand (kg/m^3)	155	(0/4, well-graded)
OPC (kg/m^3)	393	400
Superplasticizer (kg/m^3)	2.35	7.20
Retarder (kg/m^3)	–	1.0
Water (l/m^3)	172	175 ± 5[a]
w/c	0.43	0.44
CA/FA	1.32	1.40
Air (%)	1.3	–
Specific gravity	2.41	–
Yield stress (Pa)	1246	–
Viscosity after vibration (Pa.s)	94	–
Slump (mm)	150	100–150 mm[a]
Placeability	6.5	–
Segregation potential	0.81	–
Compr. strength at 1 day (MPa)	30	19[b]
Compr. strength at 28 days (MPa)	53	55[a]
Obtained characteristic strength (MPa)	–	48[a]
Elastic modulus (GPa)	41	–
Adiabatic temperature rise (°C)	55	–
Autogenous shrinkage (10^{-6})	80	–
Total shrinkage (10^{-6})	572	–
Specific basic creep (10^{-6}/MPa)	29	–
Specific total creep (10^{-6}/MPa)	56	–
Cost (FF/m^3)	508.39	–

[a] Site data, from quality control tests performed during construction.
[b] At 24 hours, 20 °C equivalent time, which means an earlier age for a cylinder cured at the same outer temperature.

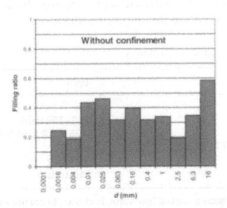

Figure 5.5 Filling diagram of the simulated C40 mix.

The simulated mixture is now compared with the C40 concrete used for the Viaduc de Carrières sur Seine,[2] built in 1993–1994 (the mixture proportions are given in Table 5.3). A compressive strength of 19 MPa at 18 hours was required for this mixture, and was finally obtained at 24 hours of 20 °C equivalent time. The two formulae are remarkably similar, although the constituents were not the same. The only marked difference lies in the dosage of superplasticizer. As already pointed out, the cement/superplasticizer affinity may vary in a large extent from one pair to another. About 1000 m^3 of the actual C40 mix was successfully pumped and placed in forms containing an amount of reinforcement close to 100 kg/m^3. The characteristic compressive strength at 28 days exceeded the design strength by about 20%, demonstrating the predominant weight of early strength requirement in this type of mixture.

5.2.2 C25 for building

Concrete for ordinary building is probably the most important market in terms of volume. These mixtures are generally produced with ordinary materials, and based upon the low level of required performance one could think that they do not deserve special attention. However, because of the intense competition that exists between the various producers, a rational optimization can have extremely beneficial consequences for the plant that makes such an effort. A typical set of specifications is given in Table 5.4. For this mix, we assume that attractive concrete facings are aimed at, with little bubbling. Moreover, the environment is supposed to be humid with moderate freeze/thaw.

Simulations and comparisons with an actual mixture

As for the materials, the standard set is maintained, except the cement strength, which is lowered to comply with current practice. In fact, blended cements are usual in Europe for this type of utilization, with a

Table 5.4 Specifications for a C25 concrete for building.

Criterion	Value
Placeability[a]	$\leqslant 6$
Slump (during one hour)	100–150 mm
Compressive strength at 28 days	$\geqslant 25 \times 1.2 = 30$ MPa
Water/cement ratio (according to NFP 18 305)	$\leqslant 0.60$
Cement content (ditto)	$\geqslant 280$ kg/m^3
Cost	Minimum

[a] In spite of a small amount of superplasticizer, such a low value is chosen to ensure easy placement, with a satisfactory aspect to concrete vertical surfaces after demoulding.

mean strength of about 45 MPa at 28 days. Alternatively, lower dosages of high-strength cement are used with supplementary cementitious materials.

The optimization procedure is similar to that used for the previous concrete. Preliminary simulations lead to a cement dosage around 350 kg/m³. So this value is taken tentatively, and the proportions of aggregate fractions are optimized to obtain a minimal K' value (maximum placeability). Finally, the paste composition is adjusted, accounting for the specifications. The obtained mixture appears in Table 5.5.

The real mixture examined was cast in 1996–1997 on a building site in Paris.[3] The cement used was a blended cement containing 19% of slag, 5% of limestone filler and 5% of class F fly ash, with a mean strength at 28 days equal to 47 MPa. The aggregate proportions and cement dosage of this concrete are similar to those of the simulated mix. However, because of differences in the nature of the materials, the active constraints are not the same in both mixtures. The compressive strength of the simulated mix is equal to the minimal value given by the specifications, while the

Table 5.5 Theoretical mixture matching the specifications, and real project concrete.

Constituents	C25 (simulations)	C25 from the Opera Victoire site, Paris
12.5/20 aggregate (kg/m³)	681	974
5/12.5 aggregate (kg/m³)	348	(4/20)
0/4 River sand (kg/m³)	590	749
0/1 Correcting sand (kg/m³)	224	(0/4)
OPC (kg/m³)	358	345
Superplasticizer (kg/m³)	1.29	2.07
Water (1/m³)	184	211 ± 5
w/c	0.51	0.60
CA/FA	1.26	1.30
Air (%)	1.4	–
Specific gravity	2.38	–
Yield stress (Pa)	1235	–
Viscosity after vibration (Pa.s)	57	–
Slump (mm)	150	159 ± 18[a]
Placeability	6	Satisfactory
Compr. strength at 28 days (MPa)	30	34.0 ± 2.6[a]
Obtained characteristic strength (MPa)	–	28.8
Elastic modulus (GPa)	32.1	–
Adiabatic temperature rise (°C)	53	–
Autogenous shrinkage (10⁻⁶)	82	–
Total shrinkage (10⁻⁶)	601	–
Specific basic creep (10⁻⁶/MPa)	49	–
Specific total creep (10⁻⁶/MPa)	98	–
Cost (FF/m³)	482.41	–

[a] Site data from quality control tests performed during construction.

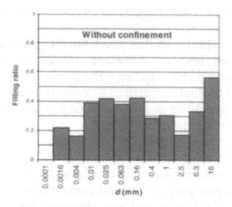

Figure 5.6 Filling diagram of the simulated C25 mix.

water/cement ratio is lower than the maximum threshold. The opposite situation is found in the real mix.

When we look at the filling diagram of the simulated mixture (Fig. 5.6), no special trend for bleeding or segregation is observed, although there is a slight gap between coarse and fine aggregate. But the cement peaks are quite high, because of the cement dosage. If a high-strength cement had been used without supplementary material, the dosage should have been lower, with more likelihood of bleeding. This is an example of the value of having different binder qualities to cover a wide range of strength, even if a high-strength cement alone permits us to obtain such a strength range.

5.3   HIGH-PERFORMANCE CONCRETE

Several definitions have been proposed for high-performance concrete (HPC). The author prefers the FIP definition (FIP, 1990), which has the advantage of being simple and objective. This definition assumes that an HPC has either a (cylinder) compressive strength at 28 days greater than 60 MPa, or a water/binder ratio lower than 0.40. In the last 10 years this range of concrete has spread in various sectors of the construction industry, mostly for durability purposes (CEB, 1994). However, it is still commonly designated by the specified compressive strength. We shall deal first with 'basic' HPC: that is, a conventional structural concrete in which the water/binder ratio has been lowered because of the action of superplasticizer. Then three special HPCs will be highlighted, showing the versatile character of this new generation of concrete.

5.3.1 'Basic' high-performance concrete

In this section, we shall concentrate on HPC in the range 60–80 MPa (design compressive strength at 28 days). After presenting a typical set of specifications, we shall proceed with simulations, which will be compared with real formulae used in two bridge projects.

Specifications

Let us assume that an HPC is devoted to cast-in-place elements of a bridge. To ensure easy placement, a placeability of 8 can be specified. As far as the slump is concerned, we have seen that HPC generally has a low yield stress, but a high plastic viscosity (see section 2.1). Therefore, if easy handling of fresh concrete is aimed for, experience has shown that a slump higher than 200 mm is desirable (Malier and de Larrard, 1993). However, this constraint could be relaxed with the use of certain new-generation superplasticizers. For an easy pumping process, it is also advisable to limit the plastic viscosity[4] to 200 Pa.

As far as compressive strength at 28 days is concerned, a margin of 20% between design and target strength is a good rule of thumb. However, in any specific case where the production conditions are known, this margin can be evaluated more precisely (see section 4.5.3). We shall then simulate the cases of 60 and 80 MPa design strength. Finally, as far as durability is concerned, we shall stick to the FIP definition, which consists in limiting the water/binder ratio to a maximum of 0.40. A summary of specifications for both HPCs is given in Table 5.6.

Simulations

Concretes matching the preceding specifications were simulated. The same constituents as those presented in section 4.2.2 were used. The C60 mixture was proportioned without silica fume (which is the present state

Table 5.6 Specifications of HPCs.

Criterion	C60	C80
Placeability	⩽ 8	⩽ 8
Slump	⩾ 200 mm	⩾ 200 mm
Plastic viscosity	⩽ 200 Pa.s	⩽ 200 Pa.s
Compressive strength	⩾ 72 MPa	⩾ 96 MPa
Water/binder ratio	⩽ 0.40	⩽ 0.40
Silica fume/cement ratio	0%	8%
Superplasticizer dosage	Saturation	Saturation
Cost	Minimum	Minimum

of the art in France), while the C80 contained a proportion of silica fume equal to 8% of the cement weight. Obtained mixture proportions appear in Table 5.7. As shown in section 4.4.3, cost considerations lead to the introduction of silica fume for compressive strength around 70–80 MPa, even if this product is beneficial at lower grades, for all secondary properties (except the colour of concrete facings). For the two mixtures, the saturation dosage of superplasticizer was taken. This choice is not the best as far as economy is concerned. However, the saturation dosage not only guarantees the least deformability of hardened concrete, but also limits the slump loss hazards. Therefore, it is the author's opinion that a

Table 5.7 Theoretical mixtures matching the specifications and real project concretes.

Constituents	C60 (simulations)	Joigny C60	C80 (simulations)	Iroise C80
12.5/20 aggregate (kg/m³)	722	1030	728	634 (10/16)
5/12.5 aggregate (kg/m³)	334	(5/20)	337	423 (4/10)
0/4 River sand (kg/m³)	644	649	650	744
0/1 Correcting sand (kg/m³)	120	105	121	(well-graded)
OPC (kg/m³)	464	450	434	450
Silica fume (kg/m³)	0	0	34.7	36
Superplasticizer (kg/m³)	9.3	11.25	12.1	17.7
Retarder (kg/m³)	–	4.5	–	1.57
Water (l/m³)	141	165	126	132
w/c	0.32	0.35	0.31	0.32
CA/FA	1.38	1.36	1.38	1.45
Specific gravity	2.43	–	2.44	2.47
Yield stress (Pa)	461	–	609	–
Viscosity after vibration (Pa.s)	200	–	200	–
Slump (mm)	265	190–250[a]	245	195–255[a]
Placeability	7.39	excellent	7.46	excellent
Compressive strength (MPa)	72	78[a]	96	95.7[a]
Obtained characteristic strength (MPa)	–	68	–	83.5
Elastic modulus (GPa)	45.4	–	50.7	47
Adiabatic temperature rise (°C)	58	57[b]	58	52[b]
Autogenous shrinkage (10⁻⁶)	175	171	212	–
Total shrinkage (10⁻⁶)	565	609	532	–
Specific basic creep (10⁻⁶/MPa)	23.2	24	13.2	–
Specific total creep (10⁻⁶/MPa)	44.8	39	26	–
Cost (FF/m³)	598	–	658	–

[a] Site data, from quality control tests performed during construction.
[b] Value calculated by correction of heat losses in a semi-adiabatic test (Schaller *et al.*, 1992).

dosage close to the saturation one is advisable for most cast-in-place HPC. This rule has to be relaxed only when a *maximum* slump (or minimum yield stress) is specified, e.g. for slope stability purposes (see section 4.1.1).

Real examples

The C60 mixture comes from the Joigny project. This was an experimental bridge cast in 1988, built within the frame of a national project (Malier *et al.*, 1992; Cadoret and Richard, 1992; Schaller *et al.*, 1992). The aim was to demonstrate the feasibility of using a central-plant-batched, non-silica-fume 60 MPa HPC for a cast-in-place bridge. The formula of the Joigny C60 concrete appears in Table 5.7. This concrete was of a very fluid consistency (no rheometer was available at this time to provide a more complete characterization), and could be pumped without difficulty. Strangely enough, the obtained characteristic strength was higher than the strength obtained in the laboratory study. The reason probably lies in the addition of a retarder, which might increase the final strength, together with the fact that the bridge was cast in about 24 hours during a very cold day. A cold early curing of cylinders is also a factor of higher final strength. The characteristics of the simulated C60 and Joigny concretes are not directly comparable, because the computer has not been fed with the Joigny material data. However, it can be seen that the mixture composition and characteristics, as obtained by simulations with relatively standard materials, are very close to a concrete selected by experienced practitioners (Cadoret and Richard, 1992).

In the Iroize project a C80 concrete was specified and fully utilized in the stability calculations (Le Bris *et al.*, 1993). At the time of its completion, and as a full concrete, cable-stayed bridge with central suspension, the 400 m main span of the Iroize bridge made it a world record. The bridge was made up with several kinds of concrete: C35/C45 normal strength, normal weight concretes for the main piers and abutments, C32 lightweight concrete for the central part of the main span (see section 5.5.1), and C60/C80 HPCs for the pylons, which were subject to heavy loading because of the wind effects. The C80 HPC mixture composition is given in Table 5.7. Again, the main proportions are very similar to those of the simulated C80. This concrete exhibited a superior placeability, which was necessitated by the casting mode (over-reinforced pylons, 4 m high lifts). Although this concrete was virtually non-vibrated, the facings at demoulding were very satisfactory, with no bubbling or honeycombing.

5.3.2 Low-heat HPC for nuclear power plant

In French nuclear technology, power production reactors are surrounded by a double-wall concrete containment. The inner wall is prestressed, while the outer wall is reinforced. The main function of the inner wall is to confine radioactive release in the case of a severe accident. This wall must be at the same time strong, gas-tight and, as much as possible, crack free in spite of its thickness of 1.2 m. The optimum design of concrete for this type of structure is therefore of great importance (Ithurralde and Costaz, 1993).

Specifications

In case of accident, a given overpressure (0.4 MPa) must be sustained by the inner wall, which corresponds to a given load per unit surface. Without internal pressure, the prestress applies this load to the concrete wall: therefore the higher the strength, the thinner the wall. On the other hand, everything being constant, we have seen that the cracking risk in massive conditions increases with the concrete strength (see section 5.1); but a lower strength means a higher air permeability (see section 2.6). Thus writing strictly rational specifications in such a case appears rather complicated: the optimization process should ideally be performed at the structure level, not at the material level. This is beyond present engineering capabilities.

A more modest approach is to fix a moderately high strength (which guarantees negligible air leakage through the concrete porosity), and to try to minimize the adiabatic temperature rise, which is the main cause of cracking. Clearly, looking for a minimum heat production leads us to take maximum advantage of the binders. This means a complete deflocculation of the fine grains, which pushes us towards high-performance concrete technology. According to French nuclear engineering practice, the concrete is pumped and placed between a reinforcement that is light in most parts, but can be dense in some special critical parts. So the fresh concrete specifications are the same as those taken in the 'basic' HPCs (see section 5.3.1). The final set of specifications is given in Table 5.8.

Simulations

In section 4.4.3 we saw that silica fume allows us to reduce the heat of hydration for a given strength level. This is a positive consequence of the

Table 5.8 Set of specifications for a special power-plant C60 HPC.

Criterion	Value
Slump	$\geqslant 200$ mm
Placeability	$\leqslant 8$
Plastic viscosity	$\leqslant 200$ Pa.s
Compressive strength	$\geqslant 72$ MPa
Adiabatic temperature rise	Minimum

filler effect, which leads to a lower water content and a lower quantity of hydrates between densely packed particles. However, there may be some limits in the use of silica fume when dealing with durability (or at least when the ultra-high-strength range is not being addressed; see section 5.3.4). A 0.15 silica fume/cement ratio, given the protected environment of a inner wall, seems reasonable.

If a cement content sufficient for preventing bleeding and segregation is taken together with a full use of superplasticizer, the obtained strength will be around 100 MPa or more. However, we have seen in section 5.1 that heat of hydration is always, for a particular type of concrete, an increasing function of compressive strength. Therefore we have to find a means of limiting the compressive strength in the desired range without disturbing the granular structure of the concrete. The most natural way is then to replace a part of the Portland cement by a limestone filler, the heat contribution of which is likely to be negligible (see section 2.2.4). We have finally asked the computer to solve the optimization problem as described in Table 5.8, for a concrete having a Portland cement/limestone filler/silica fume ternary binder system. The result appears in Table 5.9. Note that both temperature rise and autogenous shrinkage are reduced, as compared with C60 'basic' HPC (see section 5.3.1).

Real case

The Civaux II containment (near Poitiers, France) was built in a special HPC for improving the performance of this critical part of the structure (de Larrard and Acker, 1990; de Larrard *et al.*, 1990; CEB, 1994). The mix-design study was conducted with the same specifications (except for plastic viscosity, as no rheometer was available at this time). The obtained formula is given in Table 5.9. It is very similar to the theoretical formula. The water content was higher because of the shape of the particles (only crushed materials were used). This led to a lower strength; the 60 MPa requirement was just matched.

The only defect of this concrete was its poor pumpability, which originated in its high plastic viscosity (de Larrard *et al.*, 1997a). In fact the Civaux II project was one of the site experiences that pushed LCPC to launch the development of a fresh concrete rheometer. Concerning heat production, a temperature rise of 30 °C was measured during construction, instead of 40 °C in the Civaux I containment (CEB, 1994). The latter was made up with ordinary C45 concrete, and located nearby. At the end of construction, both containments were tested by inflating them with pressurized air, and then measuring the global air-leakage at the whole structure level. The amount of air lost in 24 hours was 0.38 and 0.19% of the total mass for Civaux I and II respectively. For comparison purposes the maximum limit specified by nuclear authorities was 1%. The value obtained for Civaux I was an absolute record over more than 40

Table 5.9 Theoretical mixture matching the specifications and real project concrete.

Constituents	Power plant special C60 (simulations)	Civaux II HPC
12.5/20 aggregate (kg/m³)	759	815
5/12.5 aggregate (kg/m³)	351	318
0/4 River sand (kg/m³)	677	818
0/1 Correcting sand (kg/m³)	127	(0/5)
OPC (kg/m³)	262	266
Limestone filler (kg/m³)	69	57
Silica fume (kg/m³)	39	40
Superplasticizer (kg/m³)	13.1	9.1
Retarder (kg/m³)	–	0.93
Water (l/m³)	121	161
w/c	0.49	0.55
CA/FA	1.38	1.39
Specific gravity	2.41	–
Yield stress (Pa)	900	–
Viscosity after vibration (Pa.s)	200	–
Slump (mm)	200	180–230
Placeability	7.55	Satisfactory
Compressive strength (MPa)	72	66.1
Obtained characteristic strength (MPa)	–	60.7
Elastic modulus (GPa)	48	–
Adiabatic temperature rise (°C)	46.9	–
Autogenous shrinkage (10^{-6})	112	
Total shrinkage (10^{-6})	470	
Specific basic creep (10^{-6}/MPa)	11.3	
Specific total creep (10^{-6}/MPa)	23.1	
Cost (FF/m³)	598	–

Site data, from quality control tests performed during construction.

containments built in France in the last 30 years, but the use of an optimized HPC in Civaux II led to a doubling of this record (that is, a halving of the air losses).

5.3.3   Ultra-stable HPC for composite bridge deck

The application addressed in this section deals with composite bridge decks, in which a concrete slab is supported by two steel beams. This type of bridge is popular in France, in the span range around 100 m. The steel beams are first launched, and then the concrete slab is cast in place, which makes a fast and simple construction process (no scaffolding is needed). The weakness of this technology lies in the difference in delayed deformations between the two materials. Concrete shrinkage is partially restrained by the beams, so that severe transverse cracking may occur in the parts of the bridge where traffic loads provoke additional tensions. A classic technique to overcome this cracking process is to lift the extreme abutments at the end of the construction, so that a permanent positive moment is introduced in the structure. But this 'free prestress' tends to vanish by creep/relaxation mechanism in the concrete slab (Kretz and Eymard, 1993).

In addition to its durability properties, HPC can be attractive because of its low creep, combined with its high compressive strength at early age, which is a factor for fast construction. But as far as the cracking phenomenon under restraint is concerned, we have seen that the higher the strength, the more likely the occurrence of cracking, especially at early age (see section 5.1). Considering that cracking is inevitable in the long run, an optimization strategy can be to try to limit thermal and total shrinkage, which govern the crack opening. In fact, thin cracks are not detrimental to the structure behaviour as long as shear forces can be transmitted through the crack (the 'dowel effect').[5]

To minimize the total shrinkage, the use of some special admixtures could open new avenues in the future (Folliard and Berke, 1997). However, keeping the 'traditional' components, the idea is to try to limit the paste volume as much as possible, while keeping an acceptable placeability for the fresh mixture. Then comes the concept of using a very fine binder, which exerts virtually no loosening effect (see section 1.1.3) on the aggregate skeleton. Moreover, the coarse part of cement is of little use in HPC, because the final degree of hydration is around 60%, which means that 40% of the clinker volume, mainly located in the coarse fraction, remains unhydrated.

Simulations

A C70 HPC was simulated by using a fine Portland cement (90% of particles being finer than 8 μm) together with a selected fly ash of similar size. No silica fume was used, since for medium-thickness structures where thermal effects may take place it is preferable to use pozzolans having a slow reactivity. The cement is a commercial product mainly used in injection grout technology. As the emphasis was placed on preventing cracking, it was chosen not to require any particular flowing property, but rather a minimum placeability corresponding to a thorough vibration. Therefore a maximum K' value of 9 was specified, with a target strength of 84 MPa. As far as shrinkage limitation is concerned, optimizing the mixture by looking for minimal shrinkage may lead to a pure Portland cement binder (see section 4.4.2) in such a low quantity that excessive bleeding takes place. To overcome this problem another strategy was adopted, which is to minimize the water content. This leads to maximization of the filler effect played by the binder combination in relation to the aggregate skeleton. The various specifications are summarized in Table 5.10.

Using the same aggregate fractions as in the previous simulations, the mixture displayed in Table 5.11 was obtained. Compared with 'standard C70 HPC' (such as those simulated in section 5.3.1), total shrinkage and

Table 5.10 Specifications for an ultra-stable HPC.

Criterion	Value
Placeability	$\leqslant 9$
Compressive strength	$\geqslant 72$ MPa
Water content	Minimum

Table 5.11 Theoretical mixture matching the specifications and trial mixture.

Constituents	USHPC (simulations)	Trial mixture
12.5/20 aggregate (kg/m³)	759	716
5/12.5 aggregate (kg/m³)	388	568
0/4 River sand (kg/m³)	605	663
0/1 Correcting sand (kg/m³)	274	118
Fine Portland cement (kg/m³)	217	213
Fine fly ash (kg/m³)	87	95
Superplasticizer (kg/m³)	6.37	20
Water (l/m³)	112	103
w/c	0.52	0.55
CA/FA	1.31	1.64
Air (%)	1.7	0.6
Specific gravity	2.45	–
Yield stress (Pa)	1383	–
Viscosity after vibration (Pa.s)	497	–
Slump (mm)	130	215
Flowing time (s)	–	9.4
Placeability	9	Acceptable
Compressive strength (MPa)	84	87
Elastic modulus (GPa)	54.3	58.0
Adiabatic temperature rise (°C)	39	42
Autogenous shrinkage (10^{-6})	62	91
Total shrinkage (10^{-6})	348	255
Specific basic creep (10^{-6}/MPa)	6.3	–
Specific total creep (10^{-6}/MPa)	12.5	–
Cost (FF/m³)	1031	–

Source: de Larrard *et al.* (1997b).

adiabatic temperature rise were reduced by 32 and 31% for a total binder content of only 304 kg/m^3. However, the filling diagram did not suggest too high a bleeding risk, as the cement peaks were sufficiently high (Fig. 5.7).

Trial batch and experimental characterization

A similar mixture has been produced and characterized, based upon the same specifications (de Larrard *et al.*, 1997b). The mixture composition was optimized with an earlier version of the packing model, without regard to the flowing properties (see Table 5.11, column 3). Its composition is very close to that of the simulated mixture. Quite a viscous concrete was produced, but showing a fair placeability, as illustrated by the very low measured air content. No excessive bleeding was noted. Of course, this concrete did not look pumpable, but this property was out of the specifications.

Obtained total shrinkage is even lower than expected, and adiabatic temperature rise is comparable to that of a conventional 20 MPa mixture (see section 5.1). Paying attention to the very low predicted creep values, one must admit that this concrete would be very suitable for the intended use, the only limitation coming from the limited availability of the binders, together with their high cost. Also, a rather slow strength development is to be expected for this mixture, because of its high water/cement ratio and fly ash content.

5.3.4 Ultra-high-performance mortar

Having dealt with realistic applications, it is interesting to investigate the ultimate capabilities of HPC technology. We shall then try to develop a structural[6] cementitious material having a maximum compressive strength.

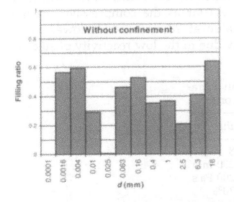

Figure 5.7 Filling diagram of the ultra-stable HPC for composite bridge deck.

Simulations

In section 4.2.1 (rule 9), we saw that for a given workability the maximum strength is attained with a pure paste. However, it is not possible to use such a material to build structural elements, because of its high deformability and brittleness. A minimum aggregate content has to be maintained, which we shall take arbitrarily equal to 30% (in volume). For easy transport and placing, the same requirements as those specified in basic HPC will be kept. Table 5.12 summarizes the set of specifications.

Simulations have led to the mixture proportions given in Table 5.13. It turns out that the aggregate phase is reduced to fine sand. This is due to the effect of maximum paste thickness on compressive strength (see section 2.3.3): the search for maximum strength leads the computer to minimize the maximum size of aggregate (MSA). A water/cement ratio of 0.18 is suggested by the computer, together with a high silica fume content (23% of the cement weight). Note that this water/cement ratio is out of the range for which models dealing with hardened concrete properties have been calibrated. Therefore all corresponding predictions are extrapolations. With this reservation, it can be seen that a compressive strength of about 263 MPa is theoretically attainable. As expected, the fact of having a high paste content leads to a high shrinkage and a high adiabatic temperature rise: this material is supposed to boil when thermally insulated!

Tests on a similar mixture

A series of mixtures were produced with the same goal of maximizing the compressive strength (de Larrard and Sedran, 1994). An earlier version of the packing model was used to determine the optimum proportions. The best mixture obtained appears in Table 5.13. Here again, both formulae look similar, even if the simulations were based upon different materials from those used in the experiments. In particular, the water/cement ratios are equal. As far as the compressive strength is concerned, the experimental value at 28 days is far lower than the predicted value. This is probably due to the low reactivity of silica fume

Table 5.12 Specifications for an ultra-high-performance mortar.

Criterion	Value
Placeability	$\leqslant 8$
Slump	$\geqslant 200$ mm
Plastic viscosity	$\leqslant 200$ Pa.s
Aggregate content	$\geqslant 30\%$
Compressive strength	Maximum

Table 5.13 Theoretical mixture matching the specifications and laboratory-tested mixture.

Constituents	UHPM (simulations)	Trial mixture (de Larrard and Sedran, 1994)
0/1 Fine sand (kg/m³)	800	813
OPC (kg/m³)	1072	1081
Silica fume (kg/m³)	250	334
Superplasticizer (kg/m³)	52.4	51.8
Water (l/m³)	152	167
w/c	0.18	0.18
Air (%)	5.1	2.0
Specific gravity	2.32	–
Yield stress (Pa)	830	–
Viscosity after vibration (Pa.s)	200	–
Slump (mm)	200	–
Flowing time (s)	–	1.0
Compressive strength (MPa)	263	165 (236[a])
Elastic modulus (GPa)	64	50.9
Adiabatic temperature rise (°C)	102	–
Autogenous shrinkage (10^{-6})	327	–
Total shrinkage (10^{-6})	1447	–
Specific basic creep (10^{-6}/MPa)	16.3	–
Specific total creep (10^{-6}/MPa)	32.1	–
Cost (FF/m³)	1570	–
Placeability	7.63	Excellent

[a] Compressive strength at 7 days after thermal treatment at 90 °C.

at such a low water/binder ratio. However, a thermal treatment at 90 °C greatly improves the final performance. The compressive strength obtained at 7 days is closer to the predicted final strength.

Since these experiments, important developments have taken place in the field of similar materials (Richard and Cheyrezy, 1994; Richard *et al.*, 1995). These authors demonstrated that, by keeping the same type of formula, with a higher curing temperature (up to 250 °C) and a high pressure applied to the hardening mixture, compressive strengths close to 800 MPa were attainable. However, to overcome the natural brittleness of ultra-high-performance mortars, steel fibres had to be incorporated in the material. It is even claimed that such fibres may replace secondary reinforcement, opening a new avenue for prestressed concrete structures made up of only fibre-reinforced UHPM and prestressing strands. By using two types of steel fibre it is even possible to obtain a material that displays a hardening behaviour after cracking in pure tension (Rossi and Renwez, 1996). In other words, the strength after cracking is superior to that of the uncracked matrix. Otherwise, the incorporation in such a mortar of small and hard aggregate particles with a maximum size of 6 mm leads to a lower compressive strength, but also to a much lower binder demand, producing an ultra-high-strength concrete of reasonable cost (de Larrard *et al.*, 1997b).

5.4   CONCRETES WITH SPECIAL PLACING METHODS

In this section the case of concrete families inducing special placement techniques will be highlighted. Roller-compacted concretes are covered by special rules, being at least partially dealt with by soil mechanics engineers. Shotcretes (sprayed concretes) used by the wet method are also subject to special practice as materials devoted mainly to repair purposes or to prototyping. On the other side of the consistency range, self-compacting concretes are very recent in the concrete landscape, and no mix-design methods are currently standardized in any country (to the best of the author's knowledge). We shall show that all these 'cousins' of mainstream structural concrete can be re-integrated into the cementitious material family with the help of the scientific approach developed in this book.

5.4.1   Roller-compacted concrete

Roller-compacted concretes (RCC) are generally devoted to massive constructions such as dams. They can be also used for pavements. In most cases, tremendous quantities of materials are used. Therefore there is a strong incentive to decrease the unit cost as much as possible. This trend, together with the prevention of thermal cracking, leads to the use of very dry concrete, with very low binder content. As a matter of fact, we have seen that the less workable the concrete, the cheaper it is for a given strength (see rule 2 in section 4.2.1). From a theoretical viewpoint, the question of RCC proportioning is one of the purest since RCC is really a packing of particles, placed with methods inducing a high compaction energy, with little concern about flowing behaviour.

Specifications

Given the fact that RCC placement is performed with mechanical compaction and vibration, this type of concrete may be considered as a packing, with a high compaction index. Remembering the dry particle experiments reviewed in chapter 1, a K value of 9 seems reasonable, as a first approximation. Here the granular system considered is the reunion of aggregate and binders, which means that the porosity is filled with water *and* entrapped air. Other requirements deal with compressive strength (at 28 days or at later ages) and adiabatic temperature rise (which can develop fully in very massive dams). In some particular cases durability requirements (minimum binder content, maximum water/

binder ratio etc.) can be added, when the concrete surface is submitted to an aggressive environment. However, RCC is generally unreinforced, which removes the usual corrosion-related concerns.

Example

We shall review an example, extracted from a large study that involved several laboratories in France (BaCaRa, 1994). An RCC mixture including seven aggregate fractions, a limestone filler and a blended cement had to be optimized on the basis of a fixed cement content of 120 kg/m^3. The cement strength at 28 days was equal to 45 MPa. Grading curves of the products were provided, and it was indicated that all aggregate particles had a round shape. The mutual proportions of aggregate fractions were determined by reference to an 'ideal' grading curve: the 'Talbot' curve (power law with a $\frac{1}{3}$ exponent). Mass percentages are given in Table 5.14. The aim of the interlaboratory tests was to compare various methods for the determination of optimum water content and corresponding mechanical properties. For this type of dry mixture the goal is to find the water content that leads to the best particle packing. With higher dosage the water in excess loosens the particle arrangement. Conversely, with lower dosage there is a 'bulking' phenomenon, which overcomes the good compaction of the system, because of capillary forces acting at the fine particle level (Fig. 5.8).

The compressible packing model (CPM) is a suitable tool for optimizing the mutual aggregate fractions; it is more accurate and reliable than the use of any reference curve (see section 1.4). However, for demonstrating its capability to predict optimum water content in an RCC, we simulated the mixture given in Table 5.15, in order to predict the voids content. The residual packing densities of constituents were not

Table 5.14 Proportions of dry materials matching Talbot's reference curve.

Constituents	% (in mass)
CHF 45 slag cement	5.00
Limestone filler	5.07
0/3	14.32
1/3	10.15
3/8	14.59
10/20	19.02
20/31.5	11.18
31.5/40	6.60
40/63	14.07

Source: BaCaRa (1994).

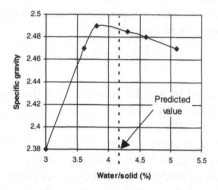

Figure 5.8 Results of the Proctor test (performed on 400 mm dia. mould) for the investigated mixture (BaCaRa, 1994).

given in the original reports. Typical values for similar materials, taken from past studies, were used in the calculations. The air content was predicted by equation (2.24) (with a slump equal to 0), and the free water content was calculated by subtraction from the voids content. Then the compressive strength was evaluated by using the general model developed in section 2.3. Obtained predictions are given in Table 5.15, to be compared with the experimental results (in Table 5.16).

The Proctor experimental curve led to a water/solid percentage of 4.0%. Following normal practice, this dosage should be matched in production with the following margin: (−0.5 to +1%), which gives the 3.5–5% interval. Then the theoretical value of 4.19% falls well in the middle of the specified interval (Fig. 5.8). As far as specific gravity is concerned, the prediction of CPM is close to the experimental values obtained by the various compaction methods. Finally, the estimated strength is somewhat lower than the measured values, while the error is close to the usual one: that is, 2–3 MPa. The compressive strength of RCC in production is usually rather scattered, as can be seen in Table 5.17.

Table 5.15 Simulation results for RCC.

Voids content (for $K = 9$, in %)	11.05
Air content (%)	1.1
Free water content (l/m³)	99.5
w/c	0.83
Water/solid ratio (%)	4.19
Specific gravity of fresh concrete	2.48
Compressive strength at 28 days (MPa)	14.3

Table 5.16 Experimental results from BaCaRa (1994) for a water/solid ratio of 4% (optimum Proctor value).

	Compaction technique			
	Proctor method	VeBe table	Kango hammer	VCPC (vibro-compaction)
Specific gravity	2.50	2.47	2.49	2.38
Compressive strength at 28 days, measured on 250 × 500 mm cylinders (MPa)	–	16.3	18.0	15.2[a]

[a] 300 × 600 mm cylinders.

Table 5.17 Strength data from real projects.

RCC dam	Obtained compressive strength (in laboratory). Mean value (MPa) ± standard deviation
Knellpoort (South Africa)	17.2 ± 7.8
Wolwedans (South Africa)	15.8 ± 4.5
Puding (China)	23 ± 2.3

Source: BaCaRa (1994).

Therefore examination of this laboratory concrete shows that the general approach developed in this book applies reasonably well to RCC.

Further interest in using a sound packing model rather than relying upon laboratory tests is provided by the possibility of dealing with concrete having a very large maximum aggregate size (MSA). RCCs with MSA up to 100 mm and more are common, but an experimental optimization of this type of mixture is very difficult to perform in the laboratory, for obvious reasons. However, one may measure the packing density of aggregate fractions, accounting for the wall effect exerted by the mould (see section 3.1.4), and then calculate the packing density of the whole mixture either in bulk or in a container of limited dimensions.

5.4.2 Shotcrete (wet process)

Shotcrete is a concrete that is sprayed with a nozzle towards a form of any orientation. The fresh mix is intended to stick on the surface (which can be either the form or a previously sprayed concrete layer). This technology is mainly used for repair works, or for prototyping. Even applications where shotcrete has a permanent structural function can be found nowadays. Shotcrete can be either premixed, pumped and sprayed (wet process) or moistened just before shooting, with a water injection in the nozzle (dry process). Only the first technology will be investigated hereafter.

Specifications

As usual, the specifications dealing with this type of concrete come logically from the different stages induced by its utilization.

First of all, the mixture has to be *pumpable*. This is obtained by trying to obtain an even filling diagram, together with a slump of at least 50 mm (see section 4.1.1). Theoretically, the plastic viscosity should be also limited. However, this constraint can be defined only for mixtures that are fluid enough: that is, which have a slump higher than 100 mm (see section 2.1.1). Moreover, the pumping length is generally low in shotcrete technology. Therefore the plastic viscosity requirement can be removed in this particular case.

Second, once concrete has been shot, it must *compact* itself on the wall. This corresponds to a certain minimum placeability, which we describe through a maximum value of the compaction index, K'. Clearly, the shooting process is less efficient than the usual placing technique, which entails the action of gravity and vibration. As a preliminary rule, we shall assume that a reduction of 0.5 has to be applied with regard to the values given for the conventional technology, where concrete is placed by gravity and vibration. Then the maximum compaction index will be equal to 5.5 and 7.5 for mixtures without and with superplasticizer respectively.

Third, *fresh concrete must stick to the support*. In order to analyse the stability of a freshly sprayed concrete layer, let us study the two following cases. In the first case (an overhead layer), concrete is sprayed on to a ceiling. Let us assume that fresh concrete is perfectly cohesive, but develops no bond with the wall. The maximum thickness is attained when the concrete weight is just balanced by the effect of atmospheric pressure (Fig. 5.9). We have

$$\rho g h = P \tag{5.4}$$

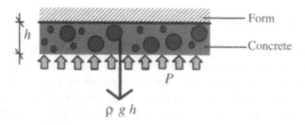

Figure 5.9 Stability of an overhead fresh concrete layer, assuming perfect cohesion.

where ρ is the concrete unit mass, g is the acceleration due to gravity, h is the thickness of the concrete layer, and P is the atmospheric pressure. For $\rho = 2400 \text{ kg/m}^3$, $g = 10 \text{ m/s}^2$ and $P = 10^5$ Pa, we find $h = 4.16$ m. This value is very large, compared with actual layer thickness. We conclude that the limiting factor for overhead stability is not the general force equilibrium, but rather the fresh concrete cohesiveness: if the coarse aggregate phase is able to sink throughout the concrete layer, gravity will provoke the fall of gravels, while the mortar will stick to the ceiling (Fig. 5.10). So the main requirement for overhead stability is the absence of segregation in the fresh mixture.

Let us now examine the stability of a vertical layer. Here it is the shear resistance of fresh concrete that must resist the effect of gravity (Fig. 5.11). This requirement gives the following equation:

$$\rho g h < \tau_0 \tag{5.5}$$

where τ_0 is the fresh concrete yield stress. Following this equation, the higher the yield stress, the thicker the fresh concrete layer will be before failure.

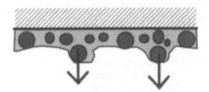

Figure 5.10 Fall of an overhead fresh concrete layer by lack of cohesiveness.

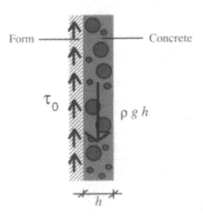

Figure 5.11 Stability of a vertical concrete layer.

A numerical application of this equation leads to very realistic orders of magnitude (see Table 5.18). At this point, one may wonder why it is possible to shoot several concrete layers before setting. The reason lies in the *thixotropic* character of fresh concrete. If a shear stress is applied to a fresh mixture which has remained at rest for some minutes, the minimum value which leads to concrete flowing, called the shear yield stress at rest (de Larrard *et al.*, 1997c), can yield several times the value of 'conventional' yield stress (Hu and de Larrard, 1996).

Finally it is concluded that the requirements for optimum 'shootability' (Beaupré, 1994) are equivalent to specifying a minimum yield stress (or a maximum slump), together with a minimum segregation potential. In addition to fresh concrete requirements, shotcrete is subject to any type of specifications dealing with hardening/hardened concrete. In particular, it is possible to specify HPC properties (high strength and/or low water/binder ratio), and to use all the ingredients used in normally cast structural concrete.

Simulations and comparison with current state of the art

We shall simulate the case of a C30 shotcrete. Since the in situ properties of these mixtures are quite scattered, we shall adopt a target compressive strength of 40 MPa. No superplasticizer will be used in the mix design. All specifications are given in Table 5.19.

Table 5.18 Maximum thickness of a vertical concrete layer as predicted from equation (5.5). The assumed unit mass of fresh concrete is 2400 kg/m^3, and the slump is evaluated from equation (2.19).

Yield stress (Pa)	Slump (mm)	Thickness (mm)
2500	0	104
2000	41	83
1500	114	63
1000	186	42

Table 5.19 Specifications of a C30 shotcrete.

Type of problem addressed	Quantitative requirement
Pumpability	Slump $\geqslant$ 50 mm and $S \leqslant 0.85$
Placeability	$K' \leqslant 5.5$
Shootability	Slump $\leqslant$ 70 mm and $S \leqslant 0.85$
Strength	$fc_{28} \geqslant 40$ MPa
Durability	$w/c \leqslant 0.5$

With the same materials as those used in the previous simulations, the mixture given in Table 5.20 is the one that matches all the specifications at minimum cost. The filling diagram of the mix appears in Fig. 5.12.

As a general rule, shotcretes sprayed by the wet process are rather rich and sandy mixtures. The cement content is generally between 400 and 450 kg/m³ (Beaupré, 1994). This condition is verified by the present mix, which also approximately matches ACI specifications in terms of aggregate grading (Fig. 5.13).

Table 5.20 Mixture proportions and properties of the C30 shotcrete (simulations).

Constituents	C30 shotcrete
12.5/20 aggregate (kg/m³)	322
5/12.5 aggregate (kg/m³)	420
0/4 River sand (kg/m³)	753
0/1 Correcting sand (kg/m³)	176
OPC (kg/m³)	433
Water (l/m³)	206.9
w/c	0.471
CA/FA	0.80
Air (%)	2.7
Specific gravity	2.314
Yield stress (Pa)	1893
Slump (mm)	50
Placeability	5.5
Segregation potential	0.817
Compressive strength (MPa)	40
Elastic modulus (GPa)	31.4
Cost (FF/m³)	501.29

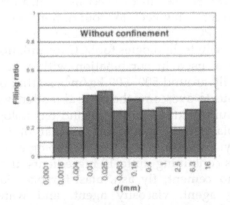

Figure 5.12 Filling diagram of the simulated C30 shotcrete.

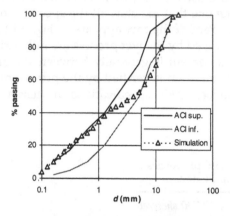

Figure 5.13 Aggregate size distribution of the simulated C30 shotcrete, compared with ACI 506 requirements.

For a given strength level, shotcretes are generally more expensive than ordinary cast concrete, because the high amount of sand entails a higher cement content. Furthermore, a part of the produced concrete is lost by rebound (between 3 and 15% of the sprayed volume; Beaupré, 1994). However, cost savings can be made since fewer forms are needed, as compared with cast concrete. Also, the construction rate is increased by the use of shotcrete technology.

5.4.3   Self-compacting concrete

By definition, self-compacting concretes (SCC) do not need vibration for being placed and consolidated. They can flow through congested reinforcement without presenting severe segregation or blockage (Sedran, 1995). These concretes can be used in a variety of applications, from low-strength, mass concrete devoted to foundation works (Tanaka *et al.*, 1993), to C80 HPC for bridges (de Larrard *et al.*, 1996a). There are currently two main technologies for SCC formulation: in the first, a superplasticizer is used in combination with a large quantity of fine materials (usually around 500–600 kg/m^3), so that the paste phase is sufficiently viscous to overcome the natural tendency of aggregate to sink in the very fluid mixture. In the second technology viscosity agents are added, playing the same role as that of fine particles in the first technology. Sometimes the two technologies are mixed. One can even find SCCs containing up to 10 ingredients: three fractions of aggregate, Portland cement, fly ash, blast-furnace slag, superplasticizer, air-entraining agent, viscosity agent, and water. Fortunately, it is possible to design simpler mixtures that have a satisfactory behaviour. In this section, we shall restrict ourselves to SCCs *without* viscosity agents.

Specifications

From the fresh concrete viewpoint SCCs are probably the mixtures that are the most difficult to design, because they have to match a plurality of criteria.

First, they have to be *placeable under their own weight*. This induces a need for a K' index lower than in usual mixtures. Based on a large experimental plan already and abundantly used in section 2.1 (Ferraris and de Larrard, 1998), a value of 7 for a superplasticized mix seems to be a good order of magnitude.

Second, for satisfactory flowing behaviour, where aggregate is borne by the fine phase, the fine particles must be packed sufficiently in the aggregate interstices. Hence it is the same type of requirement as for bleeding limitation (see section 2.1.7). If the condition of having a slump flow[7] higher than 600 mm, without visual separation, is taken as an experimental 'self-compactability' criterion, the mixtures matching this criterion in the data set (Ferraris and de Larrard, 1998) are those for which the K'_c coefficient (contribution of cement to the compaction index of de-aired concrete; see section 2.1.7) is higher than 3.4 (Fig. 5.14).

Another problem has to be prevented in SCC: the *coarse aggregate blockage* between rebars, or in the space between form and reinforcement. This concern leads the formulator to limit the amount of coarse aggregate with regard to the maximum quantity permitted by the presence of finer grains. Mathematically, this ratio is exactly expressed by the K'_g coefficient defined in section 2.1.7. We have seen that the fact that K'_g is lower than 2

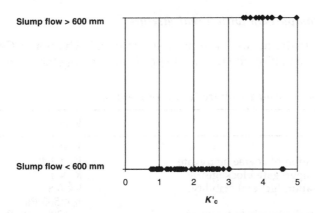

Figure 5.14 Correlation between self-compactability and K'_c value, for super-plasticized mixtures extracted from Ferraris and de Larrard (1998).

guarantees that coarse aggregate is not expelled from fresh concrete during shear (limitation of dilatancy). Adopting a conservative approach, we shall limit K'_g to a value of 1 (this condition being generally matched by SCCs taken from the literature). Let us note that this last condition applies only if the mixture is to be placed in a heavily reinforced structure.

A last condition dealing with prevention of segregation is the requirement for a continuous grading, with a well-balanced size distribution. This concept is described by the *segregation potential* (see sections 1.5.2 and 2.1.7). In the absence of a large amount of data dealing with segregation of SCCs, it is difficult to fix a precise bound for S. After examination of some mixtures taken from the literature, it seems that a maximum value of 0.8 for S leads to acceptable mixes.

Besides the stability problems, an SCC should be close to the *self-levelling* state, which means that the slope of a free surface should tend spontaneously to become horizontal. As already seen in section 4.1, this corresponds to a limitation of yield stress. Again, a maximum value of 500 Pa seems to be a good rule of thumb.

All the above requirements dealing with fresh concrete properties are summarized in Table 5.21. Note that, except for yield stress, these quantities should be evaluated in a confined medium, which means taking account of the geometry of the form and rebars. A limitation of plastic viscosity must be also added if the concrete is to be pumped (see section 4.1.1). In addition, any type of requirements can be added dealing with hardening/hardened concrete properties. One should keep in mind that, because of their high fine particles content, SCCs may exhibit a high paste volume, which is a factor for high shrinkage and creep. Also, the adiabatic temperature rise should be limited by a suitable use of supplementary cementitious materials (SCMs).

Simulations and comparison with a real mixture

Keeping the same materials as those used so far in this chapter, a C40 SCC was designed matching the previous specifications, together with a

Table 5.21 Specifications dealing with fresh SCC properties.

Criterion	Value
Self-placeability	$K' \leqslant 7$
Prevention of segregation of coarse aggregate	$K'_c \geqslant 3.4$
Prevention of coarse aggregate blockage[a]	$K'_g \leqslant 1$
Prevention of segregation (general stability)	$S \lesssim 0.8$
Self-levelling ability	$\tau_0 \leqslant 500$ Pa
Pumpability (if this technique is to be used)	$\mu \leqslant 200-300$ Pa.s
Cost	Minimum

[a] In the case of dense reinforcement.

target compressive strength of 48 MPa at 28 days. The saturation dosage of superplasticizer was adopted (see section 3.4.2). This choice is probably not the most economical. However, it tends to decrease the likelihood of early workability losses. Moreover, it is also a factor for lower delayed deformations (which can be high with SCCs, as already mentioned). The mixture proportions appear in Table 5.22, and the filling diagram is given in Fig. 5.15.

Looking at the different specified properties, it turns out that most constraints are active, with the exception of those dealing with flowing properties (yield stress and plastic viscosity). Comparing the hardened concrete properties with those of mixtures of similar compressive strength simulated in section 5.1, it is apparent that this SCC mixture keeps relatively unchanged deformability properties. This would not be the case if a lower superplasticizer dosage had been used.

Table 5.22 Theoretical mixture matching the specifications, and real project concrete.

Constituents	C40 SCC	SCC for precast tetrapods[a]
12.5/20 aggregate (kg/m^3)	410	571 (10/20)
5/12.5 aggregate (kg/m^3)	591	363 (4/10)
0/4 River sand (kg/m^3)	680	852
0/1 Correcting sand (kg/m^3)	68	(0/4)
OPC (kg/m^3)	323	350
Limestone filler (kg/m^3)	187	134
Superplasticizer (kg/m^3)	11.7	7.1
Water (l/m^3)	157	168
w/c	0.50	0.47
CA/FA	1.34	1.10
Air (%)	1.1	–
Specific gravity	2.421	–
Yield stress (Pa)	328	–
Viscosity after vibration (Pa.s)	157	⩽150
Slump (mm)	283	–
Slump flow (mm)	–	600–700
Placeability	7	–
K'_c	3.4	–
K'_g	1	–
S	0.80	–
Compressive strength (MPa)	48	50
Elastic modulus (GPa)	36.4	–
Adiabatic temperature rise (°C)	49	–
Autogenous shrinkage (10^{-6})	70	–
Total shrinkage (10^{-6})	733	–
Specific basic creep (10^{-6}/MPa)	40	–
Specific total creep (10^{-6}/MPa)	78	–
Cost (FF/m^3)	561.67	–

[a] Sedran *et al.* (1996).

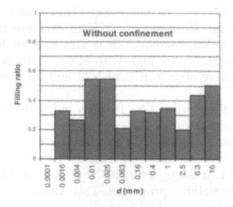

Figure 5.15 Filling diagram of the simulated C40 SCC.

Let us now compare these simulations with a real mixture. An SCC has been designed with local materials for construction of precast tetrapodes (Sedran *et al.*, 1996). These large concrete pieces (4 m^3 each) were previously cast with ordinary concrete, and the aim was to accelerate the filling of the moulds in order to increase the productivity of the factory. The mixture was designed using an earlier version of the packing model, together with the use of the BTRHEOM rheometer for checking the rheological properties. The mixture proportions appear in Table 5.21. They are quite similar to the result of simulations. Thanks to this new concrete, the filling time could be cut from 15 to 2 minutes, showing the great economic potential of SCCs.

5.5   CONCRETES WITH SPECIAL COMPOSITION

In this last series of concretes we shall review mixtures in which at least one component is unusual, in terms of nature or proportion. In lightweight aggregate concrete, the coarse aggregate phase is replaced by particles made up with a special, lightweight material. High-volume fly ash concretes have the peculiarity of incorporating a supplementary cementitious material (fly ash) in higher quantity than Portland cement. Finally, coarse aggregates are simply omitted in sand concretes.

5.5.1   Lightweight aggregate concrete

Lightweight aggregates (LWA) have long been used in structural concretes. Thanks to these materials it is possible to obtain concretes that are lighter than normal-weight concretes, while providing a better thermal insulation. In terms of general material properties, lightweight aggregate concretes (LWAC) are generally less rigid at equal strength, but their durability is excellent, thanks to the unique bond developed between porous LWA and cement paste.

Departing from an ordinary mixture, a substitution of the coarse aggregate by LWA gives concretes with specific gravity ranging from 1.5 to 2, and a compressive strength between 20 and 80 MPa at 28 days. Further substitution of the fine aggregate by LWA produces very light materials that are no longer structural, with compressive strength hardly reaching some MPa. This last type of material is not covered in this section.

After a short presentation of available LWAs, the effect of LWA incorporation on concrete mechanical properties will be emphasized. A modification of the models developed in Chapter 2 will be proposed, leading us to apply the general procedure to this particular concrete category. Then typical specifications of LWAC will be presented, and the case of an LWAC for a bridge will be simulated and compared with the mixture used in an actual project. Finally, correlations between hardened concrete properties will be discussed through a series of additional simulations. Most developments presented here are extracted from de Larrard (1996).

Lightweight aggregates

Some rocks, such as pumice, have low specific densities (less than 2) in their natural condition. Crushing them therefore yields lightweight aggregates. This constitutes a first category of LWA. This is followed by the by-products of other sectors of industry. One of them, foamed slag, is produced by putting molten blast-furnace slag in contact with water. The cooling of the mineral phase is concurrent with the evaporation of the water, yielding a cellular material, which must generally be crushed to control its grading. Finally, artificial LWAs (basically expanded clays or shales), specially produced, are made by first forming a dense paste, mixing a powder with water. This paste is then divided into small pellets measuring a few millimetres, which are put through a furnace at very high temperature. The expansion of the water and the vitrification of the mineral lead to the formation of more or less spherical aggregates, consisting of a shell enclosing gas bubbles scattered in a matrix. The operation of the furnaces can be adjusted to favour more or less expansion; there is then a correlation between the mean size of the grains and their specific gravity (the coarsest being the least dense). The main properties of LWA are as follows.

- *Shape*: As stated above, artificial LWAs are generally rather round; expanded clays are smooth, and so the packing densities rather high, whereas the shales are sometimes rougher;
- *Absorption of water*: LWAs have total porosities that are very high (up to 50% and more); fortunately, not all of this porosity is accessible to

water under normal conditions, but the absorption is higher than with aggregates of normal density. The kinetics of absorption are relatively fast in the first hours; it is therefore essential to have pre-wetted the LWA if it is desired to have a concrete of stable consistency when placed. It is also necessary to take account of the quantity of water present in the aggregates in the mixture-proportioning calculations (in particular of density);

- *Dry specific gravity*: One of the fundamental characteristics, which ranges from 0.8 for the lightest clays to 1.5–1.6 for by-products (foamed slags). There are also much lighter aggregates such as vermiculite, or even expanded polystyrene, but these are not used in structural concretes. Furthermore, recall that the coarser an expanded aggregate, the lower its density.

- *Mechanical characteristics*: Several tests have been proposed in the past to characterize the contribution of the lightweight aggregate to the mechanical properties of the concrete. However, according to a theoretical approach explained below, it is the equivalent elastic modulus of the aggregate that is the fundamental mechanical quantity. This quantity is hardly measurable directly, but it correlates more or less well with specific gravity (Fig. 5.16). Table 5.23 groups a few characteristics of lightweight aggregates used to make LWAC, taken from the literature.

Effect of LWA on concrete properties

The main features of LWAC at the fresh state are as follows.

- Because of their high water absorption, LWAs are generally pre-wetted, to avoid rapid slump losses. However, if concrete can

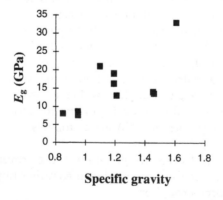

Figure 5.16 Correlation between equivalent modulus and specific gravity (de Larrard, 1996).

Table 5.23 Characteristics of a few LWAs. The equivalent elastic moduli have been calculated by fitting a model, described later, to concrete modulus measurements.

Name	Grading range	Absorption of water (% weight at 2 hours)	Specific gravity	E_g (GPa)
Argi 16	4/12	6.5	0.85	8
Isol S	3.15/8	5.2	1.21	13.1
Leca	4/10	14.6	0.95	7.6
Surex	6.3/10	3.7	1.19	16.2
Galex	3/8	2.4	1.61	33
LWC1 crushed	10/20	–	1.46	13.6
LWC2 pellet	5/20	–	1.45	14

Source: de Larrard (1996).

be placed quickly after mixing, it can be tempting to use dry LWA. In such a case all the water absorbed by the LWA phase after placement leads to a general improvement of concrete properties, because of the lowering of the free water/cement ratio. In reality, intermediate situations are the most common, which means that LWACs are more subject to slump losses than normal-weight aggregate concretes (NWACs);

- For a given yield stress an LWAC has a lower slump than an NWAC. This comes from the fact that slump is an increasing function of specific gravity (see equation (2.19) in section 2.1.4). As a result, LWACs are often produced with lower slumps than more usual structural concretes.
- Finally, the difference in specific gravity between LWA and mortar tends to promote a higher proneness to segregation. It is then generally advisable to produce concretes with stiffer consistency.

Because of their softness, LWAs exhibit a particular behaviour in hardened concrete, and deserve special modelling. As far as deformability properties are concerned, LWACs can be modelled by a triple sphere (see section 2.5.1), where LWA and mortar stand for the aggregate and matrix phases respectively. The mortar characteristics are covered by the models given in Chapter 2.

Let us now deal with the effect of LWAs on concrete compressive strength, for which a specific model is needed. The physical assumptions on which the model is based are as follows:

- The LWA develops a 'perfect' bond with the mortar, because of its porosity and roughness.
- Since the stiffness of the mortar is generally greater than that of the LWA, stress transfers occur between the two phases. It follows that

for a given external loading level the mortar 'works' more in an LWAC than in a traditional concrete. It is the failure of the mortar, and not that of the LWA, that governs the failure of the concrete.

Therefore stress transfers are governed by the mechanical properties of the mortar (compressive strength, elastic modulus), and by the LWA[8] equivalent elastic modulus. The latter will be determined by using, 'backwards', the triple-sphere model. Let us recall that this model is based on calculations for a basic cell of the material subjected to a *hydrostatic* pressure. It is well suited to estimating the deformability properties of the composite, but not to describing failure in *uniaxial* compression.

The simplest rheological model illustrating the dispersion of LWAs in a mortar matrix is assuredly that of Fig. 5.17, in which the grains are assumed to be cubic (model used by Virlogeux, 1986, to estimate the modulus of the lightweight concrete). By expressing the respective volumes in terms of the volume proportion, g, of LWA, one naturally arrives at the uniaxial parallel/series model of Fig. 5.18.

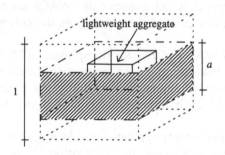

Figure 5.17 Cubic LWA of volume $a^3 = g$ (volume proportion of the lightweight aggregate in the concrete) included in a mesh of unit volume.

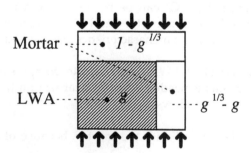

Figure 5.18 Rheological model for predicting the compressive strength of LWAC.

E_g and E_m represent the moduli of the LWA and mortar phases respectively; if a compressive stress, σ, is applied on a concrete unit cell, the stress in the part of the mortar in parallel is therefore

$$\sigma_m = \sigma \frac{E_m}{(1 - g^{2/3})E_m + g^{2/3}E_g} \qquad (5.6)$$

The compressive strength of the concrete is simply written versus that of the mortar as follows:

$$fc_{\text{concrete}} = \left[1 - \left(1 - \frac{E_g}{E_m} \right) g^{2/3} \right] fc_{\text{mortar}} \qquad (5.7)$$

if $E_g < E_m$.

Therefore it turns out that there is no strict proportionality between these two quantities. In effect, as the strength of the mortar increases, its modulus also becomes greater, and the aggregate bears a smaller share of the load. The strength of the LWAC therefore increases less quickly than that of the mortar, which has suggested to some authors that there is a limitation of the strength caused by crushing of the lightweight aggregate. In fact, a dry packing of LWA presents so high a deformability in uniaxial compression that no significant pre-peak failure may occur in this material when it is incorporated in concrete, which fails at a low compressive strain.

To validate this model, some data have been taken from the literature. The mechanical characteristics of the mortars were estimated using BETONLAB software (de Larrard and Sedran, 1996).[9] As stated above, the equivalent moduli of the LWAs were adjusted so that the triple-sphere model replicated the modulus of the experimental concrete. It was then possible to calculate a theoretical compressive strength, compared in Fig. 5.19 with the experimental value. In one case (foamed slag LWAs), the equivalent modulus is greater than that of the mortar. According to our model (Fig. 5.18), here it is the part of the mortar in series that fails first; the strength of the concrete is then simply equal to that of the mortar. The agreement of the model is satisfactory over the whole range of LWACs (mean prediction error equal to 2.2 MPa, or 6.5% in relative value).

Specifications, simulation, and comparison with a real example

Specification of an LWAC is particularly delicate, since by comparison with NWACs at least one property has to be added: the specific gravity. Furthermore, there is an important interaction between specific gravity

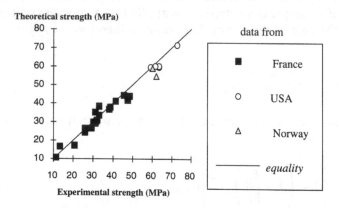

Figure 5.19 Comparison between theoretical strength and experimental strength for LWACs (de Larrard, 1996).

and compressive strength. But the correlation between these two properties depends on the nature of the particular LWA dealt with (see the last paragraph of this section). For demonstrating the capability of the system to master LWAC technology, let us address the case of a C32 LWAC for bridges, the specifications of which are given in Table 5.24.

Simulations were performed with a 4/10-mm expanded shale as LWA. This material had a water absorption of 6% in mass and a dry specific gravity of 1.1, which according to Fig. 5.16 gives an equivalent elastic modulus close to 16 GPa. Other materials were the same as those used in the previous simulations, including the superplasticizer. The models used for fresh concrete properties were the same as those already used, except the one dealing with entrapped air. Hence the slump is one of the main parameters of the model given by equation (2.23), while the real influencing physical property is the yield stress. Therefore this model is likely to overestimate air content for LWACs. In the present simulations a fixed value of 1.5% was taken for entrapped air. Obtained mixture proportions appear in Table 5.25.

Table 5.24 Specifications of an LWAC for bridges (mix containing superplasticizer).

Specified property	Value
Placeability	$K' \leqslant 8$
Slump	$\geqslant 80$ mm
Specific gravity	$\leqslant 1.78$
Compressive strength at 28 days	$\geqslant 38$ MPa
Cement dosage	$\geqslant 400$ kg/m^3

Table 5.25 Mixture proportions and properties of C32 LWAC.

Constituents	Simulations	Iroize bridge
5/12.5 aggregate (kg/m³)	424	507
River sand (kg/m³)	689	657
0/1 Correcting sand (kg/m³)	72	(0/3)
OPC (kg/m³)	409	400
Superplasticizer (kg/m³)	5.0	2
Retarder (kg/m³)	–	≤ 1
Water (l/m³)	203	212
w/c[a]	0.44	0.45
CA/FA[b]	1.34	1.85
Air (%)	1.5	–
Specific gravity	1.80	1.78
Yield stress (Pa)	1347	–
Viscosity after vibration (Pa.s)	186	–
Slump (mm)	80	–
Placeability	7.65	Satisfactory
Compressive strength (MPa)	38	44
Elastic modulus (GPa)	20.9	19.7
Adiabatic temperature rise (°C)	71	–

[a] Free water/cement ratio.
[b] In volume.

When we look at the properties of simulated concrete, it appears that the active constraints deal with slump, specific gravity and compressive strength. The yield stress necessary to obtain an 80 mm slump would lead to a 110–160 mm slump in the case of NWAC.

The filling diagram of the simulated LWAC appears in Fig. 5.20. The coarse aggregate peak is very high, which means that this mixture is probably not pumpable. However, as there is a natural tendency for coarse aggregate to rise throughout the fresh concrete volume, the fact

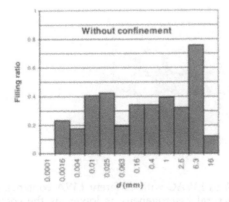

Figure 5.20 Filling diagram of the simulated C32 LWAC.

that this fraction is closely packed tends to minimize this hazard (Fig. 5.21). Furthermore, pumping of LWACs is generally detrimental to the hardened concrete properties, because water penetrates inside LWAs when the mixture is submitted to the pumping pressure. Later, when the normal pressure is restored, this water flows out of the aggregate, forming a very porous and weak transition zone.

Another feature is the high predicted adiabatic temperature rise. This comes from the lower thermal capacity per unit volume of concrete, as compared with NWACs. In reality, the situation can be even worse, for two reasons:

- The water stored in LWAs may migrate in the matrix, enhancing the level of cement hydration and leading to a heat production higher than in the simulation.
- The thermal insulation provided by LWAs makes the real temperature rise closer to the adiabatic one, for a given structure geometry.

Therefore more attention must be paid to the prevention of thermal cracking in structures made up with LWAC.

The mixture proportions of a real LWAC have been extracted from a laboratory study (Lange and Cochard, 1991). This formula (see Table 5.25) was devoted to the central span of the Iroize bridge, already reported in section 5.3.1, and designed according to the same specifications. Again, while the materials were different from those used in the simulations (except the LWA), the obtained formulae are comparable.

Correlations between properties of LWACs

Before we deal with another type of concrete, it is of interest to examine additional simulations, together with data extracted from the literature, in order to realize the relationships that link together the various

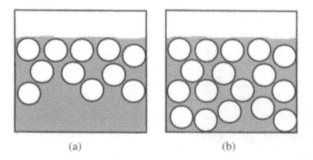

(a)   (b)

Figure 5.21 Segregation of LWA in LWAC with different LWA contents. In (b), the risk of structural heterogeneity is lower, as the coarse aggregate is densely packed.

properties of LWACs. Two series of concretes made up with the same LWA and mortars of variable quality were simulated. Expanded clay (with an equivalent elastic modulus of 8 GPa and a specific gravity of 0.85) and expanded shale ($E_g = 19$ GPa, $\rho_g = 1.20$) were used in these series. Figures 5.22 and 5.23 show compressive strength versus mortar strength and versus concrete specific gravity respectively. For the latter, experimental results taken from the literature have been plotted, together with the results of simulations.

For a given series it is found, as stated above, that the stronger the mortar, the greater the loss of strength, with respect to the mortar, due to the LWA. The larger the contrast of modulus between the phases, the larger this effect. On the other hand, it can be seen in Fig. 5.23 that the

LWAC compr. strength (MPa)

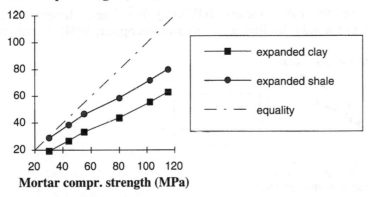

Mortar compr. strength (MPa)

Figure 5.22 Relationship between strength of LWAC and strength of mortar (de Larrard, 1996).

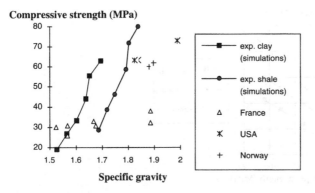

Figure 5.23 Relationship between strength and specific gravity (de Larrard, 1996).

lightest LWA, while it leads to concretes that are not as strong, allows the attainment of higher strength/specific gravity ratios.

The possible drawback then lies in the elastic modulus/strength ratio, which is lower than with stiffer LWAs (see Fig. 5.24). Finally it is possible, still using the same simulations, to examine the correlation between the compressive strength of the LWAC and the compressive strength that would be exhibited by a traditional concrete having the same composition (in volume), made up with round gravels. In the case of the expanded shale used in these simulations, it is only at a strength level greater than 40 MPa that the use of the LWA decreases the strength of the concrete. In the region 20–45 MPa, the effect of the deformability of the LWA is offset by that of its excellent bond with the mortar (Fig 5.25).

5.5.2 High-volume fly ash concrete

High-volume fly ash concrete (HVFAC) has been developed by CANMET in Canada (Malhotra and Ramezanianpour, 1994), mainly for

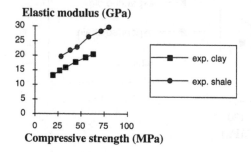

Figure 5.24 Relationship between elastic modulus and compressive strength (de Larrard, 1996).

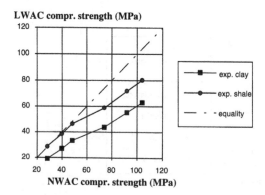

Figure 5.25 Compressive strength of LWAC vs. compressive strength of normal-weight round-aggregate concrete (de Larrard, 1996).

environmental purposes. North America produces tremendous quantities of fly ash, which must be used in construction materials rather than tipped in land fill. This type of mixture incorporates a fly ash at a higher level than Portland cement. It can be considered as a side result of high-performance concrete (HPC) technology, as it is made possible by the existence of superplasticizers. HVFAC is basically an HPC in which a maximum proportion of the cement has been substituted by fly ash. In the fresh state it behaves like a conventional HPC (see section 5.3.1), while in the hardened state the properties are closer to those of a normal-strength structural concrete (NSC).

Numerous studies have been performed on HVFAC. It appears that, from a concrete technology viewpoint, these mixtures are acceptable for many applications. The only reservations regarding the use of HVFACs could deal with their slow strength development (especially in cold weather), their colour (which is generally darker than that of conventional NSC), and their poor scaling behaviour when submitted to deicing salts. Also, the production consistency of a fly ash source must be thoroughly assessed prior to the choice of this fly ash, if constant properties are desired for the concrete to be produced. On the other hand, HVFAC offers the general advantages of concretes containing pozzolans, such as decreased permeability, better resistance to sulphate attack, and lower susceptibility to alkali–silica reaction (ASR).

Specifications

The fresh concrete properties will be specified as those of basic HPC. As far as compressive strength is concerned, it can be interesting to deal with compressive strength at a later age than for conventional NSC, because of the slow strength development of HVFAC. However, it must be kept in mind that fly ash mixtures generally require a more thorough curing than ordinary mixtures. Therefore the difference in quality between laboratory, water-cured and in situ concrete will be higher if the structural element is thinner, if the climate is drier, or if the concrete is older. An example of a specification set for a C50 HVFAC is given in Table 5.26.

Table 5.26 Specifications of a C50 HVFAC.

Criterion	Value
Placeability	$K' \leqslant 8$
Slump	$\geqslant 200$ mm
Compressive strength at 120 days	$\geqslant 60$ MPa
Cost	Minimum

Simulations and comparison with an HVFAC used in an actual project

A simulation has been performed with the standard materials used throughout this chapter, with the same process as that already used (that is, aggregate fraction optimization for maximizing the placeability, then paste optimization for matching the specifications). The saturation dosage of superplasticizer has been adopted, for the same reasons as listed in section 5.3.1. The result appears in Table 5.27, together with a real formula used in the Park Lane Complex project, Halifax, Canada (Malhotra and Ramezanianpour, 1994).

The similarity between the two formulae is noticeable, especially in terms of cement and fly ash content. This is a further proof that a sound, scientifically based approach leads directly to results comparable to those obtained after a long, empirical process based on trial and error. Looking at the predicted properties of the simulated mixture, the following

Table 5.27 Theoretical mixtures matching the specifications, and real project concretes.

Constituents	Simulations	Park Lane Complex project
12.5/20 aggregate (kg/m^3)	750	1100
5/12.5 aggregate (kg/m^3)	334	(19 mm MSA)
0/4 River sand (kg/m^3)	676	800
0/1 Correcting sand (kg/m^3)	82	(presumably well graded)
OPC (kg/m^3)	186	180
Class F fly ash (kg/m^3)	221	220
Superplasticizer (kg/m^3)	6.94	6
Water (l/m^3)	138	110
w/c	0.75	0.61
CA/FA	1.43	1.38
Air (%)	1.0	–
Specific gravity	2.39	–
Yield stress (Pa)	539	–
Viscosity after vibration (Pa.s)	297	–
Slump (mm)	250	–
Placeability	8	–
Compr. strength at 28 days (MPa)	43	32–55[a]
Compr. strength at 120 days (MPa)	60	52–74[a]
Elastic modulus (GPa)	37.8	–
Adiabatic temperature rise (°C)	38	–
Autogenous shrinkage (10^{-6})	0	–
Total shrinkage (10^{-6})	579	–
Specific basic creep (10^{-6}/MPa)	22	–
Specific total creep (10^{-6}/MPa)	46	–
Cost (FF/m^3)	458.98	–

[a] Site data, from quality control tests performed during construction.

remarks can be made:

- The adiabatic temperature rise is very low, compared with that of a low strength Portland cement concrete, especially if the high final strength is taken into account (see section 5.1). Such a quality is a general feature of HVFAC (Malhotra and Ramezanianpour, 1994).
- The autogenous shrinkage is virtually non-existent, because of the high water/cement ratio. This is another characteristic that is favourable for the prevention of early age cracking.
- The total shrinkage and creep deformations are quite low, because of the maximum use of superplasticizer.

Therefore, based upon the simulations, HVFAC would be a suitable material for large prestressed concrete structures. However, the variability of the compressive strength, as reported from the real project, could be a limitation for this type of mixture. This issue is closely related to the quality assurance plan adopted at the fly ash production plant.

5.5.3 Sand concrete

Sand concrete is a concrete formulated without coarse aggregate. It cannot be identified as a mortar, because of its limited Portland cement content (comparable to that of a conventional concrete). Moreover, it can be used as a structural material, while mortar is generally applied in thin layers, for binding bricks/blocks or for wall coating. Sand concrete technology has been developed primarily for regions where coarse aggregate was rare or non-existent. However, when gravels are available at a reasonable price, sand concrete can even be competitive in some specific markets as architectural concretes with very fine texture. Note that intermediate mixtures between conventional and sand concretes are technically feasible, and may be of interest when coarse aggregate is available, but expensive. As high-volume fly ash concretes, sand concretes are also 'cousins' of HPC, since they require the use of a superplasticizer. Basically, a sand concrete is a superplasticized mortar in which a part of the cement is replaced by a mineral admixture, which can be fly ash, slag, limestone filler (Chanvillard and Basuyaux, 1996), and/or silica fume.

One of the first utilizations of sand concrete took place in France as early as 1863, in a building that is still visible in the northern suburbs of Paris. Later, this technology spread over the former USSR, where coarse aggregate is scarce in numerous areas. More recently, a national project was carried out in France, aiming to disseminate the use of sand concrete among the various fields of concrete utilization. The main achievements of this project are summarized in a book (Sablocrete, 1994). On the whole it appears that sand concretes may replace conventional concretes in most applications, at a higher cost[10] and with different secondary properties.

In terms of specifications, there are no real differences between conventional concrete and sand concrete. Because of the small size of aggregate, sand concretes present little proneness to segregation. Therefore less attention has to be paid to fresh state stability criteria. On the other hand, because of a reduced grading range, the water and fine particle demands will be systematically higher for sand concrete than for those of ordinary mixtures containing coarse aggregate. As a result, the material deformability will become more critical, in terms of elastic modulus, shrinkage, creep, and thermal strains.

Simulation, and comparison with a mixture used in a real project

The case of a C40 sand concrete for precast prestressed beams will be investigated. The chosen specifications are given in Table 5.28. Note that a lower slump than usual for HPC is required, because of the industrial context (prefabrication). Also, no early strength requirement is formulated, since a thermal treatment is assumed to be optimized in order to obtain the desired in situ strength development.

Preliminary simulations show that silica fume is useful for attaining a sufficient strength level, while keeping a suitable dimensional stability. As for C80 HPC (see section 5.3.1), a proportion of 8% of the cement mass is taken. To avoid too high a retarding effect, the superplasticizer is *not* proportioned at the saturation dosage. This choice may lead to early slump losses, but such a phenomenon can be more easily managed in a factory than on site. The numerical optimization process, based upon the above-mentioned set of specifications, leads to the formula given in Table 5.29.

Again, it is demonstrated that the system is able to produce automatically a formula very close to that obtained after a conventional mix-design study. Examination of Table 5.29 leads to the following remarks:

- Compared with conventional superplasticized concrete, sand concrete is characterized by a low K' value, which means an excellent

Table 5.28 Specifications of a C40 sand concrete.

Criterion	Value
Placeability	$K' \leqslant 7$
Slump	$\geqslant 150$ mm
Compressive strength at 28 days	$\geqslant 50$ MPa
Cost	Minimum

Table 5.29 Theoretical mixtures matching the specifications, and real project sand concretes.

Constituents	Simulations	Sablocrete mixture[a]
0/4 River sand (kg/m^3)	1462	1600
0/1 Correcting sand (kg/m^3)	30	(fineness modulus 2.49)
OPC (kg/m^3)	390	380
Limestone filler (kg/m^3)	81	70
Silica fume (kg/m^3)	31	35
Superplasticizer (kg/m^3)	4.0	11.4 (plasticizer)
Water (l/m^3)	217	219
w/c	0.56	0.60
CA/FA	0	0
Air (%)	4.3	–
Specific gravity	2.22	–
Yield stress (Pa)	1184	–
Viscosity after vibration (Pa.s)	32	–
Slump (mm)	150	–
Placeability	5.71	–
Compr. strength at 28 days (MPa)	50	44.5
Elastic modulus (GPa)	31.1	23[b]
Adiabatic temperature rise (°C)	69	–
Autogenous shrinkage (10^{-6})	78	–
Total shrinkage (10^{-6})	958	–
Specific basic creep (10^{-6}/MPa)	38	–
Specific total creep (10^{-6}/MPa)	79	–
Cost (FF/m3)	557.89	–

[a] p. 127 in Sablocrete (1994).
[b] This value is probably wrong. In fact other C40 sand concretes cited in the same reference exhibit an elastic modulus close to 30 GPa.

placeability. This feature has always been mentioned in all sand concrete applications (Sablocrete, 1994). Hence, if we remember the model for yield stress (see section 2.1.3), we note that the sand contribution to the yield stress is much greater than that of coarse aggregate. As a result, the level of compaction is more limited by aggregate friction when all coarse particles are removed. So the requirements dealing with flowing behaviour are critical in sand concrete, compared with those dealing with compactibility.

- The plastic viscosity is equally low. In addition to the fact that segregation seldom occurs in sand concrete, it turns out that sand concretes present outstanding pumpability.
- Conversely, the hardened concrete properties are less attractive from a mechanical viewpoint: the elastic modulus is low, the adiabatic temperature rise is very high, and the total shrinkage is almost twice that of conventional concrete of similar compressive strength (see section 5.1). Again, these predictions are confirmed by the practical experience gained through the Sablocrete project.

The amount of creep deformation does not lead to sand concrete's being recommended for structural applications where the material is heavily loaded. However, sand concretes are particularly suitable for shotcrete application: their excellent pumpability and placeability, together with their high cohesiveness, allow the user to spray thick layers with a low rebound.

NOTES

1 Such as packing density parameters.
2 Private communication from F. Cussigh, GTM Dumez, December 1997.
3 Private communication from P. Laplante, RMC Technical Management, Wissous, January 1998.
4 As measured with the BTRHEOM rheometer, after a cycle under vibration (see section 2.1.2).
5 Note that all bridge decks are covered with a watertight membrane in France. Thus water leakage through the cracks is not a problem.
6 That is, a material that can be used for construction of large elements.
7 Mean diameter of concrete spread after a conventional slump test.
8 It would be wrong to speak of a *true elastic modulus*, because of the impossibility of measuring this quantity on an aggregate of any shape, and because of the large heterogeneity of the grains.
9 The models incorporated in this software and devoted to the prediction of hardened concrete properties are close to those related in Chapter 2.
10 At least if the cost of binders and admixtures is addressed.

Conclusion

THE CONCRETE SYSTEM

At the end of this work, it is the author's hope that the reader has gained or improved his or her understanding of concrete as a *dynamic system*. To a certain extent, the versatility of concrete, coming from the multiplicity of available components, is one of the most recent and important trends in all civil-engineering-related technologies. Rather than being a 'petrified' material, concrete has become a system able to fit a wide range of different needs, in a wide range of changing contexts. Facing this reality, an empirical approach is still possible. When a concrete utilization is sufficiently long-lived an acceptable mix design is eventually found, whatever the method. The only thing is that all errors and successive trials undergone before stabilization of the process have a cost, which is supported by all the participants involved (owner, concrete producer, contractor etc.). What can be expected from a scientific approach is much faster convergence towards a good solution from both the technical and the economic viewpoint. Also, a wider number of possibilities can be envisaged through simulations. Tests are still necessary, but can be managed and optimized much more rigorously. Finally, a scientific optimization can sometimes lead to innovative formulae (see section 5.3.2), which would not be suggested by past experience.

A further step of material science implementation in concrete technology would be to monitor concrete production in order to perform in real time mixture-proportion adjustments for the sake of maximum uniformity in terms of the main concrete properties. These adjustments would be based upon numerical optimizations accounting for measured changes in the characteristics of components. To reach this level of technology, two conditions must be fulfilled:

- the mathematical models should be sufficiently accurate, which means that their error should be lower than the accepted fluctuations of the corresponding property;

- the constituent parameters responsible for the changes in the property should be measured continuously, or on a regular basis, assuming monotonic variations between the tests.

As far as compressive strength is concerned, we have seen that available models can predict experimental results with a 2–3 MPa error, which is satisfactory, especially at high strength levels. Unfortunately, the two main relevant material parameters – the aggregate water content and the cement strength – are difficult to obtain in due time. They can be evaluated *a posteriori*: the former can be estimated from changes in slump (Day, 1995), while other factors, especially in the presence of admixtures, can provoke slump changes. In mixtures where the entrapped air content is very low, as in fluid mixtures, the fresh concrete specific gravity is a better way of assessing water content. As for cement, the strength of a cement delivery is generally measured when the concrete is already at the hardened state; but with the collaboration of material providers it is possible to anticipate and account for changes in the constituent deliveries.

Concerning fresh concrete properties, the precision of the models still must be improved. However, for partially unclear reasons, it is not uncommon to see a discrepancy between predictions and measurements that is reasonably constant for a given set of materials, making it possible to perform successfully an empirical correction (see the predictions of slump for unplasticized concrete in Fig. 2.21). In production, the critical factors inducing changes in rheology are again changes in aggregate water content, in aggregate size distribution (with a special emphasis on fine content), and – especially in the presence of superplasticizer – changes in cement water-demand. Apart from the first factor, the rest can be easily monitored and controlled.

To conclude, it is not unrealistic to dream of a process in which a concrete plant would be connected with a computer, which was in charge of managing the mixture proportions on the basis of periodically refreshed material parameters. This organization would lead to a material that was more uniform in terms of properties, but more variable in terms of compositions. In this context, prescription-based standards are no longer suitable, and must be replaced by performance-based contracts.

RESEARCH NEEDS

The main content of this book is a collection of mathematical models. While the current state of development already permits practical use of the system for proportioning a wide variety of concretes (see Appendix), further progress is still needed to improve these models.

First, an open question is to know whether *analytical* models, such as those given in this book, will continue to be sufficient, or whether cellular models, in which the geometrical, three-dimensional structure of a concrete sample is modelled in a computer, followed by calculations (e.g. finite elements), will become necessary (Garboczi and Bentz, 1996, 1997). It is the author's opinion that for mechanical properties, including fresh concrete rheological parameters, analytical techniques are capable of great achievements; but for durability, where local heterogeneity plays a major role, it may be necessary to move towards more sophisticated techniques, which require more powerful computation tools.

Let us now review the various concrete properties that in our view most deserve additional research:

- In terms of *fresh concrete rheology*, we have shown the critical role of admixtures, especially superplasticizer. More knowledge should be acquired regarding the adsorption of these products on supplementary cementitious materials, for a better prediction of the behaviour of mixtures formulated from a binary or ternary system of binders. The effect of intermediate doses should also be better quantified. Likewise, the role of entrained air, fibres and viscosity agents in the fundamental fresh concrete properties is not well documented, to the best of the author's knowledge.

- A very important issue is constituted by *fresh concrete stability*. Whatever the power of the filling diagram concept (see sections 1.5.2 and 2.1.7) to predict bleeding and segregation, these phenomena remain imperfectly understood, and are dealt with by a surprisingly low number of researchers worldwide.

- The author and his colleagues have put much effort into trying to build a comprehensive model for *compressive strength*. But this model still remains to be improved for a better prediction power at early age – a very important issue for many practical applications (Chanvillard and Laplante, 1996) – and also for a correct account of blast-furnace slag and class C fly ash.

- The *deformability properties*, especially creep and shrinkage, are covered by a set of new models, some of which need further validation, and completion by the modelling of creep and relaxation at early age. In this respect, recent findings about the creep mechanisms open new avenues (Guenot-Delahaie, 1997; Ulm and Bazant, 1998).

- In addition to their contribution to fresh concrete rheology, the *role of fibres in the post-cracking behaviour of hardened concrete* must be clarified and quantified, as a function of the matrix/fibre properties (Rossi, 1998). When this stage is reached, a scientific mix design of steel/organic fibre concrete will become feasible.

- In the hypothesis in which fracture-mechanics-based methods will eventually find international acceptance for the design of concrete structures, it will be important to select real material properties[1] (fracture toughness, fracture energy, brittleness number etc.) as inputs for these models. Then the question of integrating fracture mechanics in the system will emerge, and there will be a need to model the related properties as functions of the mixture proportions (Lange-Kornbak and Karihaloo, 1996).

- However, the wider gap in the current set of models is of course that dealing with *durability*. In coming years one may expect to see the emergence of sound, scientifically based and quantitative models for predicting properties such as gas permeability and diffusivity, carbonation depth vs. time, and alkali–silica reaction development (ASR). Eventually it will be possible to predict the induction time for steel corrosion as a function of concrete mix design, cover thickness and environment. The connection of these models, which are nowadays insufficiently developed and validated, with the successors of those given in this book will complete the construction of a global system for the mastery of concrete design and production. At this time, the scientific approach, defined by René Descartes in 1637, will reach a state of full completion as regards concrete mixture proportioning.

---

[1] That is, properties that are not too much affected by size effects.

References

Abrams, D. (1919) Cited by Neville (1995).

ACI 211.1-91 (1994) Standard practice for selecting proportions for normal, heavyweight and mass concrete, *ACI Manual of Concrete Practice*, Part I: *Materials and General Properties of Concrete*, ACI, Detroit, Michigan.

ACI 304 (1982) Placing concrete by pumping methods, *ACI, Manual of Concrete Practice*, Part 3.

Acker, P. (1982) Drying of concrete: consequences for the evaluation of creep tests, in *Fundamental Research on Creep and Shrinkage of Concrete*, Martinus Nijhoff Publishers, pp. 149–169.

Addis, B.J. and Alexander, M.G. (1994) Cement-saturation and its effects on the compressive strength and stiffness of concrete. *Cement and Concrete Research*, 24(5), 975–986.

Aitcin, P.C. and Albinger, J.M. (1989) Les BHP – expérience nord-américaine et française. *Annales de l'Institut Technique du Bâtiment et des Travaux Publics*, No. 473, March–April.

Aitcin, P.C. and Mehta, P.K. (1990) Effect of coarse-aggregate characteristics on mechanical properties of high-strength concrete. *ACI, Materials*, 87(2), 103–107.

Aitcin, P.C., Jolicoeur, C. and McGregor, J.G. (1994) Superplasticizers: how do they work and why they sometimes don't. *Concrete International*, 16(5), 45–52.

Al-Amoudi, O.S., Maslehuddin, M. and Asi, I.M. (1996) Performance and correlation of the properties of fly ash cement concrete. *Cement, Concrete and Aggregates*, CCAGDP, 18(2), 71–77.

Alexander, K.M. *et al.* (1961) Discussion of Walker, S. and Bloem, D.L. (1960) Effects of aggregate size on properties of concrete *ACI Journal*, 32(9), 1201–1258.

Alou, F., Ferraris, C.F. and Wittmann, F.H. (1987) Etude expérimentale du retrait du béton. *Materials and Structures*, 20, 323–333.

Auperin, M., de Larrard, F., Richard, P. and Acker, P. (1989) Retrait et fluage de bétons à hautes performances aux fumées de silice – influence de l'âge du chargement. *Annales de l'Institut Technique du Bâtiment et des Travaux Publics*, No. 474, May–June.

Baalbaki, W. (1990) Bétons à haute performance à matrice constante: influence de la nature des granulats sur le comportement mécanique. Mémoire de Maîtrise ès Sciences Appliquées, University of Sherbrooke.

BaCaRa (1994) Programme inter-laboratoire de fabrication des éprouvettes en BCR, Projet National Barrages en béton compacté au rouleau. *10 Rapports d'étude Tranche 5*, poste 6.4, IREX (Institut pour la Recherche et l'Expérimentation en Génie Civil), Paris.

Bache, H.H. (1981) Densified cement/ultrafine particle-based materials, in *Second CANMET/ACI International Conference on Superplasticizers in Concrete*, Ottawa (ed. V.M. Malhotra), pp. 185–213.

Baroghel-Bouny, V. (1994) *Caractérisation des pâtes de ciment et des bétons – Méthodes, analyse et interprétations*, Laboratoire Central des Ponts et Chaussées, Paris.

Baron, J. and Lesage, R. (1976) *La composition du béton hydraulique, du laboratoire au chantier*. Rapport de recherche des Laboratoires des Ponts et Chaussées No. 64, December.

Baron, J., Bascoul, A., Escadeillas, G. and Chaudouard, G. (1993) From clinker to concrete – an overall modelling. *Materials and Structures*, RILEM, **26**, 319–327.

Bartos, P.J.M. and Hoy, C.W. (1996) Interactions of particles in fibre-reinforced concrete, in *Production Methods and Workability of Concrete*, Proceedings of the International RILEM Conference (eds P.J.M. Bartos, D.L. Marrs and D.J. Cleland), E & FN Spon, London, pp. 451–462.

Bazant, Z.P., Kim, J.K. and Panula, M. (1991–92) Improved prediction model for time-dependent deformations of concrete – Part 1-6. *Materials and Structures*, **24**, 327–344; 409–421; **25**, 21–28, 84–94, 219–223.

Beaupré, D. (1994) Rheology of high-performance shotcrete. PhD thesis, University of British Columbia, Vancouver.

Ben-Aïm, R. (1970) Etude de la texture des empilements de grains. Application à la détermination de la perméabilité des mélanges binaires en régime moléculaire, intermédiaire, laminaire. Thèse d'Etat de l'Université de Nancy.

Bentz, D.P., Waller, V. and de Larrard, F., (1998) Prediction of adiabatic temperature rise in conventional and high-performance concretes using a 3-D microstructural model. *Cement and Concrete Research*, **28**(2), 285–297.

Blot, G. and Nissoux, J.-L. (1994) Les nouveaux outils de contrôle de la granulométrie et de la forme. *Mines et Carrières, Les Techniques*, **94**, suppl., December.

Bolomey, J. (1935) Granulation et prévision de la résistance probable des bétons. *Travaux*, **19**(30), 228–232.

BRE (1988) *Design of Normal Concrete Mixes*, Building Research Establishment, Watford, UK.

British Standards Institution (1975) BS 812 *Methods for Testing Aggregate. Part 2: Methods for Determination of Physical Properties*, BSI, London.

Buil, M., Witier, P., de Larrard, F., Detrez, M. and Paillere, A.-M. (1986) Physicochemical mechanism of the action of the naphthalene-sulfonate based superplasticizers on silica fume concretes, in *Proceedings of 2nd International ACI-CANMET, Conference on Fly Ash, Silica Fume, Slag and Natural Pozzolans in Concrete*, Madrid (ed. V.M. Malhotra), ACI SP 91, pp. 959–972.

Cadoret, G. and Richard, P. (1992) Full scale use of high-performance concrete in building and public works, in *High-Performance Concrete – From Material to Structure* (ed. Y. Malier), E & FN Spon, London, pp. 379–411.

Caquot, A. (1937) Rôle des matériaux inertes dans le béton. *Mémoire de la Société des Ingénieurs Civils de France*, July–August.

Caré, S., Baroghel, V., Linder, R. and de Larrard, F. (1998) Effet des fillers sur les propriétés d'usage des bétons – Plan d'expérience et analyse statistique. Internal report, LCPC, Paris.

Carino, N.J., Knab, L.I. and Clifton, J.R. (1992) Applicability of the maturity method to high-performance concrete. Building and Fire Research Laboratory, National Institute of Standards and Technology, Gaithersburg, MD, NISTIR 4819, May.

Cariou, B. (1988) Bétons à très hautes performances. Caractérisation de huit fumées de silice. Lafarge Corporation, internal report.

CEB (1994) Application of high-performance concrete. *CEB, Bulletin d'information* No. 222, November, Comité Euro-international du Béton.

Chang, C. and Powell, R.L. (1994) Effect of particle size distributions on the rheology of concentrated bimodal suspensions. *Journal of Rheology*, 38(1), 85–98.

Chanvillard, G. and Basuyaux, O. (1996) Une méthode de formulation des bétons de sable à maniabilité et résistance fixées. *Bulletin des Laboratoires des Ponts et Chaussées*, No. 205, September–October, pp. 49–64.

Chanvillard, G. and Laplante, P. (1996) Viser une résistance à court terme pour tenir les délais de fabrication, in *Les bétons – Bases et données pour leur formulation* (ed. J. Baron and J.P. Ollivier), Eyrolles, Paris.

Chaudorge, L. (1990) Compacité des mélanges sable + rubans Fibraflex. Etude expérimentale et modélisation. Rapport de stage de 1° année, Ecole nationale des ponts et chaussées, September.

Christensen, R.M. and Lo, K.H. (1979) Solutions for effective shear properties in three-phase sphere and cylinder models. *Journal of the Mechanics and Physics of Solids*, 27(4), Erratum (1986) 34.

Cintre, M. (1988) Recherche d'un mode opératoire de mesure de compacité de mélanges vibrés à sec de classes élémentaires de granulats. Rapport du LRPC de Blois, January.

Coquillat, G. (1992) Fluage des bétons HP chargés au jeune âge. *Voies Nouvelles du Matériau Béton*, Rapport de Recherche No. 41 101, CEBTP, St Rémy lès Chevreuses.

Cordon, W.A. and Gillespie, H.A. (1963) Variables in concrete aggregates and Portland cement paste which influence the strength of concrete. *Journal of the American Concrete Institute*, August.

Coussot, P. and Piau, J.-M. (1995) A large-scale field coaxial cylinder rheometer for the study of the rheology of natural coarse suspensions. *Journal of Rheology*, 39, 104–123.

Cumberland, D.J. and Crawford, R.J. (1987) *Handbook of Powder Technology*, Vol. 6, *The Packing of Particles* (eds J.C. Williams and T. Allen), Elsevier, Amsterdam.

Day, K.W. (1995) *Concrete Mix-Design, Quality Control and Specification*, E & FN Spon, London.

de Champs, J.F. and Monachon, P. (1992) Une application remarquable: l'arc du pont sur la Rance, in *Les bétons à hautes performances – Caractérisation, durabilité, applications* (ed. Y. Malier), Presses de l'Ecole Nationale des Ponts et Chaussées, Paris, pp. 601–614.

de Larrard, F. (1988) Formulation et propriétés des bétons à très hautes performances. Thèse de Doctorat de l'Ecole nationale des ponts et chaussées, Rapport de Recherche des LPC, No. 149, March.

de Larrard, F. (1989) Ultrafine particles for the making of very high-strength concretes. *Cement and Concrete Research*, **19**, 161–172.

de Larrard, F. (1995) Modelling the compressive strength of structural fly ash concrete, in *Proceedings of the 5th International ACI-CANMET, Conference on Fly Ash, Silica Fume, Slag and Natural Pozzolans in Concrete* (ed. V.M. Malhotra), Milwaukee, May, ACI, SP 153-6, pp. 99–108.

de Larrard, F. (1996) Mixture proportioning of lightweight aggregate concrete, in *Proceedings of the International Conference on High-Performance Concrete and Performance and Quality of Concrete Structures* (ed. L.R. Prudencio, P.R. Helene and D.C.C. Dal Molin), Florianopolis, Brazil, June, pp. 154–166.

de Larrard, F. and Acker, P. (1990) Un exemple d'ingéniérie du matériau: amélioration de l'étanchéité à l'air des enceintes internes de centrales nucléaires – Intérêt de l'emploi d'un béton à hautes performances de formulation spéciale. Rapport des Laboratoires, Série Ouvrage d'Art, OA 7, LCPC, October.

de Larrard, F. and Belloc, A. (1992) Are small aggregates really better than coarser ones for making high-strength concrete? *Cement, Concrete and Aggregates*, CCAGDP, **14**(1), 62–66.

de Larrard, F. and Belloc, A. (1997) The influence of aggregate on the compressive strength of normal and high-strength concrete. *ACI Materials Journal*, **94**(5), 417–426.

de Larrard, F. and Le Roy, R. (1992) Relations entre formulation et quelques propriétés mécaniques des bétons à hautes performances. *Materials and Structures*, RILEM, **25**, 464–475.

de Larrard, F. and Marchand, J. (1995) Discussion on the paper by B.J. Addis and M.G. Alexander: 'Cement-saturation and its effects on the compressive strength and stiffness of concrete'. *Cement and Concrete Research*, **25**(5), 1124–1128.

de Larrard, F. and Sedran, T. (1994) Optimization of ultra-high performance concrete by using a packing model. *Cement and Concrete Research*, **24**(6), 997–1009.

de Larrard, F. and Sedran, T. (1996) Computer-aided mix-design: predicting final results. *Concrete International*, **18**(12), 39–41.

de Larrard, F. and Tondat, P. (1993) Sur la contribution de la topologie du squelette granulaire à la résistance en compression du béton. *Materials and Structures*, RILEM, **26**, 505–516.

de Larrard, F., Torrenti, J.M. and Rossi, P. (1988) Le flambement à deux échelles dans la rupture du béton en compression. *Bulletin de Liaison des Laboratoires des Ponts et Chaussées*, no. 154, March–April, 51–55.

de Larrard, F., Ithurralde, G., Acker, P. and Chauvel, D. (1990) High-performance concrete for a nuclear containment. *Proceedings of the 2nd International Conference on Utilization of High-Strength Concrete, Berkeley* (ed. Weston T. Hester), ACI SP 121–27, May, pp. 549–576.

de Larrard, F., Gorse, J.F. and Puch, C. (1992) Comparative study of various silica fumes as additives in high-performance cementitious materials. *Materials and Structures*, RILEM, **25**, 265–272.

de Larrard, F., Sedran, T. and Angot, D. (1994a) Prévision de la compacité des mélanges granulaires par le modèle de suspension solide. I: Fondements théoriques et calibration du modèle. *Bulletin de Liaison des Laboratoires des Ponts et Chaussées*, no. 194, November–December.

de Larrard, F., Sedran, T. and Angot, D. (1994b) Prévision de la compacité des mélanges granulaires par le modèle de suspension solide. II: Validations et cas des mélanges confinés. *Bulletin de Liaison des Laboratoires des Ponts et Chaussées*, no. 194, November–December.

de Larrard, F., Belloc, A., Renwez, S. and Boulay, C. (1994c) Is the cube test suitable for high-performance concrete. *Materials and Structures*, 27, 580–583.

de Larrard, F., Acker, P. and Le Roy, R. (1994d) Shrinkage, creep and thermal effects, in *High-Performance Concrete and Applications* (eds S.P. Shah and S.H. Ahmad), Edward Arnold, London, pp. 75–114.

de Larrard, F., Szitkar, J.C., Hu, C. and Joly, M. (1994e) Design of a rheometer for fluid concretes, in *Proceedings of the International RILEM workshop Special Concretes: Workability and Mixing* (ed. P.J.M. Bartos), E & FN Spon, London, pp. 201–208.

de Larrard, F., Hu, C., Sedran, T. (1995) Best packing and specified rheology: two key concepts in high-performance concrete mixture-proportioning. CANMET, Adam Neville Symposium, Las Vegas, June.

de Larrard, F., Gillet, G. and Canitrot, B. (1996a) Preliminary HPC mix-design study for the 'Grand Viaduc de Millau': an example of LCPC's approach, in *Proceedings of the 4th International Symposium on the Utilization of High-Strength/High-Performance Concrete* (eds F. de Larrard and R. Lacroix), Paris, May, pp. 1323–1332.

de Larrard, F., Belloc, A., Boulay, C., Kaplan, D., Renwez, S. and Sedran, T. (1996b) Formulations de référence – Propriétés mécaniques jusqu'à l'âge de 90 jours. Rapport Projet National BHP 2000, December.

de Larrard, F., Sedran, T., Hu, C., Sitzkar, J.C., Joly, M. and Derkx, F. (1996c) Evolution of the workability of superplasticized concretes: assessment with BTRHEOM rheometer in *Proceedings of the International RILEM Conference on Production Methods and Workability of Concrete*, Paisley, June (eds P.J.M. Bartos, D.L. Marrs and D.J. Cleland), pp. 377–388.

de Larrard, F., Bosc, F., Catherine, C. and Deflorenne, F. (1997a) The AFREM method for the mix-design of high-performance concrete. *Materials and Structures*, no. 198, May.

de Larrard, F., Belloc, A. and Boulay, C. (1997b) Variations sur le thème des BHP. *Annales des Ponts et Chaussées*, no. 83, 4–14.

de Larrard, F., Hu, C., Sedran, T., Sitzkar, J.C., Joly, M., Claux, F. and Derkx, F. (1997c) A new rheometer for soft-to-fluid fresh concrete. *ACI Materials Journal*, 94(3), 234–243.

de Larrard, F., Ferraris, C.F. and Sedran, T. (1998) Fresh concrete: a Herschel–Bulkley material. Technical note, to appear in *Materials and Structures*.

Detwiler, R.J. and Tennis, P. (1996) The use of limestone in Portland cement: a state of the art review. RP 118, *Portland Cement Association*, Skokie, IL.

Dewar, J.D. (1986) The structure of fresh concrete. First Sir Frederick Lea Memorial Lecture, Institute of Concrete Technology, reprinted by British Ready Mixed Concrete Association.

Dewar, J. (1993) *Questjay – Mixism*, computer program, issue 3.0, January.

Dreux, G. (1970) *Guide pratique du béton*, Collection de l'ITBTP.

Dreux, G. and Gorisse, F. (1970) Vibration, ségrégation et ségrégabilité des bétons. *Annales de l'Institut Technique du Bâtiment et des Travaux Publics*, no. 265, Janvier, 37–86.

Farris, R.J. (1968) Prediction of the viscosity of multimodal suspensions from unimodal viscosity data. *Transactions of the Society of Rheology*, 12(2), 281–301.

Fascicule 65 A (1992) Exécution des ouvrages en béton armé ou en béton précontraint par post-tension. Cahier des clauses techniques générales applicables aux marchés publics de travaux, Décret n° 92–72, January.

Faury, J. (1944) *Le béton – Influence de ses constituants inertes – Règles à adopter pour sa meilleure composition, sa confection et son transport sur les chantiers*. 3rd edn, Dunod, Paris.

Féret, R. (1892) Sur la compacité des mortiers hydrauliques. *Annales des Ponts et Chaussées, Série 7*, 4, 5–164.

Ferraris, C.F. (1986) Mécanismes du retrait de la pâte de ciment durcie. PhD thesis no. 621, Ecole polytechnique fédérale de Lausanne.

Ferraris, C.F. and de Larrard, F. (1998) Testing and modelling of fresh concrete rheology. NISTIR 6094, National Institute of Standards and Technology, February.

FIP (1990) State of the art report on high-strength concrete. Fédération Internationale de la Précontrainte, June.

Foissy, A. (1989) Les effets fluidifiants. Séminaire Connaissance générale du béton, Ecole nationale des ponts et chaussées, Novembre.

Folliard, K.J. and Berke, N.S. (1997) Properties of high-performance concrete containing shrinkage-reducing admixture. *Cement and Concrete Research*, 27(9), 1357–1364.

Foster, B.E. and Blaine, R.L. (1968) A comparison of ISO and ASTM tests for cement strength, in *Cement, Comparison of Standards and Significance of Particular Tests*, ASTM STP 441, American Society for Testing and Materials, Philadelphia.

Garboczi, E. (1997) Stress, displacement, and expansive cracking around a single spherical aggregate under different expansive conditions. *Cement and Concrete Research*, 27(4), 495–500.

Garboczi, E. and Bentz, D. (1996) Modelling of the microstructure and transport properties of concrete. *Construction and Building Materials*, 10(5), 293–300.

Garboczi, E. and Bentz, D. (1997) An electronic monograph: Modelling the structure and properties of cement-based materials. Internet address http://ciks.cbt.nist.gov/garboczi/.

Giordano, R. and Guillelmet, J.P. (1993) Etude de l'activité des fines calcaires – Mesure des indices d'activité – Influence du ciment et des fines – Récapitulatif des résultats. Rapports CEBTP, CEMEREX-Marseille, Dossier 4122.3.600, October.

Granger, L. (1996) Comportement différé du béton dans les enceintes de centrales nucléaires: analyse et modélisation. Thèse de doctorat de l'ENPC and Etudes et recherche des LPC, OA 21, available at LCPC Paris.

Granger, L., Torrenti, J.M. and Acker, P. (1997a) Thoughts about drying shrinkage – scale effect and modelling. *Materials and Structures*, 30(196), 96–105.

Granger, L., Torrenti, J.M. and Acker, P. (1997b) Thoughts about drying shrinkage – experimental results and quantification of structural drying creep. *Materials and Structures*, 30(204), 588–598.

Guenot-Delahaie, I. (1997) Contribution à l'analyse physique et à la modélisation du fluage propre du béton. Etudes et recherches des Laboratoires des Ponts et Chaussées, OA 25, April.

Guyon, E. and Troadec, J.P. (1994) *Du sac de billes au tas de sable*. Editions Odile Jacob, Paris.

Hashin, Z. (1962) The elastic moduli of heterogeneous materials. *Journal of Applied Mechanics*, **29**, 143–150.

Hashin, Z. and Shtrikman, S. (1963) A variational approach to the theory of the elastic behaviour of multiphase materials. *Journal of Mechanics and Physics of Solids*, **11**, 127–140.

Helmuth, R., Hills, L.M., Whiting, D.A. and Bhattacharja, S. (1995) Abnormal concrete performance in the presence of admixtures. PCA report RP333, Portland Cement Association, Skokie, IL.

Hirsh, T.J. (1962) Modulus of elasticity of concrete as affected by elastic moduli of cement paste matrix and aggregate. *Journal of the ACI*, **59**, 427–451.

Hobbs, D.W. (1972) The compressive strength of concrete: a statistical approach to failure. *Magazine of Concrete Research*, **24**(80), 127–138.

Holland, T.C., Krysa, A., Luther, M.D. and Liu, T.C. (1986) Use of silica-fume concrete to repair abrasion-erosion damage in the Kinzua Dam stilling basin, in *Proceedings of 2nd International ACI-CANMET Conference on Fly Ash, Silica Fume, Slag and Natural Pozzolans in Concrete*, Madrid (ed. V.M. Malhotra), ACI SP 91–40, pp. 841–864.

Hu, C. (1995) *Rhéologie des bétons fluides*, Etudes et recherches des Laboratoires des Ponts et Chaussées, OA 16.

Hu, C. and de Larrard, F. (1996) The rheology of fresh high-performance concrete. *Cement and Concrete Research*, **26**(2), 283–294.

Hu, C., de Larrard, F. and Gjørv, O.E. (1995) Rheological testing and modelling of fresh high-performance concrete. *Materials and Structures*, **28**, 1–7.

Hu, C., de Larrard, F., Sedran, T., Boulay, C., Bosc, F. and Deflorenne, F. (1996) Validation of BTRHEOM, the new rheometer for soft-to-fluid concrete. *Materials and Structures*, RILEM, **29**(194), 620–631.

Hua, C. (1992) Analyse et modélisation du retrait d'autodessiccation de la pâte de ciment durcissante. Doctoral Thesis of Ecole Nationale des Ponts et Chaussées.

Ithurralde, G. and Costaz, J.L. (1993) HPC for the improvement of tightness of nuclear reactor containment in case of severe accidents, in *HPC in Severe Environments* (ed. P. Zia), ACI SP 140-11, November, pp. 227–238.

Johansen, V. and Andersen, P.J. (1991) Particle packing and concrete properties. *Materials Science of Concrete II* (eds J. Skalny and S. Mindess), The American Ceramic Society, Westerville, OH, pp. 111–148.

Johnston, C.D. (1990) Influence of aggregate void conditions and particle size on the workability and water requirement of single-sized aggregate, in *Properties of Fresh Concrete*, Proceedings of RILEM Colloquium, Hanover, October (ed. H.J. Wienig), pp. 67–76.

Joisel, A. (1952) Composition des bétons hydrauliques. Annales de l'ITBTP, 5ème année, No. 58, Série: Béton. Béton armé (XXI), October.

Justnes, H., Sellevold, E. and Lundevall, G., (1992a) High-strength concrete binders, part A: Composition of cement pastes with and without silica fume, in *Proceedings of the 4th International Symposium on Fly Ash, Silica Fume, Slag and Natural Pozzolans in Concrete*, Istanbul (ed. V.M. Malhotra), ACI SP 132-47.

Justnes, H., Meland, I., Bjoergum, J. and Krane, J. (1992b) The mechanisms of silica fume action in concrete studied by solid state ^{29}Si NMR, in Seminar on Nuclear Magnetic Resonance, Guerville, France, CTG Ital-Cementi.

Kaetzel, L.J. and Clifton, J.R. (1995) Expert/knowledge based systems for materials in the construction industry: state-of-the-art report. *Materials and Structures*, **28**, 160–174.

Kantha Rao, V.V.L. and Krishnamoothy, S. (1993) Aggregate mixtures for least-void content for use in polymer concrete. *Cement, Concrete and Aggregates*, CCAGDP, **15**(2), 97–107.

Kheirbek, A. (1994) Retrait et séchage des bétons. Mémoire de diplôme approfondi, LCPC/Ens, Cachan, June.

Kim, J.K., Park, Y.D., Sung, K.Y. and Lee, S.G. (1992) The production of high-strength fly ash concrete in Korea, in *Proceedings of the 4th CANMET/ACI International Conference on Fly Ash, Silica Fume, Slag and Natural Pozzolans in Concrete*, Istanbul, May, Supplementary papers.

Kretz, T. and Eymard, R. (1993) Fluage des dalles de pont en ossature mixte. *Bulletin Ponts Métalliques*, no. 16, 111–118.

Krieger, I.M. and Dougherty, T.J. (1959) *Transactions of the Society of Rheology*, **III**, 137–152.

Kronlof, A. (1994) Effect of very fine aggregate on concrete strength. *Materials and Structures*, RILEM, **27**(165), 15–25.

Lange, C. and Cochard, A. (1991) Pont de l'Elorn – Béton léger BL 32 pour le béton précontraint du tablier. Report from CEBTP Rennes, File # 8132 6 923, July.

Lange-Kornbak, D. and Karihaloo, B.L. (1996) Design of concrete mixes for minimum brittleness. *Advanced Cement-Based Materials*, **3**, 124–132.

Laplante, P. (1993) Comportement mécanique du béton durcissant. Analyse comparée du béton ordinaire et à très hautes performances. *Etudes et Recherches des Laboratoire des Ponts et Chaussées*, OA 13, LCPC, Paris.

Le Bris, J., Redoulez, P., Augustin, V., Torrenti, J.M. and de Larrard, F. (1993) High-performance concrete at the Elorn Bridge, in *High Performance Concrete in Severe Environments* (ed. P. Zia), SP 140-4, American Concrete Institute, Detroit, November.

Lee, D.J. (1970) Packing of spheres and its effects on the viscosity of suspensions. *Journal of Paint Technology*, **42**, 579–587.

Lecomte, A. and Zennir, A. (1997) Modèle de suspension solide et formulation de bétons calcaires en Lorraine. *Bulletin des Laboratoires des Ponts et Chaussées*, no. 211, September–October, 41–52.

Le Roy, R. (1996) Déformations instantanées et différées des bétons à hautes performances. *Etudes et Recherches des Laboratoires des Ponts et Chaussées*, September.

Le Roy, R., de Larrard, F. and Duval, D. (1995) *Retrait et fluage du béton du viaduc sur le Rhône*, Rapport de l'étude no. 324, 47, Février, LCPC, Paris.

Le Roy, R., de Larrard, F. and Pons, G. (1996) Calcul des déformations instantanées et différées des BHP. *Bulletin des Laboratoires des Ponts et Chaussées*, Numéro spécial XIX.

Lyse, I. (1932) Tests on consistency and strength of concrete having constant water content. *Proceedings of the ASTM*, **32**, Part II, 629–636.

Malhotra, V.M. (1986) Mechanical properties and freezing-and-thawing resistance of non-air-entrained and air-entrained condensed silica-fume concrete using ASTM Test C 666, Procedure A and B, *Second CANMET/ACI/International Conference on Fly Ash, Silica Fume, Slag and Natural Pozzolans in Concrete*, Madrid, (ed. V.M. Malhotra), SP 91, 1–53.

Malhotra, V.M. and Ramezanianpour, A.A. (1994) *Fly Ash in Concrete*. 2nd edn, MSL 94-45(IR), CANMET, Ottawa, Canada.

Malier, Y. and de Larrard, F. (1993) French bridges in high-performance concrete, in *Proceedings of the 3rd International Conference on the Utilization of High-Strength Concrete*, June, Lillehammer, (eds I. Holland and E. Sellevold), pp. 534–544.

Malier, Y., Brazillier, D. and Roi, S. (1992) The Joigny bridge: an experimental high-performance concrete bridge, in *High-Performance Concrete – From Material to Structure* (ed. Y. Malier), E & FN Spon, London, pp. 424–431.

Marchand, J. (1992) Pâtes de ciment. Résultat des essais mécaniques à 28 jours. Internal report, LCPC, June.

Markestad, S.A. (1976) The Gukild–Carlsen method for making stabilized pastes of cements. *Materials and Structures*, 9(50), 115–117.

Marusin, S.L. and Bradfor Shotwell, L. (1995) ASR in concrete caused by densified SF lumps – a case study, in *5th International Conference on Fly Ash, Silica Fume, Slag and Natural Pozzolans in Concrete*, Milwaukee, June, Supplementary papers, pp. 45–60.

Marzouk, H. (1991) Creep of high-strength and normal-strength concrete. *Magazine of Concrete Research*, 43(155), 121–126.

Maso, J.C. (1980) L'étude expérimentale du comportement du béton sous sollicitations monoaxiales et pluriaxiales, in *Le béton hydraulique* (eds J. Baron and R. Sauterey), Presse de l'Ecole nationale des ponts et chaussées, pp. 275–294.

Matsuyama, H. and Young, F. (1997) Superplasticizer effects on cement rheology, ACBM Semi-annual Meeting, Northwestern University, March.

Mehta, P.K. (1975) Evaluation of sulfate-resisting cement by a new test method. *ACI Materials Journal*, 72(10), 573–575.

Mehta, P.K. (1990) Durability of high-strength concrete. Paul Klieger Symposium on Performance of Concrete, ACI SP 122-2, pp. 19–28.

Mehta, P.K. (1997) Durability – critical issues for the future. *Concrete International*, 19(7), 27–33.

Monachon, P. and Gaumy, A. (1996) The Normandy bridge and the Société Générale tower – HSC grade 60, in *Proceedings of the 4th International Symposium on the Utilization of High-Strength/High-Performance Concrete*, Paris, May, (eds F. de Larrard and R. Lacroix), pp. 1525–1536.

Mooney, M. (1951) The viscosity of concentrated suspensions of spherical particles. *Journal of Colloids*, 6, 162.

Mori, T. and Tanaka, K. (1973) Average stress in matrix and average elastic energy of materials with misfitting inclusions. *Acta Metallurgica*, 21(571).

Mørtsell, E., Maage, M. and Smeplass, S. (1996) A particle-matrix model for prediction of workability of concrete. *Production Methods and Workability of Concrete, Proceedings of the international RILEM conference* (eds P.J.M. Bartos, D.L. Marrs and D.J. Cleland), E & FN Spon, London, pp. 429–438.

Naproux, P. (1994) Les micro-cendres (cendres volantes traitées) et leur emploi dans les bétons hydrauliques, Doctoral thesis of Institut National des Sciences Appliquées de Toulouse.

Neville, A.M. (1995) *Properties of Concrete*. Longman, Harlow.

Nilsen, A. and Monteiro, P.J.M. (1993) Concrete: a three-phase material. *Cement and Concrete Research*, **23**, 147–151.

Oger, L. (1987) Etude des corrélations structure-propriétés des mélanges granulaires (Study of correlations between structure and properties of granular mixtures), Thèse d'Etat, Université de Rennes.

Ohta, T., Yamazaki, N. and Nishida, A. (1996) Prediction of the adiabatic temperature rise of HSC using low heat Portland cement, ACI Fall meeting, New Orleans, November.

Oluokun, F.A. (1991) Prediction of concrete tensile strength from its compressive strength: evaluation of existing relations for normal-weight concrete. *ACI Materials Journal*, **88**(3).

Oluokun, F.A., Burdette, E.G. and Deatherage, J.H. (1991) Splitting tensile strength and compressive strength relationship at early ages. *ACI Materials Journal*, **88**(2), 115–121.

Paillère, A.M., Buil, M. and Serrano, J.J. (1989) Effect of fiber addition on the autogenous shrinkage of silica fume concrete. *ACI Materials Journal*, **86**(2), 139–144.

Pashias, N., Boger, D.V., Summers, J. and Glenister, D.J. (1996) A fifty cent rheometer for yield stress measurement. *Journal of Rheology*, **40**(6), 1179–1189.

Pigeon, M. and Pleau, R. (1995) *Durability of Concrete in Cold Climates*. E & FN Spon, London.

Popovics, S. (1982) *Fundamentals of Portland Cement Concrete: A Quantitative Approach*, Vol. 1: *Fresh Concrete*, John Wiley & Sons.

Popovics, S. (1990) Analysis of concrete strength vs. water-cement ratio relationship. *ACI Materials Journal*, **87**(5), 517–528.

Popovics, S. and Popovics, J. (1995) Computerization of the strength versus W/C relationship. *Concrete International*, **17**(4), 37–40.

Powers, T.C. (1968) *Properties of Fresh Concrete*, John Wiley & Sons, New York.

Powers, T.C. and Brownyard, T.L. (1946–47) Studies of the physical properties of hardened Portland cement paste. *Journal of the American Concrete Institute*, **18**, 101–132, 249–336, 469–504, 549–602, 669–712, 845–880, 933–992.

Raphael, J.M. (1984) Tensile strength of concrete. *Concrete International*, **81**(2), 158–165.

Richard, P. and Cheyrezy, M. (1994) Reactive powder concretes with high ductility and 200–800 MPa compressive strength. *ACI Spring Convention*, Washington, ACI SP-144, pp. 507–519.

Richard, P. *et al.* (1995) Les bétons de poudres réactives (BPR) à ultra-haute résistance (200 à 800 MPa). *Annales de l'ITBTP*, no. 532, 81–143.

Rollet, M., Levy, C. and Chabanis, F. (1992) Physical, mechanical and durability characteristics of various compressive strength concretes (from 20 to 120 MPa). CANMET/ACI International Conference on Advances in Concrete Technology, May, Research Papers, pp. 207–226.

Rosato, A., Strandburg, R., Prinz, F. and Swendsen, R.H. (1987) Why the Brazil nuts are on the top: size segregation of particulate matter by shaking. *Physical Review Letters*, **58**(10) 1038–1040.

Rossi, P. (1998) *Les bétons de fibres métalliques*, Presses de l'Ecole Nationale des Ponts et Chaussées, Paris.

Rossi, P. and Renwez, S. (1996) High performance multimodal fibre reinforced cement composites (HPMFRCC), in *Proceedings of the 4th International Symposium on the Utilization of High-Strength/High-Performance Concrete*, May, Paris (eds F. de Larrard and R. Lacroix), pp. 687–694.

Rossi, P., Wu X., Le Maou, F. and Belloc, A. (1994) Scale effect on concrete in tension. *Materials and Structures*, **27**, 437–444.

Sablocrete (1994) *Bétons de sable – Caractéristiques et pratiques d'utilisation*, Projet national de recherche/développement Sablocrete, Presses de l'Ecole Nationale des Ponts et Chaussées, Paris.

Schaller, I., de Larrard, F. and Sudret, J.P. (1992) L'instrumentation du pont de Joigny, in *Les Bétons à hautes performances* (ed. Y. Malier), Presse de l'Ecole Nationale des Ponts et Chaussées, pp. 483–520.

Schiessl, P. *et al.* (1997) New approach to durability design – An example for carbonation induced corrosion. *CEB Bulletin*, no. 238.

Schrage, I. and Springenschmid, R. (1996) Creep and shrinkage data of high-strength concrete, in *Proceedings of the 4th International Symposium on the Utilization of High-Strength/High-Performance Concrete*, Paris, May, (eds F. de Larrard and R. Lacroix) pp. 331–348.

Sedran, T. (1995) Les bétons autonivelants (BAN) – Etude bibliographique. *Bulletin de Liaison des Laboratoires des Ponts et Chaussées*, no. 196.

Sedran, T. (1997) Mise au point de béton autonivelant à hautes performances pour tuyaux d'assainissement. Contrat SABLA, LCPC Report, September.

Sedran, T. (1999) Rhéologie et rhéométrie des bétons. Applications à la formulation des bétons autonivelants. Doctoral thesis of Ecole Nationale des ponts et Chaussées.

Sedran, T. and de Larrard, F. (1994) RENÉ-LCPC – Un logiciel pour optimiser la granularité des matériaux de génie civil, Technical Note, *Bulletin de Liaison des Laboratoires des Ponts et Chaussées*, no. 194, November–December.

Sedran, T., de Larrard, F., Hourst, F. and Contamines, C. (1996) Mix-design of self-compacting concrete (SCC) in *Production Methods and Workability of Concrete, Proceedings of the international RILEM conference* (eds P.J.M. Bartos, D.L. Marrs and D.J. Cleland), E & FN Spon, London, pp. 439–450.

Shah, S.P. (1997) An overview of the fracture mechanics of concrete. *Cement, Concrete and Aggregates*, CCAGDP, **19**(2), 79–86.

Sellevold, E. (1992) Shrinkage of concrete: effect of binder composition and aggregate volume fraction from 0 to 60%. *Nordic Concrete Research*, no. 11, 139–152.

Shilstone, J.M. (1993) High-performance concrete mixtures for durability, in *High Performance Concrete in Severe Environments* (ed. P. Zia), SP 140-14, pp. 281–305.

Sicard, V., François, R., Ringot, E. and Pons, G. (1992) Influence of creep and shrinkage on cracking in high-strength concrete. *Cement and Concrete Research*, **22**, 159–168.

Stock, A.F., Hannant, D.J. and Williams, R.I.T. (1979) The effect of aggregate concentration upon the strength and modulus of elasticity of concrete. *Magazine of Concrete Research*, **31**(109), 225–234.

Stovall, T., de Larrard, F., Buil, M. (1986) Linear Packing Density Model of Grain Mixtures. *Powder Technology*, **48**, (1), September.

Struble, L. and Sun, G.K. (1995) Viscosity of Portland cement paste as a function of concentration. *Journal of Advanced Cement-Based Materials*, **2**, 62–69.

Tanaka, K., Sato, K., Watanabe, S., Arima, I. and Suenaga, K. (1993) Development and utilization of high-performance concrete for the construction of the Akashi Kaikyo Bridge, in *High-Performance Concrete in Severe Environments* (ed. P. Zia), ACI SP 140, pp. 25–51.

Tatersall, G.H. (1991) *Workability and Quality Control of Concrete*. E & FN Spon, London.

Tazawa, E. and Miyazawa, S. (1993) Autogenous shrinkage of concrete and its importance in concrete technology, in *Proceedings of the 5th RILEM Symposium on Creep and Shrinkage of Concrete*, Barcelona, September, pp. 158–168.

Tazawa, E. and Miyazawa, S. (1996) Influence of autogenous shrinkage on cracking in high-strength concrete, in *Proceedings of the 4th International Symposium on the Utilization of High-Strength/High-Performance Concrete*, Paris, May (eds F. de Larrard and R. Lacroix), pp. 321–330.

Toralles-Carbonari, B., Gettu, R., Agullo, L., Aguado, A. and Acena, V. (1996) A synthetical approach for the experimental optimization of high-strength concrete, in *Proceedings of the 4th International Symposium on the Utilization of High-Strength/High-Performance Concrete*, Paris, May, (eds F. de Larrard and R. Lacroix), pp. 161–167.

Torrenti, J.M., Guenot, I., Laplante, P., Acker, P. and de Larrard, F. (1994) Numerical simulation of temperatures and stresses in concrete at early ages. Euro-C 1994, Innsbruck, March.

Toutlemonde, F. and le Maou, F. (1996) Protection des éprouvettes de béton vis-à-vis de la dessiccation. Le point sur quelques techniques de laboratoire. *Bulletin des Laboratoires des Ponts et Chaussées*, no. 203, May–June, 105–119.

Troxell, G.E., Raphael, J.M. and Davis, R.E. (1958) Long-time creep and shrinkage tests of plain and reinforced concrete. *Proceedings of the ASTM*, **58**, 1101–1120.

Ulm, F.-J. and Bazant, Z.P. (1998) Modeling of creep and viscous flow mechanisms in early-age concrete, proposed to *Journal of Engineering Mechanics, ASCE*.

Van Breugel, K. (1991) Simulation of hydration and formation of structure in hardening cement-based materials, Doctoral thesis, Delft Technical University.

Virlogeux, M. (1986) Généralités sur les caractères des bétons légers, in *Granulats et bétons légers – bilan de 10 ans de recherche*, Presses de l'Ecole Nationale des Ponts et Chaussées, Paris, pp. 111–246.

Walker, S. and Bloem, D.L. (1960) Effect of aggregate size on properties of concrete. *Journal of the American Concrete Institute*, **32**(3), 283–298.

Waller, V. (1998) Relations entre composition, exothermie en cours de prise et résistance en compression des bétons, submitted as a Doctoral Thesis to Ecole Nationale des Ponts et Chaussées, Paris.

Waller, V., de Larrard, F. and Roussel, P. (1996) Modelling the temperature rise in massive HPC structures, in *Proceedings of the 4th International Symposium on the Utilization of High-Strength/High-Performance Concrete*, Paris, May, (eds F. de Larrard and R. Lacroix), pp. 415–422.

Waller, V., Naproux, P. and de Larrard, F. (1997) Contribution des fumées de silice et des cendres volantes silico-alumineuses à la résistance en compression du béton – Quantification. *Bulletin des Laboratoires des Ponts et Chaussées*, no. 208, March–April, 53–65.

Weber, S. and Reinhardt, H.W. (1996) Various curing methods applied to high performance concrete with natural and blended aggregates, in *Proceedings of the 4th International Symposium on the Utilization of High-Strength/High-Performance Concrete*, Paris, May (eds F. de Larrard and R. Lacroix), pp. 1295–1304.

Wiens, U., Breit, W. and Schiessl, P. (1995) Influence of high silica fume and high fly ash contents on alkalinity of pore solution and protection of steel against corrosion, in *Proceedings of the 5th International Conference on Fly Ash, Silica Fume, Slag and Natural Pozzolans in Concrete*, Milwaukee, (ed. V.M. Malhotra), ACI SP 153-39, June.

Yamato, T. Emoto, Y. and Soedn, M. (1986) Strength and freezing-and-thawing resistance of concrete incorporating condensed silica-fume, in *Second CANMET/ACI International Conference on Fly Ash, Silica Fume, Slag and Natural Pozzolans in Concrete*, Madrid (ed. V.M. Malhotra), SP 91-54.

Yamato, T., Soedn, M. and Emoto, Y. (1989) Chemical resistance of concrete containing condensed silica fume, in *Third CANMET/ACI International Conference on Fly Ash, Silica Fume, Slag and Natural Pozzolans in Concrete*, Trondheim (ed. V.M. Malhotra), SP 114-44.

Yssorche, M.P. and Ollivier, J.P. (1996) Relationship between air permeability and compressive strength of concrete, in *Proceedings of the 4th International Symposium on the Utilization of High-Strength/High-Performance Concrete*, Paris, May (eds F. de Larrard and R. Lacroix), pp. 463–470.

List of symbols

A list of all symbols used in the book is presented here. A given parameter appears in the list of the chapter in which it is first used. Unfortunately, it has not been possible to avoid completely the use of several meanings for some symbols. In such cases the reader should look for the most recent definition, with regard to the place where the symbol is used.

Chapter 1

a_{ij}	Parameter describing the loosening effect exerted by class j on the dominant class i
$a(x)$	Same parameter, considered as a function of the fine/coarse diameter ratio
a_F	Transverse dimension (width) of a prismatic fibre
b_{ij}	Parameter describing the wall effect exerted by class j on the dominant class i
$b(x)$	Same parameter, considered as a function of the fine/coarse diameter ratio
b_F	Transverse dimension (height) of a prismatic fibre
D	Maximum diameter in a polydisperse mix
d	Diameter[1] of particles in a monodisperse mix, or minimal diameter in a polydisperse mix
d_i	Diameter of ith class of particles. When $i > j$, $d_i < d_j$.
e	Voids index of a granular mix
$H(x)$	Function describing the contribution of an aggregate fraction to the compaction index
K	Compaction index
K_i	Contribution of class i to the compaction index
k_F	Ratio between the distance at which the fibre perturbation propagates and the grain size

---

[1] For non-spherical particles, d is assumed to be the diameter found by sieving.

k_w	Ratio between local and overall packing density (in the wall effect zone)
l_F	Length of a fibre
N_F	Number of fibres by unit volume
n	Number of grain classes
p_{min}	Minimal (actual) porosity of a granular mix having a given grading span
S	Segregation potential of a granular mix
S_i	Proportion of the container in which no i grain would occur in the case of full segregation
V_p	In a total unit volume, the volume in which the packing of particles is perturbed by the container wall
v_p	Elementary volume perturbed by a single fibre
y_i	Volume fraction of the ith class, related to the total *solid* volume
α	Actual packing density of a monodisperse mix
α_i	Actual packing density of a monodisperse fraction having a diameter equal to d_i (for a given value of K index)
$\overline{\alpha_i}$	Mean value of α_i, in a mix affected by either a container or a fibre perturbing effect
β	Residual packing density (virtual packing density of a monodisperse mix)
β_i	Residual packing density of a monodisperse fraction having a diameter equal to d_i
$\overline{\beta_i}$	Mean value of β_i, in a mix affected by either a container or a fibre perturbing effect
γ	Virtual packing density of a polydisperse mix
γ_i	Virtual packing density of a polydisperse mix, when the i fraction is dominant
Φ	Solid volume of a granular mix, in a unit total volume
Φ_F	Proportion (in volume) of fibres, related to the total volume of the container
Φ_i	Proportion (in volume) of the i fraction, related to the total volume of the container
Φ_i^*	Maximum proportion of the i fraction, for fixed values of the other Φ_j ($j \neq i$)
λ	Ratio between successive diameters in an appolonian mixture ($\lambda < 1$)
$\lambda_{i \to j}$	Parameter describing the interaction exerted by class i on the dominant class j
λ^*	Critical percolation ratio
π	Porosity of a granular mix
π_{min}	Minimal (virtual) porosity of a granular mix having a given grading span

Chapter 2

A	Constant in the hardening function of a concrete
a	Volume of entrapped air in a unit volume of fresh cement paste
a_i	Multiplicative coefficient for the contribution of the i fraction to the yield stress of fresh concrete
B	Constant expressing the acceleration effect of a limestone filler
b	Dimension of the basic cell in the model of hardened cement paste
C^{th}	Heat capacity of fresh concrete
CA	Mass of coarse aggregate per unit volume of concrete
c	Mass of cement per unit volume of concrete
c_i^{th}	Heat capacity of constituent i
c_{eq}	Equivalent mass of cement (in the compressive strength model)
c_i	Concentration of phase i in a composite material
D	Particle diameter in Hashin's model, and maximum size of aggregate (MSA) in a polydisperse mixture
D'	Thickness of the matrix crust in Hashin's model
$d(t)$	Cement kinetics term expressing the compressive strength development at age t
E	Elastic modulus of concrete
E_d	Delayed elastic modulus of concrete (including the effect of basic creep)
E_d'	Delayed elastic modulus of concrete (including the effect of total creep)
E_{max}	Maximum value of E when the aggregate concentration equals its packing density
E_s	Elastic modulus of the solid phase
E_g	Elastic modulus of the aggregate phase
E_m	Elastic modulus of the matrix phase
E_{md}	Delayed modulus of the matrix phase (in sealed conditions)
E_{md}'	Delayed modulus of the matrix phase (in drying conditions)
E_p	Elastic modulus of cement paste
E_{pd}	Delayed modulus of the cement paste
e	Transverse dimension of beam or plate in the model of hardened cement paste
e_c	Thickness of the external layer in the triple-sphere model
FA	Mass of fine aggregate per unit volume of concrete
fa	Mass of fly ash per unit volume of concrete
fc	Compressive strength (of concrete or mortar)
fc_m	Compressive strength of the matrix
fc_p	Compressive strength of cement paste
fi	Mass of limestone filler per unit volume of concrete
ft	Tensile (splitting) strength of concrete or mortar
G_i	Shear modulus of phase i in a composite material

Gh_i	Hashin–Shtrikman bound for the shear modulus of a composite material, relative to the ith phase
g	Aggregate volume in a unit volume of concrete
g^*	Packing density of the aggregate used for making a concrete
h_c	Final degree of hydration of Portland cement
h_{FA}	Final degree of consumption of fly ash (ASTM Class F)
h_{SF}	Final degree of consumption of silica fume
K'	Compaction index of the (de-aired) fresh concrete (or placeability)
K^*	Compaction index dealing with a placing process
K_c	Autogenous shrinkage coefficient (cement parameter)
K'_c	Contribution of the cement to the compaction index of the (de-aired) fresh concrete
K_{cr}	Creep coefficient (sealed conditions)
K'_{cr}	Creep coefficient (drying conditions)
K_i	Bulk modulus of phase i in a composite material
K'_i	Contribution of fraction i to the compaction index of the (de-aired) fresh concrete
K_{FI}	Activity coefficient describing the contribution of a limestone filler to the compressive strength (binding effect)
K_g	Coefficient describing the effect of aggregate on the compressive strength
K'_g	Contribution of the coarse aggregate to the compaction index of the (de-aired) fresh concrete
Kh_i	Hashin–Shtrikman bound for the bulk modulus of a composite material, relative to phase i
K_p	Activity coefficient describing the contribution of a pozzolan to the compressive strength, irrespective of its dosage
k_t	Aggregate coefficient appearing in the tensile vs. compressive strength relationship (tensile strength coefficient)
MPT	Maximum paste thickness
m	Multiplicative coefficient in the Herschel–Bulkley model for fresh concrete
m_i	Mass of component i per unit volume of concrete
n	Exponent in the Herschel–Bulkley model for fresh concrete
P	Proportion of plasticizer/superplasticizer, as a percentage of the mass of binder
P^*	Value of P corresponding to the saturation amount
p	Coefficient describing the contribution of paste–aggregate bond to concrete compressive strength (bond coefficient)
pl	Mass of plasticizer (in dry extract) per unit volume of concrete
pz	Mass of pozzolan per unit volume of concrete
Q	Total heat released in adiabatic conditions during concrete hardening

q	Coefficient describing the limitation of concrete compressive strength by that of aggregate (ceiling effect coefficient)
Rc_{28}	Compressive strength of the cement at 28 days, following ENV 196-1 standard
r	Exponent of the MPT in the compressive strength model
SL	Slump of fresh concrete (in mm)
sf	Mass of silica fume per unit volume of concrete
sp	Mass of superplasticizer (in dry extract) per unit volume of concrete
t	Time (in days)
t_{C_3A}	Percentage of tricalcium aluminate in the cement (following the Bogue composition)
t_{C_4AF}	Percentage of tetracalcium ferro-aluminate in the cement
t_{C_2S}	Percentage of dicalcium silicate in the cement
t_{C_3S}	Percentage of tricalcium silicate in the cement
v_a	Volume of entrapped air in a unit volume of fresh cement paste
v_c	Volume of cement in a unit volume of fresh cement paste
v_w	Volume of water in a unit volume of fresh cement paste
w	Volume of water in a unit volume of fresh cement paste or concrete
α, β	Constants linking the delayed modulus (in sealed conditions) and the compressive strength of cement paste
α', β'	Constants linking the delayed modulus (in drying conditions) and the compressive strength of cement paste
α_c	Maximum cement concentration in a unit volume of cement paste (or matrix)
$\Delta\theta$	Final adiabatic temperature rise, for a concrete having an initial temperature of 20 °C
γ	Constant expressing the contribution of pozzolans to the hydraulic stress
$\dot{\gamma}$	Strain gradient of fresh concrete (or shear rate)
ε_{as}	Concrete autogenous shrinkage
ε_{as}^m	Matrix autogenous shrinkage
ε_{bc}^s	Specific basic creep of concrete (creep of a sealed specimen submitted to a unit stress)
ε_{ts}	Concrete total shrinkage
ε_{ts}^m	Matrix total shrinkage
ε_{ts}^g	Aggregate total shrinkage
η_a	Apparent viscosity of fresh concrete (for a strain gradient of $3\ \mathrm{s}^{-1}$)
Φ	Solid volume in a unit volume of fresh concrete or hardened cement paste
Φ_0	Solid volume in a unit volume of fresh cement paste
Φ_i	Proportion (in volume) of the i fraction, related to the total volume of fresh concrete

Φ^* Packing density of the squeezed fresh concrete

Φ_i^* Maximum proportion of the i fraction, for fixed values of the other Φ_j ($j \neq i$)

λ Dilatation ratio (from dry packing of aggregate to aggregate diluted in cement paste)

Ψ Contribution of a pozzolan to the compressive strength of concrete

Ψ' Contribution of a limestone filler (binding effect) to the compressive strength of concrete

Ψ_{max} Maximum value of Ψ

Ψ'_{max} Maximum value of Ψ'

μ Plastic viscosity of fresh concrete

ρ Specific gravity of concrete (dimensionless parameter)

ρ_c Specific gravity of cement (dimensionless parameter)

σ Stress applied on the whole material (macroscopic scale)

σ_s Stress applied on the solid phase (microscopic scale)

τ Shear stress applied to fresh concrete

τ_0 Yield stress of fresh concrete

Chapter 3

d Diameter of particles in a monodisperse mix

$d(t)$ Kinetics coefficient in the compressive strength model

E_g Elastic modulus of aggregate

fc Compressive strength of concrete (measured on cylinder)

fc_g Compressive strength of the parent rock of aggregate

fc_m Matrix compressive strength

ft Tensile splitting strength of concrete

g^* Actual packing density of the aggregate used in a concrete mix

h Final height of the specimen in the measurement of packing density

$i(t)$ Activity index at age t for a pozzolanic admixture

K Compaction index

K_c Autogenous shrinkage coefficient (cement parameter)

$Kp(t)$ Activity coefficient at age t for a pozzolanic admixture

k_t Aggregate coefficient appearing in the tensile vs. compressive strength relationship (tensile strength coefficient)

k_w Ratio between overall and local packing density (in the wall effect zone)

M_D Mass of dry aggregate used in the determination of specific gravity and/or water absorption

M_{SSD} Mass of saturated surface-dry aggregate used in the determination of specific gravity and/or water absorption

n Porosity of aggregate particles accessible to water

p	Coefficient describing the contribution of paste–aggregate bond to concrete compressive strength (bond coefficient)
q	Coefficient describing the limitation of concrete compressive strength by that of aggregate (ceiling effect coefficient)
Rc_t	Compressive strength of ISO mortar at age t
Rc_{28}	Compressive strength of ISO mortar at 28 days
V_{SSD}	Volume of saturated surface-dry aggregate used in the determination of specific gravity and/or water absorption
WA	Water absorption of aggregate, in percent
β_i	Virtual packing density of a monodisperse fraction having a diameter equal to d_i
$\bar{\beta_i}$	Mean value of β_i, in a mix affected by a container wall effect
Φ	Solid volume of a granular mix, in a unit total volume
$\varnothing$	Cylinder diameter (for packing density measurement)
ρ_D	Specific gravity of aggregate in dry conditions
ρ_{SSD}	Specific gravity of aggregate in saturated surface-dry conditions

Chapter 4

C	Volume of cement per unit volume of concrete
CA	Mass of coarse aggregate per unit volume of concrete
F	Volume of inert filler per unit volume of concrete
FA	Mass of fine aggregate per unit volume of concrete
f_c	Compressive strength of concrete at a given age
g	Volume of aggregate in a unit volume of concrete
g^*	Packing density of the aggregate (when it is separately packed as a dry granular mixture)
g^*_{max}	Value of g^* obtained with the best combination of the aggregate fractions
h	Height between the top surface and the bottom of a deck
MSA	Maximum size of aggregate
P	Dosage of plasticizer, as a proportion of the cement volume
W	Volume of water per unit volume of concrete
x	Optimal coarse/fine aggregate ratio (leading to the maximum packing density g^*_{max})
Φ	Solid volume (aggregate + filler + cement) per unit volume of concrete
Φ^*	Packing density of the solid materials (when they are separately packed as dry granular materials)
Φ^*_{max}	Value of Φ^* obtained with the best combination of all available solid materials
φ	Function linking the water/cement ratio with the compressive strength of concrete
η_a	Apparent viscosity of the fresh concrete, used as a workability index

Ψ_1	Contribution of the cement paste to the apparent viscosity (in the paste/aggregate model)
Ψ_2	Contribution of the aggregate to the apparent viscosity (in the paste/aggregate model)
Ψ_3	Contribution of the plasticizer to the apparent viscosity (in the solid concentration model)
Ψ_4	Contribution of the solid materials to the apparent viscosity (in the solid concentration model)
ρ	Mass of concrete per unit volume
τ_0^{crit}	Yield stress above which the fresh concrete will flow

Chapter 5

c_{te}	Coefficient of thermal expansion
E_g	Elastic modulus of lightweight aggregate
E_m	Elastic modulus of mortar
g	Volume of lightweight aggregate in a unit concrete volume
Ic_{sealed}	Cracking index of concrete in sealed conditions
Ic_{drying}	Cracking index of concrete in dry conditions
Ic_{massive}	Cracking index of concrete in massive conditions
σ	Compressive stress applied on a concrete unit cell
σ_m	Compressive stress in the parallel phase of mortar, in the parallel/series model

Appendix: Flowchart for mixture simulation

In this part, a logical order for using the various models developed in Chapter 2 is proposed. The aim is to predict the properties of a mixture, the proportions of which are given by the user. This flowchart is suitable for programming the models in a spreadsheet. If a solver is available and linked with the simulation system, a calibration of some material parameters becomes possible. Likewise an optimization, based on a set of specifications, is equally feasible.

It is assumed that all parameters dealing with the properties of constituents have been previously determined (see Chapter 3) and implemented in a data sheet. The mixture composition is therefore given under the following form: percentage of each aggregate fraction with respect to the total aggregate volume, masses of binders and free water per unit volume of fresh de-aired concrete, dosage of superplasticizer (dry extract), in percentage of the cement weight.

Step	Name of the parameter/property	Symbol	Granular system[1]	Equation number	Comments
1	Water/cement ratio	w/c			
2	Coarse/fine aggregate ratio	CA/FA			This parameter can be calculated in mass or in volume
3	Solid volume	Φ	3	−	Total volume minus the free water volume
4	Fresh concrete specific gravity	ρ	3	−	The value is that of de-aired concrete, which is slightly higher than that of fresh concrete
5	Saturation dose of superplasticizer	P^*	−	−	Sum of SP demands (by each binder)/mass of cement, with account of the fine particles coming from the aggregate
6	Volume fractions of the n classes	y_i	3	−	The volumes of each component and then the overall size distribution are calculated

(continued)

Step	Name of the parameter/property	Symbol	Granular system[1]	Equation number	Comments
7	Residual packing densities of the n classes	β_i	3	2.4 then 1.16	Calculation of the 'equivalent' residual packing density, where the size distributions of several constituents overlap, and accounting for the presence of superplasticizer (for the fine particles)
8	Virtual packing density when the ith class is dominant	γ_i	3	1.33	–
9	Contributions to compaction index of the n classes (in bulk)	K_i'	3	1.47	–
10	Yield stress	τ_0	3	2.17	A distinction is made between particles bigger and smaller than 0.08 mm in the presence of SP
11	**Slump**	SL	3	2.19	ρ is slightly overestimated, but this error has a negligible effect, given the precision of the slump formula
12	Mass of fine aggregate $(0.08 \leqslant d_i \leqslant 5 \text{ mm})$	FA	3	–	
13	**Air content**	a	1	2.24	
14	**Mixture proportions**	–	1	–	Calculated with account of entrapped air and water absorption by the aggregate fractions
15	**Fresh concrete specific gravity**		1	–	Calculated from ρ with account of entrapped air
16	Packing density of the squeezed fresh concrete	Φ^*	4	1.47	Given by an implicit equation, with $K=9$. The unique real root for which $\Phi^* < \gamma_i \atop 1 \leqslant i \leqslant n$ can be obtained by various numerical methods, such as Newton's or Lagrange's.
17	Plastic viscosity before vibration	μ	3	2.7	
18	**Plastic viscosity after vibration**		3	2.8	
19	Residual packing densities of the n classes, taking account of the form/rebars wall effect	$\overline{\beta}_i$	5	1.56	'Confined' values calculated from the already determined β_i, taking account of the structure geometry

(continued)

Step	Name of the parameter/property	Symbol	Granular system[1]	Equation number	Comments
20	Virtual packing density when the *i*th class is dominant	γ_i	5	1.33	–
21	Contributions to the compaction index of the *n* classes	K_i	5	1.47	–
22	**Filling diagram**	$\dfrac{K_j}{1+K_j}$	5	–	Diagram giving the filling ratio versus the mean size of the clustered aggregate fractions (the K_i are summed up within each clustered fraction)
23	**Contribution of cement to placeability**	K_c'	5	–	Used to assess the bleeding tendency, and as a criterion for designing self-compacting concrete
24	**Contribution of coarse aggregate to placeability**	K_g'	5	–	Used to assess the risk of coarse aggregate blockage in self-compacting concrete
25	**Placeability of in-place fresh concrete**	K'	5	1.47	–
26	**Segregation potential**	S	5	2.25	
27	Heat capacity	C^{th}		2.31	
28	Final degree of hydration of Portland cement	h_c		2.39–2.42	This parameter and the two following ones are calculated by solving a system of four equations (see section 2.2.2)
29	Final degree of transformation of fly ash	h_{FA}		2.39–2.42	
30	Final degree of transformation of silica fume	h_{SF}		2.39–2.42	
31	Concrete heat of hydration	Q		2.43	
32	**Adiabatic temperature rise**	$\Delta\theta$		2.44	
33	Aggregate volume	g	6	–	Volume of aggregate fractions with $d_i \geqslant 0.08$ mm
34	Volume fractions of the *n*-classes in the aggregate fraction	y_i	2	–	

(continued)

Step	Name of the parameter/property	Symbol	Granular system[1]	Equation number	Comments
35	Virtual packing density when the ith class is dominant	γ_i	2	1.33	–
36	Contributions to compaction index of the n aggregate classes	K_i	2	1.47	–
37	Packing density of the aggregate phase	g^*	2	1.47	Implicit equation, with $K = 9$
38	Maximum size of aggregate	D	2	–	Size corresponding to 90% of total aggregate passing
39	Maximum paste thickness	MPT	6	2.64	–
40	Equivalent cement content vs. age	$c_{eq}(t)$	–	2.82	–
41	Matrix compressive strength vs. age	$fc_m(t)$	–	2.83	–
42	**Concrete compressive strength vs. age**	$fc(t)$	–	2.84	–
43	**Concrete tensile strength vs. age**	$ft(t)$	–	2.89	–
44	Matrix elastic modulus vs. age	$E_m(t)$	–	2.99	–
45	**Concrete elastic modulus vs. age**	$E(t)$	6	2.97	–
46	Matrix delayed modulus	E_{md}	–	2.102	–
47	Concrete delayed modulus	E_d	6	2.101	–
48	**Specific basic creep**	ε_{bc}^s	–	2.103	–
49	Matrix delayed modulus in drying conditions	E'_{md}	–	2.106	–
50	Concrete delayed modulus in drying conditions	E'_d	6	2.105	–
51	**Specific total creep**	ε_{tc}^s	–	2.103	–
52	Hydraulic stress	σ_H	–	2.112	–
53	Matrix autogenous shrinkage	ε_{as}^m	–	2.111	–

(*continued*)

Step	Name of the parameter/property	Symbol	Granular system[1]	Equation number	Comments
54	**Concrete autogenous shrinkage**	ε_{as}	6	2.110	–
55	Matrix total shrinkage	ε_{ts}^{m}	–	2.115	–
56	**Concrete total shrinkage**	ε_{ts}	6	2.114	–
57	Unit cost of concrete				Calculated from the mixture proportions, by adding the costs of each component plus an overhead cost

[1] See section 2.7.

Index

Milton Keynes UK
Ingram Content Group UK Ltd.
UKHW021838071024
449327UK00021B/1516